Phase Change, Interphase Coupling, and Multiphase Transport in Porous Structures

Phase Change, Interphase Coupling, and Multiphase Transport in Porous Structures

Guest Editors

Liu Yang
Haitao Zhang
Yanjun Zhang
Bo Wang

Basel • Beijing • Wuhan • Barcelona • Belgrade • Novi Sad • Cluj • Manchester

Guest Editors

Liu Yang
State Key Laboratory for
Tunnel Engineering
China University of Mining
and Technology
Beijing
China

Haitao Zhang
School of Energy Resources
China University of
Geosciences (Beijing)
Beijing
China

Yanjun Zhang
College of Petroleum
Engineering
Xi'an Shiyou University
Xi'an
China

Bo Wang
School of Petroleum
China University of
Petroleum-Beijing at Karamay
Karamay
China

Editorial Office
MDPI AG
Grosspeteranlage 5
4052 Basel, Switzerland

This is a reprint of the Special Issue, published open access by the journal *Materials* (ISSN 1996-1944), freely accessible at: https://www.mdpi.com/journal/materials/special_issues/1F25Y08LNT.

For citation purposes, cite each article independently as indicated on the article page online and as indicated below:

Lastname, A.A.; Lastname, B.B. Article Title. *Journal Name* **Year**, *Volume Number*, Page Range.

ISBN 978-3-7258-4413-5 (Hbk)
ISBN 978-3-7258-4414-2 (PDF)
https://doi.org/10.3390/books978-3-7258-4414-2

Contents

materials

MDPI

Review

Fabrication and Tribological Performance of Self-Lubricating Porous Materials and Composites: A Review

Ashish K. Kasar, Subin Antony Jose, Brian D'Souza and Pradeep L. Menezes *

Department of Mechanical Engineering, University of Nevada, Reno, NV 89557, USA; ashishkasar91@gmail.com (A.K.K.); subinj@unr.edu (S.A.J.); brian39dsouza@gmail.com (B.D.)
* Correspondence: pmenezes@unr.edu

Abstract: Porous materials have recently attracted significant attention in the aerospace and biomedical fields for addressing issues related to friction and wear. Porous materials are beneficial in applications where continuous lubrication is not feasible or for components that operate under extreme conditions, such as high speeds, elevated temperatures, and heavy loads. The pores can serve as reservoirs for liquid lubricants, which are gradually released during the operation of the components. The tribological properties of these materials depend on their porosity, the lubricants used, and any additional additives incorporated into the porous materials. This review article provides insight into common fabrication techniques for porous materials and examines their tribological performance for all three classes of materials—polymers, metals, and ceramics. Additionally, it discusses design criteria for porous self-lubricating materials by highlighting the critical properties of both the substrate and lubricants.

Keywords: porosity; coefficient of friction; wear rate

1. Introduction

Porous materials are engineered with voids or open spaces within their structure to provide several distinct benefits. Porous materials can be classified into two types based on their porosity—open pores (through porosity), which can provide the passage of fluids, and closed pores (independent voids). The pores can be further classified based on their size. Pores 2 nm or smaller are called micropores, those ranging from 2 nm to 50 nm are known as mesopores, and pores larger than 50 nm are referred to as macropores [1]. Porous materials have a high surface area due to their internal structure, making them useful for various applications for all three classes of materials—polymer, metal, and ceramic. Porous polymers are often used as a drug delivery system, gas storage, sensing devices, and membranes of filtration [2,3]. Porous metals and alloys are frequently employed for their lightweight structural strength, catalytic capabilities, and roles in the filtration process and heat exchangers [4]. Porous ceramic materials find applications in filtration, separation processes, thermal insulation, and more [5]. Both porous metals and ceramic materials are extensively utilized in biomedical implants, serving as scaffolds for bone and tissue growth [6,7]. Porous materials offer superior thermal properties, including an increased convective heat transfer coefficient and higher thermal conductivity, relative to their base materials. It makes porous materials highly suitable for solar energy applications [8].

The application of porous materials in tribology is independent of the aforementioned application and has evolved over time. One notable early use of porous materials in tribology is the use of porous bronze bearing, which dates back to the early to mid-20th century by the automobile industry [9,10]. The voids present in the sintered bronze allow for the retention of lubricating oil and release of it to the surface during sliding, providing a self-lubricating capacity. The application of porous material in the field of tribology helps to overcome the limitations of conventional lubricating strategies that involve the continuous application of oil/grease on moving machinery. The continuous supply of lubricants is challenging in extreme conditions, especially in hard-to-reach places and when continuous lubrication is not possible. For such conditions, porous self-lubricating materials have gained a lot of attention in recent years to solve friction, wear, and lubrication problems [11,12]. These materials are capable of storing lubricants in their pores, which can help to reduce the friction of the components, hence making them useful in extreme conditions [13]. The lubricants stored in the pores can be released during sliding when a load is applied, and their effectiveness can be influenced by the temperature of the surrounding environment [14,15]. In addition, depending on the material–lubricant pair and pore size, the reabsorption of the lubricant can be realized through capillary action under static conditions. This enhances the recyclability of lubricants [16]. These self-lubricating porous materials find applications in equipment operating at high speeds, elevated temperatures, and heavy loads [17]. For example, Wang et al. [18]. demonstrated that polyimide as a retainer for a roller bearing failed in the first 1000 s of the test at 800 RPM under the axial load of 500 N, whereas a porous polyimide with filled oil did not show any sign of lubrication failure, even after 12 h of testing under the same testing conditions.

Engineering porous material for self-lubrication can be challenging. Different parameters need to be considered while designing a porous self-lubricating material. The first important factor is the type of material based on the application. In this review article, three classes of matrix materials are covered: (1) polymers, (2) metals, and (3) ceramics. Each class of self-lubricating porous materials has its applications and is specifically designed for certain environments and operating conditions. For example, ceramic substrates can have higher chemical and thermal resistance due to their structural stability compared with polymers or metals.

The pore size, pore concentration, and pore distribution of all the materials can be controlled by fabrication techniques, optimizing the type and concentration of additives and pore-forming agents. A wide range of fabrication methods, ranging from powder metallurgy to 3D printing, are discussed, with a focus on each method's capability to control pore size and distribution. The matrix material plays a crucial role in impregnating and supplying lubricant in porous materials. The significance of considering surface energy is extensively discussed in the design criteria for self-lubricating materials. The tribological properties of various porous self-lubricating materials are discussed in terms of lubrication regimes and longevity.

2. Fabrication of Porous Materials

Porous polymeric materials are fabricated using various techniques, with some manufacturing processes being tailored to specific classes of porous materials. The initial step in the fabrication process involves selecting the matrix materials. Various types of materials are utilized as the matrix, chosen according to the specific application requirements. Some commonly used matrix materials for porous polymers are polyetheretherketone (PEEK) [19,20], polyphenylene sulfide (PPS) [21,22], and polyimide (PI) [23,24] due to their exceptional mechanical strength, chemical resistance, and high-temperature stability compared with other polymers.

One of the reasons for pore formation in materials is the defects that arise during fabrication processes, such as the combination of compression and heat treatment. For example, polyimide porous substrates were prepared by compressing the material within a fixed volume to achieve a 1.1 g/cm^3 density. Subsequently, the compressed substrate was heat treated at 350 °C for 1 h to promote densification [25]. This compression and heat treatment fabrication route often results in closed surface pores as the surface stays in contact with die/mold walls. For the infiltration of lubricants, post-processing methods are required to remove the top surface of closed pores, which can vary from simple processes such as grinding or polishing [23,26] to laser etching. For example, Ye et al. [25] generated different pore sizes on the polyimide substrate using different sets of laser power (5 and 6 W) and a different number of scanning cycles (1 to 30 times). The surface of the substrate resulted in different pore sizes, as shown in the Figure 1. However, though this study did not report pore size values, the surface roughness (R_a) increased from 0.21 to 0.36, which indicates an increase in pore size.

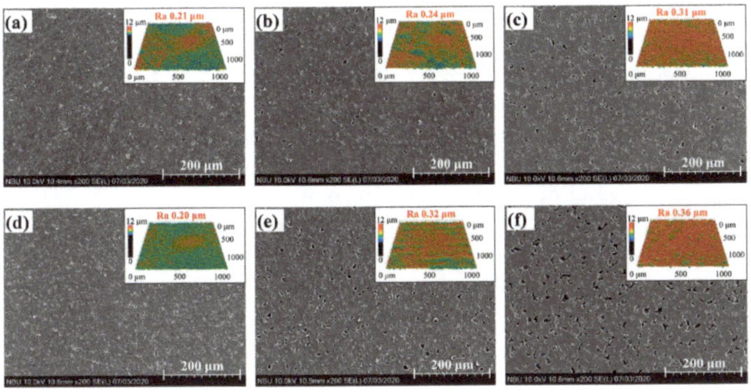

Figure 1. Surface micrographs of the polyimide substrate after laser scanning: (**a**) 5 W, 1 cycle, (**b**) 5 W, 10 times, (**c**) 5 W, 30 times, (**d**) 6 W, once, (**e**) 6 W, 10 times, and (**f**) 6 W, 30 times [25].

Another technique for creating pores is through the addition of pore-forming agents such as sodium chloride (NaCl) [15,27]. Three-dimensional (3D) printing is another newer process that is used to fabricate porous polymers. Several techniques within the 3D printing process can produce porous polymers, such as designing pores in the print path [28,29], polymer matrices with sacrificial fillers [30,31], pores formed by solvent evaporation, and printing polymers with intrinsic porosity.

Metals and ceramics materials have similar fabrication techniques for porous materials that can carry lubricants. The most common technique is powder metallurgy, which uses powders that are pressed in a die or molds to produce green samples. Then, these green samples are heated to a suitable temperature, time, and environment to produce the desired part. The powder metallurgy process provides control over pores at both stages—pressing to produce a green sample—and also during sintering [32]. Apart from this, the addition of reinforcement particles can also limit the sintering and result in porous [33,34] materials. A typical powder metallurgy method shown in Figure 2 includes mixing, cold compaction, and sintering followed by lubrication impregnation. This method has been used in various studies. For example, porous metallic Cu-Sn-Ti composites [35] were reinforced with cubic boron nitride (CBN) and graphene nanoplatelets, where these powders were blended with carbamide particles using the ball milling technique for 30 min. Carbamide is a water-soluble organic compound, also known as urea, that is used as a pore-forming agent. These mixtures are subsequently pressed unidirectionally in a single-ended mold at 300 MPa for 30 s. To leach carbamide particles, these green samples were placed in distilled water at room temperature for 4 h. After drying, the samples were sintered at 880 °C with a holding time of 30 min under a high vacuum environment.

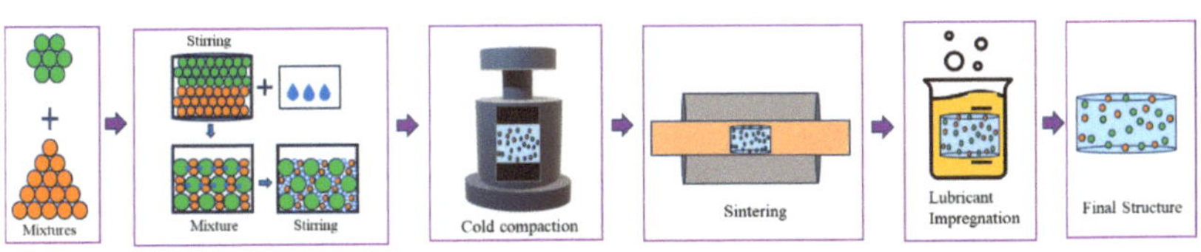

Figure 2. Schematic of a typical powder metallurgy method for self-lubricating composites starting from individual powder, mixing of powders, cold compaction followed by sintering to achieve porous material and liquid impregnation to achieve final structure of self-lubricating composite. The arrow shows the directionality in the process.

This inherent porosity is due to limited sintering that occurs uniformly by the formation of necks between powders. In addition to inherent pores during sintering, larger secondary pores can be obtained by space holders and gaseous blowing agents [36]. Space holders are typically the materials that can be burnt during sintering or washed away after sintering, e.g., NaCl [37], ammonium hydrogen bicarbonate [38], and polymeric materials [39]. These space holders are mixed with metal or ceramic powders before compaction, which are subsequently removed during or after sintering. Moreover, lower melting point metals can also be used as a space holder to generate higher porosity. For example, Mg has been used in Ti alloys to generate a porosity of up to 50–65%, where Mg evaporates during sintering without reacting with Ti [40]. Using a gaseous blowing agent, secondary large pores can be obtained. This technique is widely used to generate foam materials with a porosity of 90–95% for all three classes of materials: polymers [41,42], metals [43], and ceramics [44].

The additive manufacturing (AM) of metals and ceramics provides the most attractive methods for fabricating porous materials, which is useful for creating complex functional materials while retaining good properties. The AM process is a layer-by-layer material deposition technique using an energy source such as an electron or laser beam that can yield a precisely controlled internal and external structure [45]. These porous structures are widely used in implants for bone–tissue engineering. One such porous Ti-6Al-4V alloy's diamond and hatched structure as produced by a selective electron beam is shown in Figure 3 [46]. These porous Ti alloys exhibited maximum compressive strength in the range of 20–30 MPa for the diamond structure and 125–148 MPa for the hatched structure. These studies on additive manufacturing show that an optimal porous material can be prepared to control the volume porosity, corrosion resistance, and wear resistance [47,48]. The cold spray (CS) process is another coating/additive manufacturing technique capable of producing porous structures [49]. By adjusting the spraying parameters, the porosity level within the deposit can be controlled [50]. Wathanyu et al. [51] utilized the CS process to fabricate Ti porous coating on 316L stainless steel substrate by controlling the process gas temperature and pressure. However, the CS process is limited by material selection, as sufficient deformation of the spraying particles is required for effective bonding. The other methods for generating porosity in metals and ceramics are fugitive scaffolds of polymers [52], sol-gel processing [53], gel casting [54], etc.

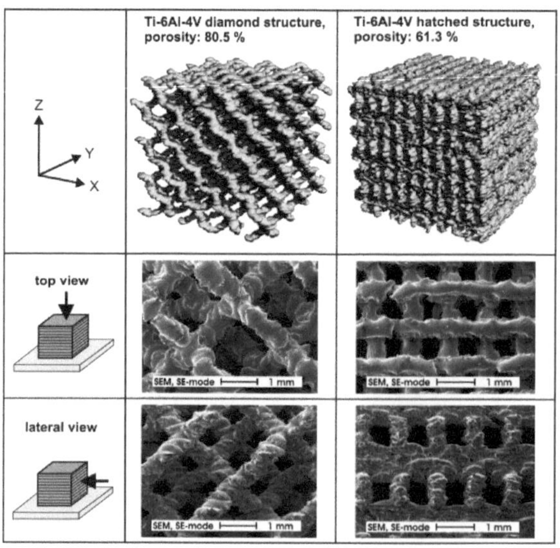

Figure 3. Porous diamond and hatched structure of Ti-6Al-4V [46].

3. Designing the Criteria and Lubrication Mechanism for Porous Self-Lubricating Composites

Reducing friction and wear are two very important objectives when designing porous self-lubricating composites. Friction is defined as the resistance to relative motion between two solid surfaces. The coefficient of friction (COF) is the ratio of the force of friction to the normal force applied, and it is a dimensionless number. Wear is defined as the progressive loss of material during sliding between two bodies. The lubrication mechanism of the porous composites can vary depending on various factors such as the type of material used as the matrix, the fabrication process, the pore-forming agents, and the lubricants. Firstly, the design criteria and lubrication mechanism are discussed in the following section.

The first step in the development of a porous lubricating material is the impregnation and storage of lubricants. The impregnation of the liquid lubricants depends on both the substrate and liquid lubricant properties. The critical properties of the substrate are the pore size and surface energy, whereas the critical parameters for the liquid lubricants are the surface tension (γ) and viscosity (η). On a smooth substrate, the spreading of liquid as measured in terms of the contact angle can be calculated by Young's equation [55] and schematically shown in Figure 4.

$$\cos\theta = \frac{\gamma_{sv} - \gamma_{sl}}{\gamma_{lv}} \tag{1}$$

where θ is the contact angle between substrate and liquid and γ_{sv}, γ_{sl}, γ_{lv} are the surface energy of the solid, interfacial surface tension, and surface tension of the liquid, respectively. For the liquid impregnation rate, Waldner and Hirn [56] reported that the Lucas–Washburn (LW) equation [57] can be used to represent the flow of liquid under the assumption of circular pores and capillary pressure as a driving force. The LW equation is given as

$$\frac{dh}{dt} = \frac{r\gamma_{lv}\cos\theta}{4\eta h} \tag{2}$$

where r and h are the pore radius and impregnation depth, respectively. By replacing the contact angle term from Equations (1) and (2), impregnation depth can be estimated using surface energy values. However, the measurement of surface energies (γ_{sv} and γ_{sl}) is difficult to perform. Therefore, surface energy modeling approaches can be used. The most widely known surface energy estimation approaches are Owens, Wendt, Rabel, and Kaelble (OWRK) [58], van Oss, Good, Chaudhury (vOGC) [59], and Wu [60]. The OWRK is based on the dispersive and polar components and vOGC and Wu are based on the acid–base components. These approaches are widely used for surface energy measurement using suitable liquids. Waldner and Hirn [56] developed this approach by combining the Young equation, LW equation, and these surface energy estimations to test the penetration depth for mixed liquids of water, glycerol, hexanediol, and naphthol blue black dye on a porous media, which yielded a higher accuracy ($R^2 > 0.75$) concerning experimental results. This combined approach of liquid flow and surface energy suggests that the most important parameters to design a porous self-lubricating system are the viscosity for the liquid lubricant, whereas it should be surface energy and pore size for the substrate material. The most technique for liquid impregnation inside the porous substrate is vacuum impregnation at higher temperature. The application of vacuum helps in speeding the impregnation process by taking out the entrapped air inside the voids, whereas higher temperature reduces the viscosity of the liquid lubricants to flow inside the voids.

Once the liquid lubricant is impregnated in the porous substrate, the lubrication mechanism during sliding against a counterbody is explained by the sweating mechanism. In this mechanism, as the material is subjected to friction during sliding, it releases the liquid stored in the pores under capillary force [61,62]. The coupled effect of load and sliding increases temperature, which assists the liquid lubricant to flow and form a layer on the surface of the material, preventing direct contact between the two materials. The increase in temperature at the asperity level (flash temperature) and bulk temperature at the surface can be estimated by the model developed by Ashby et al. [63]. Also, the liquid lubricant is restored in the pores when the material is not in operating conditions [64]. In this sweating mechanism, the release of liquid lubricants under capillary force is not the only factor; the deformation of the lubricant-carrying substrate is also a contributing factor. Zhan et al. [65] considered the porous lubricating system as a solid–liquid biphase model and analyzed the system numerically under different loading conditions to estimate the amount of stress on the solid (substrate) and fluid pressure (lubricant). The deformation of the substrate causes an increase in fluid pressure that is the driving force for the seepage of lubricants and spread over the sliding interface.

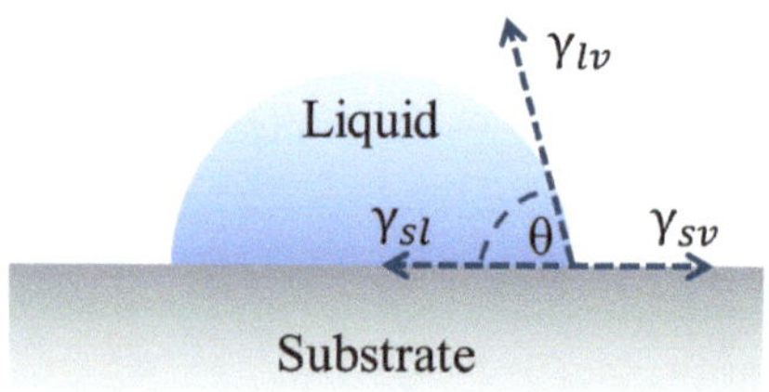

Figure 4. Schematic of a liquid droplet on the substrate with surface energy components.

Pores play an important role in lubrication mechanisms for porous materials. Pore generation relates to the fabrication of the porous substrate. Controlling pores is related to the relative density of the matrix materials. Relative density is defined as the ratio of the measured porous substrate density to its theoretical density. In general, substrates with higher relative density (>90%) will lead to closed pores that are not connected, which limits the lubrication carrying capacity of the materials, and starved lubrication yields poor lubrication performance. On the other hand, lower density to achieve connecting open pores will have poor mechanical strength due to weak particle-to-particle strength. During sliding, the weak joints can break and cause higher wear. Additionally, the increase in wear may lead to the closing of pores or a reduction in pore size that limits lubrication and increases friction, as shown in Figure 5. Therefore, optimized pore conditions or density are required to achieve consistent lubrication performance in the impregnated porous substrate materials. Close pores on the substrate can limit the lubrication carrying capacity of the materials, whereas higher pores reduce the overall strength of materials. Therefore, the pore size is a critical design parameter for self-lubricating porous material.

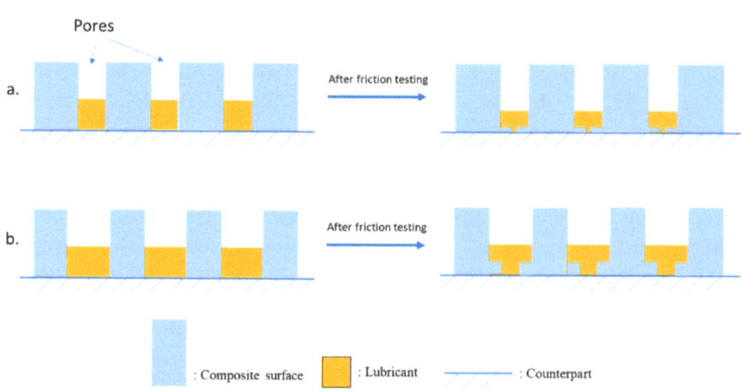

Figure 5. (**a**) Higher relative density substrates have closed pores that hold less lubrication, providing poor lubrication performance. (**b**) Lower relative density substrates are more prone to wear, which can break weak joints and close pores, leading to less lubrication.

3.1. Polymer Matrix Porous Composites

In polymers, there are three principal classes—thermoplastics, thermosets, and elastomers. Among these three, thermoplastics and thermosets are often known as polymers and are most suitable for porous matrix material due to their higher stiffness [66]. The Young's modulus of thermosets and thermoplastics is typically one to two orders of magnitude higher than those of elastomers, which makes them ideal candidates because contact pressure during sliding can be in the order of 1 GPa [67]. Meanwhile, elastomers are widely suitable, wherein structure rigidity is not required due to their extremely high flexibility.

In a study conducted by Wang et al. [61], porous PEEK composites were filled with ionic liquid (1-butyl-3-methylimidazolium hexafluorophosphate), and tribological tests were performed. The study found that PEEK composites with nano/micro porous structures (23.6% porosity) impregnated with ionic liquid showed excellent tribological properties compared with pure PEEK composites. The impregnation was carried out at 80 °C for 2 h under vacuum. The tribological test was conducted using a ring-on-ring setup at 250 N and 0.69 m/s against a 1045 steel ring. The ionic liquid-impregnated PEEK yielded COF = 0.05, which is 65% less than dry PEEK, and a wear rate of 2.0×10^{-14} m^3/Nm, which is 10^3 times less than that of dry PEEK. A similar study by Liu et al. [62] showed that the micro/nano porous carbon fiber/polytetrafluoroethylene/PEEK composites impregnated with 1-hexyl-3-methylimidazolium tetrafluoroborate ionic liquid yielded the minimum COF of 0.0197 and wear rate of 4.14×10^{-15} m^3/Nm. For comparison, the COF of PEEK against 100Cr6 Stainless steel under dry sliding conditions is ~0.6, and the wear rate is ~3.0×10^{-6} m^3/Nm, which is much higher than that of ionic liquid-impregnated PEEK. The addition of solid lubricants such as nano-to-micron WS_2 and MoS_2 to PEEK can help in reducing the COF to 0.4 and wear rate by ~50%, which are still higher [26]. These enormous differences in the COF and wear rate of dry PEEK versus liquid-impregnated PEEK suggest that impregnated porous materials are the solution to increase the life of the machine components.

In the porous self-lubricating composites, lubrication regimes can also be altered by using different porosities. Wang et al. [23] synthesized porous polyimide (PI) composites with different porosities ranging from 16 to 20%, which assisted in retaining different oil (poly-α-olefin-PAO4) content. The tribological tests were conducted using a pin-on-disk setup at room temperature under loads of 100, 200, and 400 N with increasing sliding velocity from 0.025 to 1.875 m/s to understand lubrication regimes. The results were also compared with polyimide without any pores (PI0) being tested in fully flooded conditions (oil bath). For PI0, the friction coefficient increased with an increase in load, whereas the friction coefficient decreased for PI13 (polyimide with 13% oil content and 17.28% porosity) with the increase in load. The difference is more significant between mixed and boundary lubrication regimes. The higher load flushes away the oil and increases solid-to-solid contacts for PI0. In the case of PI13, the higher load caused the release of oil from the pores to the interacting surface and decreased friction. It is also important to notice that the electrohydrodynamic lubrication region occurred earlier under a higher load for PI13 due to the greater release of oil from pores. The other tribological studies on polyimide porous material are listed in Table 1.

For porous elastomers, limited tribological studies have been carried out. In a study performed by Yamaguchi et al. [68], the authors investigated the dry sliding friction behavior of non-porous and porous ethylene-vinyl acetate (EVA) blocks with various porosity (0 to 90%) values against a stainless steel plate. Figure 6 depicts the results that showed that the highly porous EVA block provided larger contact areas due to the elastic collapse at the interior part of the block, which resulted in a high COF. In contrast, low porosity reduced the COF through a reduction in the contact area resulting from an increase in surface pores. Additionally, a decrease in velocity increased the COF, which was due to increase in the length of the apparent contact area at a lower sliding velocity. These results suggest that soft, porous polymers can be utilized as light high-friction materials, such as lightweight shoe sole materials with high cushioning and high grip, under dry conditions.

Table 1. Tribological properties of porous materials impregnated with lubricants.

Matrix	Lubricant and Impregnation Condition	Testing Condition	% COF and Wear	Ref.
Polyimide—20% porosity	Silicone oil (32 h at room temperature in vacuum)	Reciprocating sliding test against GCr15 steel ball at 20 N	COF = 0.03 (~10 times lesser than that of dry non-porous polyimide)	[11]
Polyphenylene sulfide (PPS) + PTFE with TiO_2 whisker	Lithium-based grease (2 h at 120 °C in vacuum)		Wear resistance improved by 6.45×10^3 times compared with pure PPS	[21]
PPS with 1% zeolite (20% porosity)	Lithium grease (2 h at 120 °C in vacuum)	Ring-on-ring test at 150 N with 1.4 m/s	COF = 0.024 and 1.79×10^{-16} m^3/Nm (COF and wear were 90% and 4.67×10^4 lesser compared with dry PPS, respectively)	[15]
Polyimide (PPI) with 33.5% porosity	Diesel engine oil (12 h at room temperature in vacuum)	Ring-on-disk setup at 100 N and 0.22 m/s against stainless steel ring	COF = 0.1 (70% lesser than dry porous PI)	[18]
PEEK (16.8% porosity)	Ionic liquid: 1-butyl-3-methylimidazolium hexafluorophosphate (2 h at 80 °C in vacuum)	Ring-on-ring setup at 250 N and 0.69 m/s against 1045 steel ring	COF = 0.05 and wear rate 2.0×10^{-14} m^3/Nm (COF and wear rate were 65% and 10^3 and were lower)	[61]
PPI (17.48%)	poly-α-olefin oil (20 h at 100 °C in vacuum)	Ball-on-disk (GCr15 steel ball) at 10 N and 0.2 m/s	PPI sample with the smallest pore diameter (1.48 μm) yielded the lowest COF ~0.035	[25]

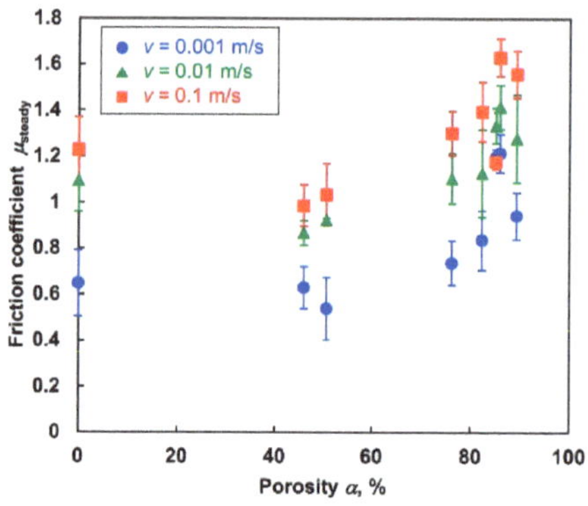

Figure 6. Effect of porosity on the mean value of the COF for different sliding velocities [68].

As seen from many studies performed, porosity is inversely related to the tribological properties of a composite [32,69]. Increasing porosity decreases the hardness of the composites, which leads to a decrease in the friction and wear resistance of the materials. Hence, the material needs to have an optimum amount of porosity that does not significantly reduce the mechanical properties of the material.

Lubricants are added to porous composites to improve their friction and wear resistance. The type of lubricants incorporated in the pores can vary from solid to liquid lubricants. The latter is commonly used in porous materials as lubricants can be effectively stored in the pores. Solid lubricants are used in conditions where liquid lubricants may be lost during working conditions, such as gear and chain lubrication and reciprocating motion. In these cases, solid lubricants are helpful as they can provide continuous lubrication and there is no loss of lubricants, thereby preventing fretting corrosion [70].

3.2. Porous Metal and Ceramic Matrix Self-Lubricating Composites

Only a few studies have been carried out on porous metal and ceramic matrices infiltrated with lubricating agents. Among porous metal matrix materials, the most famous and widely utilized is porous bearing material, which has been in use since 1925 [71]. A review article by Kumar [72] published in 1980 discusses the various porous metal bearings and their applications. The porous aluminum bearing was high in demand due to its lower cost and high thermal conductivity. However, poor galling behavior limits aluminum bearings. Later on, sintered bronze and iron bearings (porous) were used, which showed a higher eccentricity ratio than solid bearings (non-porous). The eccentricity ratio of a bearing is defined as the ratio of eccentricity and radial clearance. It is also important to notice that the oil for such bearings under a high load-carrying capacity should have a higher oxidation resistance. The commonly used oil for porous materials was silicon oil. Moreover, oxidation inhibitors can also be added to enhance the oil life in the porous material. The amount of porosity is the key factor in metals and ceramic substrates as well. Guo et al. [73] utilized the powder metallurgy method to fabricate Fe-Cu-Ni-Sn graphite porous substrate sintered at 1080 °C with a 40% porosity and 182 μm surface pore size. The authors provided a secondary treatment of rolling at 900 °C followed by another sintering at 1080 °C to decrease the porosity to ~30% and pore size to 158 μm. It was observed that the secondary treatment was beneficial in reducing the COF from ~0.95 (without secondary treatment) to 0.55 during the pin-on-disk test. The wear was measured in terms of weight loss, which also reduced from 6.5 mg to 2.75 mg by secondary treatment. The improvement in the COF and wear performance is mainly due to the reduction in porosity and pore size by the secondary treatment. Higher porosity and larger pore size can decrease the substrate strength and yield higher wear. The resulting wear debris can block the pores and affect the continuous supply of lubricants. Moreover, the addition of graphite was also helpful in reducing the COF and wear with lubricating oil, which has also been shown by other porous lubricating metallic systems such as Cu-Sn [74], Ni [17], etc. The stored lubricating oil and graphite not only reduce the friction but also protect the oxidation of the substrate and form a carbon-based tribofilm. A typical microstructural of Cu-Sn-Zn-graphite composite for bearing material is shown in Figure 7, which shows increase in pore size from 25 to 37% due to the addition of graphite ranging from 0.5 to 2.0 wt.%.

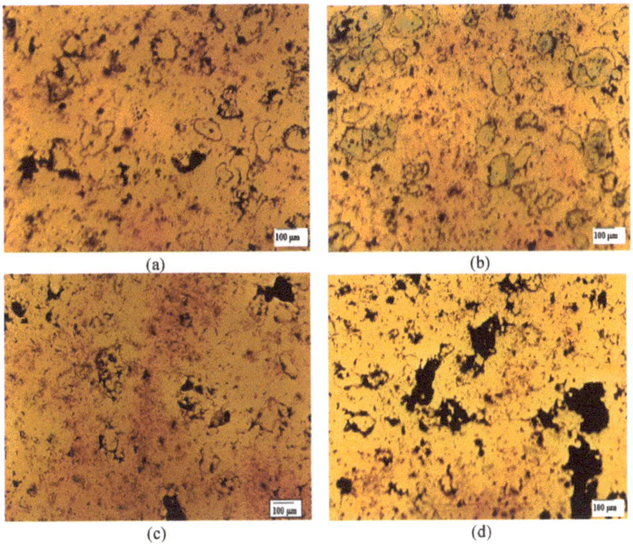

Figure 7. Microstructure of Cu-Sn-Zn composite with graphite at (**a**) 0.5, (**b**) 1.0, (**c**) 1.5 and (**d**) 2.0 wt.% [75].

While copper-based materials have been studied in the majority of papers as the self-lubricating porous substrate, austenitic stainless steel has also demonstrated similar benefits. Martin et al. [76] prepared 316 SS porous substrates by the powder metallurgy of different % pores by selecting the different initial powder sizes. The particle size < 150 μm yielded a 7.6% porosity, whereas the larger particle size (500–750 μm) provided a 28.8% porosity. The prepared samples were vacuum-impregnated at room temperature for 12 h with 0.5% aqueous solution of carboxymethyl cellulose sodium salt medium viscosity as a lubricant for biomedical applications, followed by tribological testing. The scar depth of the samples with 28.8% porosity remained consistent at ~0.05 mm during 200 m of sliding distance, whereas the scar depth continuously increased for the sample with a 7.6% porosity, with a final value of 0.2 mm at the end of the test, which was four times higher. The COF values also showed a similar trend with respect to porosity. It was found that the pores on the samples with lower porosity were closed after the initial sliding distance (<50 m) that restricted the lubricant supply.

Among ceramics, the widely used ceramic material is Al_2O_3, which has also been utilized for self-lubricating purposes. Rowthu and Hoffman [77] studied perfluoropolyether-impregnated alumina matrix materials where porous alumina matrix was prepared by slip casting followed by pre-sintering at 600 °C to remove the binding materials. The pre-sintered alumina matrix showed a relative density of ~59%, which was further sintered at different temperatures from 1150 to 1500 °C to achieve different levels of density, or in other words, different porosity. The porous alumina matrix substrates were impregnated with perfluoropolyether oil by submerging at 150 °C for 2 h. The authors observed that the composites with 90% relative density yielded a lower COF of ~0.13 at a 220 N load. The composites with lesser density had slightly higher COF values because of the weak particle-to-particle strength that caused higher wear and subsequently higher COF values. Meanwhile, the composites denser than 90% had closed pores that yielded higher COF values due to the absence of a continuous supply of lubricants. Based on Al_2O_3 ceramic, Kasar et al. [78] developed an $Al_2O_3^-$ 20 wt.% B_2O_3 ceramic composite that generates inherent porosity due to the formation of the aluminum borate phase during pressureless sintering. The composite was impregnated with phosphonium-based ionic liquids under vacuum at 60 °C for 30 min. The authors suggested that the high-temperature impregnation was needed due to the high room temperature viscosity of the phosphonium ionic liquids, which also suggests the importance of viscosity during impregnation, as discussed in Section 3. After impregnation, pin-on-disc sliding tests were performed at four different temperatures from 5 to 100 °C and also at three different sliding velocities. At lower temperatures (5 °C) and lower sliding velocity (2 mm/s), slightly higher friction was observed compared with room temperature. This study suggests the effect of the increase in flash and bulk temperature at sliding surfaces that assist in ionic liquid supply at sliding interfaces. In addition, the suitability of ionic liquid for high-temperature applications was also discussed due to thermal stability.

Pores in materials typically deteriorate their mechanical properties, which is undesirable and can cause a decrease in friction and wear resistance. To improve these mechanical properties, additives are often used during the fabrication of the materials [79]. Like the pore-forming agents, the concentration of strengthening agents must also be optimized. Adding a high concentration of strengthening agents can have an adverse effect on the mechanical and tribological properties of the materials.

Unlike liquid lubricants, a limited study has demonstrated the use of solid lubricants in porous materials to provide continuous lubrication. In a study performed by Salam et al. [70], inorganic hollow fullerene-like (IF) MoS_2 was used as a solid lubricant in porous alumina substrates, and the tribological properties were studied. Three different specimens were tested: (1) porous alumina, (2) alumina stabilized with acrylic resins, and (3) hollow IF-MoS_2 impregnated in alumina. Ball-on-disk reciprocating tests were performed using a zirconia (ZrO_2) ball at 5 and 10 N loads for 18,000 cycles, and friction and wear results were compared between the samples. Figure 8 shows the COF results for the tribological tests. It can be concluded that the hollow IF-MoS_2 specimen showed the least COF compared with the other two samples. The stabilization of alumina with acrylic resin helped to bring down the COF but could not maintain the low COF for an extended period. However, adding hollow IF-MoS_2 significantly reduced the COF of the specimen and lasted for 36,000 cycles at 10 N load.

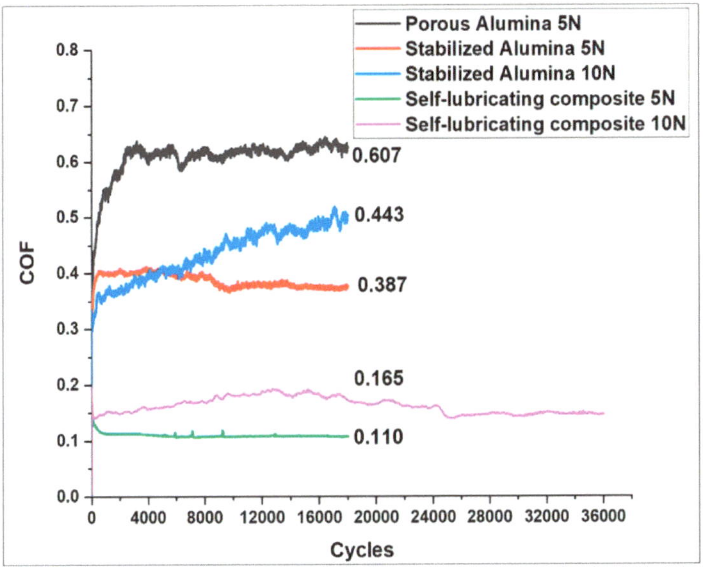

Figure 8. COF values for different alumina specimens under 5 N and 10 N loads [70].

There are multiple lubrication mechanisms for the hollow IF-MoS_2 particles. Their spherical shape acts as rollers that produce a rolling effect and reduce COF [70,80]. They also sit in between surface peaks and valleys, which help dissipate the energy produced by friction and wear during sliding. The hollow IF-MoS_2 is not destroyed immediately; rather, it slowly deteriorates, which is helpful in the long run.

A study conducted by Branco et al. [81] evaluated and compared the tribological properties of human teeth against zirconia dental prosthetics produced by robocasting (RC), a 3D printing technique with prosthetics produced by traditional unidirectional compression (UC). The RC samples were 3D-printed using a ceramic paste of 3 mol% Yttria partially stabilized zirconia at room temperature using a 0.4 mm diameter nozzle at 60 mm/s with a layer height of 0.2 mm. Meanwhile, the UC samples were produced under a load of 70 MPa for 10 s. Both the samples were sintered at 1500 °C for 2 h. The results showed that the RC samples had a higher surface roughness, lower hardness, and fracture toughness compared with the UC samples. These properties were attributed to the higher porosity of 3D-printed samples. Dental cusps were tested against both pieces to simulate chewing conditions in the presence of artificial saliva, which showed a similar specific wear rate as that shown in Figure 9. Both the dental zirconia pieces underwent mild abrasive wear, fatigue, and delamination. 3D printing techniques can fabricate products with complex geometry with identical tribological properties as the conventional method. Additionally, the higher roughness of RC samples due to porosity did not negatively affect the enamel wear. The pores could be advantageous for storing dental wear debris and avoiding sharp edges.

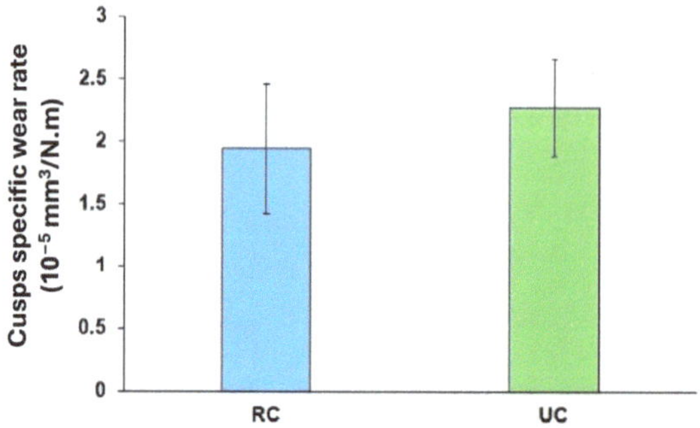

Figure 9. Specific wear rates of the dental cusps tested against 3D-printed (RC) and unidirectionally compressed (UC) plates [81].

4. Conclusions

Porous materials can be used for various applications and have been gaining more attention lately. Porous materials are used to solve friction, wear, and lubrication problems. The pores in these materials are utilized to store liquid lubricants, which come out during sliding and return to the pores after operation. Porous materials can be classified into many different categories depending on the pore size, type of materials used, and location of pores. In this review, porous materials are categorized by the type of materials used and have three categories: porous polymers, metals, and ceramics. The three different types of porous materials have different properties, and depending on the application, a certain type of material can be selected. Porous materials are fabricated by many different methods, such as sintering, molding, casting, selective laser melting, and 3D printing. Depending on the application, different materials and processing techniques are used to fabricate porous materials. Porous materials usually have lower mechanical strength. Hence, the fabrication process also involves the incorporation of strengthening additives. All of these additives have an impact on the tribological properties of the materials. The design criteria are also discussed, which suggests that the optimization of pore size is needed based on the fundamental properties of lubricants, surfaces, and contact pressure. The observation of the elastohydrodynamic lubrication regime confirms that the porous self-lubricating composites are viable routes for achieving lower friction for the difficult-to-reach components. This review article also highlights that the type of liquid lubricants is independent of the matrix materials. In different porous polymer substrates, all of the different types of liquid lubricants such as ionic liquid, lithium grease, diesel oil, etc., have been successfully used to achieve superior tribological properties. The studies discussed in this article suggest the selection of the liquid lubricant is mainly based on the requirements, e.g., viscous ionic liquid with high temperature stability can be utilized in porous substrate where viscous lubricant is required at a high-temperature application. In addition to the technical advantages of porous self-lubricating components, the ability of the porous material to hold back the lubricants while not in motion provides sustainability and avoids the excess use of lubricants.

Author Contributions: Conceptualization A.K.K. and B.D.; writing—original draft preparation, A.K.K. and B.D.; writing—review and editing, S.A.J. and P.L.M.; supervision, P.L.M.; project administration, P.L.M.; funding acquisition, P.L.M. All authors have read and agreed to the published version of the manuscript.

Funding: This research was funded by the National Aeronautics and Space Administration (NASA), USA, under cooperative agreement number 80NSSC20M0221.

Institutional Review Board Statement: Not applicable.

Informed Consent Statement: Not applicable.

Data Availability Statement: No new data were created.

Conflicts of Interest: The authors declare no conflicts of interest.

References

1. Davis, M.E. Ordered porous materials for emerging applications. *Nature* **2002**, *417*, 813–821. [CrossRef] [PubMed]
2. Ahuja, G.; Pathak, K. Porous carriers for controlled/modulated drug delivery. *Indian J. Pharm. Sci.* **2009**, *71*, 599. [CrossRef] [PubMed]
3. Wu, J.; Xu, F.; Li, S.; Ma, P.; Zhang, X.; Liu, Q.; Fu, R.; Wu, D. Porous polymers as multifunctional material platforms toward task-specific applications. *Adv. Mater.* **2019**, *31*, 1802922. [CrossRef] [PubMed]
4. Nakajima, H. Fabrication, properties and application of porous metals with directional pores. *Prog. Mater. Sci.* **2007**, *52*, 1091–1173. [CrossRef]
5. Chen, Y.; Wang, N.; Ola, O.; Xia, Y.; Zhu, Y. Porous ceramics: Light in weight but heavy in energy and environment technologies. *Mater. Sci. Eng. R Rep.* **2021**, *143*, 100589. [CrossRef]
6. Dabrowski, B.; Kaminski, J.; Swieszkowski, W.; Kurzydlowski, K.J. Porous titanium scaffolds for biomedical applications: Corrosion resistance and structure investigation. In *Proceedings of Materials Science Forum*; Trans Tech Publications Ltd.: Bäch, Switzerland, 2011; pp. 41–46.
7. Jodati, H.; Yılmaz, B.; Evis, Z. A review of bioceramic porous scaffolds for hard tissue applications: Effects of structural features. *Ceram. Int.* **2020**, *46*, 15725–15739. [CrossRef]
8. Rashidi, S.; Esfahani, J.A.; Rashidi, A. A review on the applications of porous materials in solar energy systems. *Renew. Sustain. Energy Rev.* **2017**, *73*, 1198–1210. [CrossRef]
9. Indian Institute of Technology Kharagpur. Vacuum Sintering of Self-Lubricating Bronze Bearings. Ph.D. Thesis, Indian Institute of Technology Kharagpur, Kharagpur, India, 1981.
10. Pound, M.; Rowley, A.; Elliott, J. Aspects of the Volume Production of Green Porous Bearing Compacts. *Powder Metall.* **1960**, *3*, 129–149. [CrossRef]
11. Xu, X.; Shu, X.; Pei, Q.; Qin, H.; Guo, R.; Wang, X.; Wang, Q. Effects of porosity on the tribological and mechanical properties of oil-impregnated polyimide. *Tribol. Int.* **2022**, *170*, 107502. [CrossRef]
12. Zhang, D.; Wang, T.; Wang, Q.; Wang, C. Selectively enhanced oil retention of porous polyimide bearing materials by direct chemical modification. *J. Appl. Polym. Sci.* **2017**, *134*, 45106. [CrossRef]
13. Li, X.; Olofsson, U. A study on friction and wear reduction due to porosity in powder metallurgic gear materials. *Tribol. Int.* **2017**, *110*, 86–95. [CrossRef]
14. Zhang, G.; Tong, B.; Zhang, X.; Wang, T.; Yin, Y. Thermo-hydrodynamic modeling and analysis for circular bilayer self-lubricating bearing. *J. Braz. Soc. Mech. Sci. Eng.* **2020**, *42*, 18. [CrossRef]
15. Wang, H.; Wang, G.; Zhang, S.; Yang, S.; Zhu, Y. Tribological performances on porous polyphenylene sulfide self-lubricating composites with super wear resistance. *J. Thermoplast. Compos. Mater.* **2014**, *27*, 82–92. [CrossRef]
16. Mu, B.; Yu, Z.; Cui, J.; Yan, W.; He, H.; Wang, X.; Guo, J.; Tian, L.; Yang, B.; Cui, J. Tribological Properties of Oil-Containing Polyimide Aerogels as a New Type of Porous Self-Lubricating Material. *Ind. Eng. Chem. Res.* **2022**, *61*, 14222–14231. [CrossRef]
17. Lei, Y.; Du, J.; Pang, X.; Wang, H.; Yang, H.; Jiang, J. Tribological properties and lubrication mechanism of in situ graphene-nickel matrix composite impregnated with lubricating oil. *Mater. Res. Express* **2018**, *5*, 056512. [CrossRef]
18. Wang, J.; Zhao, H.; Huang, W.; Wang, X. Investigation of porous polyimide lubricant retainers to improve the performance of rolling bearings under conditions of starved lubrication. *Wear* **2017**, *380*, 52–58. [CrossRef]
19. Shang, Y.; Zhao, Y.; Liu, Y.; Zhu, Y.; Jiang, Z.; Zhang, H. The effect of micron-graphite particle size on the mechanical and tribological properties of PEEK composites. *High Perform. Polym.* **2018**, *30*, 153–160. [CrossRef]
20. Wang, Q.; Wang, Y.; Wang, H.; Fan, N.; Wang, M.; Liu, H.; Yan, F. Comparative study of the effects of nano-sized and micro-sized CF and PTFE on the thermal and tribological properties of PEEK composites. *Polym. Adv. Technol.* **2018**, *29*, 896–905. [CrossRef]
21. Zhu, Y.; Wang, G.; Wang, H.; Zhang, S.; Yang, S. Tribological properties of porous PPS/PTFE composite filled with mesopore titanium oxide whisker. *J. Appl. Polym. Sci.* **2013**, *129*, 2321–2327. [CrossRef]
22. Bora, M.Ö.; Coban, O.; Avcu, E.; Fidan, S.; Sınmazçelik, T. The effect of TiO_2 filler content on the mechanical, thermal, and tribological properties of TiO2/PPS composites. *Polym. Compos.* **2013**, *34*, 1591–1599. [CrossRef]
23. Wang, C.; Zhang, D.; Wang, Q.; Ruan, H.; Wang, T. Effect of Porosity on the Friction Properties of Porous Polyimide Impregnated with Poly-α-Olefin in Different Lubrication Regimes. *Tribol. Lett.* **2020**, *68*, 102. [CrossRef]
24. Ruan, H.; Zhang, Y.; Li, S.; Yang, L.; Wang, C.; Wang, T.; Wang, Q. Effect of temperature on the friction and wear performance of porous oil-containing polyimide. *Tribol. Int.* **2021**, *157*, 106891. [CrossRef]
25. Ye, J.; Li, J.; Qing, T.; Huang, H.; Zhou, N. Effects of surface pore size on the tribological properties of oil-impregnated porous polyimide material. *Wear* **2021**, *484*, 204042. [CrossRef]
26. Zalaznik, M.; Kalin, M.; Novak, S.; Jakša, G. Effect of the type, size and concentration of solid lubricants on the tribological properties of the polymer PEEK. *Wear* **2016**, *364*, 31–39. [CrossRef]
27. Huang, X.; Wu, J.; Lu, X.; Feng, X.; Shi, Y. Tribological properties of porous PEEK composites containing ionic liquid under dry friction condition. *Lubricants* **2017**, *5*, 19. [CrossRef]

28. Senatov, F.S.; Niaza, K.V.; Zadorozhnyy, M.Y.; Maksimkin, A.V.; Kaloshkin, S.D.; Estrin, Y.Z. Mechanical properties and shape memory effect of 3D-printed PLA-based porous scaffolds. *J. Mech. Behav. Biomed. Mater.* **2016**, *57*, 139–148. [CrossRef] [PubMed]
29. Chen, Q.; Cao, P.F.; Advincula, R.C. Mechanically robust, ultraelastic hierarchical foam with tunable properties via 3D printing. *Adv. Funct. Mater.* **2018**, *28*, 1800631. [CrossRef]
30. Wei, P.; Leng, H.; Chen, Q.; Advincula, R.C.; Pentzer, E.B. Reprocessable 3D-printed conductive elastomeric composite foams for strain and gas sensing. *ACS Appl. Polym. Mater.* **2019**, *1*, 885–892. [CrossRef]
31. Mu, X.; Bertron, T.; Dunn, C.; Qiao, H.; Wu, J.; Zhao, Z.; Saldana, C.; Qi, H.J. Porous polymeric materials by 3D printing of photocurable resin. *Mater. Horiz.* **2017**, *4*, 442–449. [CrossRef]
32. Sinha, A.; Farhat, Z. Effect of surface porosity on tribological properties of sintered pure Al and Al 6061. *Mater. Sci. Appl.* **2015**, *6*, 549. [CrossRef]
33. Kovalčíková, A.; Balko, J.; Balázsi, C.; Hvizdoš, P.; Dusza, J. Influence of hBN content on mechanical and tribological properties of Si_3N_4/BN ceramic composites. *J. Eur. Ceram. Soc.* **2014**, *34*, 3319–3328. [CrossRef]
34. Hvizdoš, P.; Dusza, J.; Balázsi, C. Tribological properties of Si_3N_4–graphene nanocomposites. *J. Eur. Ceram. Soc.* **2013**, *33*, 2359–2364. [CrossRef]
35. Xiao, G.; Zhao, B.; Ding, W. Mechanical and tribological properties of porous metallic CBN composites reinforced by graphene nanoparticles. *Int. J. Adv. Manuf. Technol.* **2021**, *114*, 397–405. [CrossRef]
36. Manoj, A.; Kasar, A.K.; Menezes, P.L. Tribocorrosion of Porous Titanium Used in Biomedical Applications. *J. Bio-Tribo-Corros.* **2019**, *5*, 3. [CrossRef]
37. Zhao, X.; Sun, H.; Lan, L.; Huang, J.; Zhang, H.; Wang, Y. Pore structures of high-porosity NiTi alloys made from elemental powders with NaCl temporary space-holders. *Mater. Lett.* **2009**, *63*, 2402–2404. [CrossRef]
38. Wen, C.; Mabuchi, M.; Yamada, Y.; Shimojima, K.; Chino, Y.; Asahina, T. Processing of biocompatible porous Ti and Mg. *Scr. Mater.* **2001**, *45*, 1147–1153. [CrossRef]
39. Andersen, O.; Waag, U.; Schneider, L.; Stephani, G.; Kieback, B. Novel metallic hollow sphere structures. *Adv. Eng. Mater.* **2000**, *2*, 192–195. [CrossRef]
40. Aşık, E.E.; Bor, Ş. Fatigue behavior of Ti–6Al–4V foams processed by magnesium space holder technique. *Mater. Sci. Eng. A* **2015**, *621*, 157–165. [CrossRef]
41. Zhao, J.; Wang, G.; Zhang, A.; Zhao, G.; Park, C.B. Nanocellular TPU composite foams achieved by stretch-assisted microcellular foaming with low-pressure gaseous CO_2 as blowing agent. *J. CO2 Util.* **2021**, *53*, 101708. [CrossRef]
42. Grassberger, L.; Koch, K.; Oberhoffer, R.; Müller, A.; Klemmer, H.F.; Strey, R. Blowing agent free generation of nanoporous poly (methylmethacrylate) materials. *Colloid Polym. Sci.* **2017**, *295*, 379–389. [CrossRef]
43. Jee, C.; Guo, Z.; Evans, J.; Özgüven, N. Preparation of high porosity metal foams. *Metall. Mater. Trans. B* **2000**, *31*, 1345–1352. [CrossRef]
44. Li, M.; Luo, X.; Zhang, G.; Xie, Z.; Zhao, L.; Jin, H. Effect of blowing-agent addition on the structure and properties of magnesia porous material. *Refract. Ind. Ceram.* **2017**, *58*, 60–64. [CrossRef]
45. Yang, C.; Jiang, P.; Qin, H.; Wang, X.; Wang, Q. 3D printing of porous polyimide for high-performance oil impregnated self-lubricating. *Tribol. Int.* **2021**, *160*, 107009. [CrossRef]
46. Heinl, P.; Müller, L.; Körner, C.; Singer, R.F.; Müller, F.A. Cellular Ti–6Al–4V structures with interconnected macro porosity for bone implants fabricated by selective electron beam melting. *Acta Biomater.* **2008**, *4*, 1536–1544. [CrossRef]
47. Lu, P.; Wu, M.; Liu, X.; Duan, W.; Han, J. Study on corrosion resistance and bio-tribological behavior of porous structure based on the SLM manufactured medical Ti6Al4V. *Met. Mater. Int.* **2020**, *26*, 1182–1191. [CrossRef]
48. Zhang, B.; Coddet, C. Selective laser melting of iron powder: Observation of melting mechanism and densification behavior via point-track-surface-part research. *J. Manuf. Sci. Eng.* **2016**, *138*, 051001. [CrossRef]
49. Jose, S.A.; Kasar, A.K.; Menezes, P.L. Cold spray deposition of cermets: Insights into bonding mechanism and critical parameters. *Int. J. Adv. Manuf. Technol.* **2024**, *133*, 1–23. [CrossRef]
50. Sun, J.; Han, Y.; Cui, K. Innovative fabrication of porous titanium coating on titanium by cold spraying and vacuum sintering. *Mater. Lett.* **2008**, *62*, 3623–3625. [CrossRef]
51. Wathanyu, K.; Tuchinda, K.; Daopiset, S.; Sirivisoot, S.; Kondas, J.; Bauer, C. Study of the properties of titanium porous coating with different porosity gradients on 316L stainless steel by a cold spray process. *J. Therm. Spray Technol.* **2022**, *31*, 545–558. [CrossRef]
52. Kupp, D.; Claar, D.; Flemmig, K.; Waag, U.; Goehler, H. Processing of controlled porosity titanium-based materials. *Adv. Powder Metall. Partic Mater.* **2002**, *2*, 2–126.
53. Feinle, A.; Elsaesser, M.S.; Huesing, N. Sol–gel synthesis of monolithic materials with hierarchical porosity. *Chem. Soc. Rev.* **2016**, *45*, 3377–3399. [CrossRef] [PubMed]
54. Potoczek, M. Gelcasting of alumina foams using agarose solutions. *Ceram. Int.* **2008**, *34*, 661–667. [CrossRef]
55. Young, T. *III. An Essay on the Cohesion of Fluids*; Philosophical Transactions of the Royal Society of London: London, UK, 1805; pp. 65–87.
56. Waldner, C.; Hirn, U. Modeling liquid penetration into porous materials based on substrate and liquid surface energies. *J. Colloid Interface Sci.* **2023**, *640*, 445–455. [CrossRef] [PubMed]
57. Washburn, E.W. The dynamics of capillary flow. *Phys. Rev.* **1921**, *17*, 273. [CrossRef]

58. Owens, D.K.; Wendt, R. Estimation of the surface free energy of polymers. *J. Appl. Polym. Sci.* **1969**, *13*, 1741–1747. [CrossRef]
59. Van Oss, C.; Good, R.; Chaudhury, M. Additive and nonadditive surface tension components and the interpretation of contact angles. *Langmuir* **1988**, *4*, 884–891. [CrossRef]
60. Wu, S. Polar and nonpolar interactions in adhesion. *J. Adhes.* **1973**, *5*, 39–55. [CrossRef]
61. Wang, H.; Li, M.; Liu, D.; Zhao, Y.; Zhu, Y. Tribological properties tests and simulations of the nano-micro multilevel porous self-lubricating PEEK composites with ionic liquid lubrication. *J. Mater. Sci.* **2016**, *51*, 3917–3927. [CrossRef]
62. Zhu, Y.; Lin, S.; Wang, H.; Liu, D. Study on the tribological properties of porous sweating PEEK composites under ionic liquid lubricated condition. *J. Appl. Polym. Sci.* **2014**, *131*, 40989. [CrossRef]
63. Ashby, M.; Abulawi, J.; Kong, H. Temperature maps for frictional heating in dry sliding. *Tribol. Trans.* **1991**, *34*, 577–587. [CrossRef]
64. Yan, Z.; Jiang, D.; Fu, Y.; Qiao, D.; Gao, X.; Feng, D.; Sun, J.; Weng, L.; Wang, H. Vacuum tribological performance of WS_2–MoS_2 composite film against oil-impregnated porous polyimide: Influence of oil viscosity. *Tribol. Lett.* **2019**, *67*, 2. [CrossRef]
65. Zhang, G.; Cai, W.; Wei, X.; Yin, Y. Percolation and Supply Behavior of Lubricant on Porous Self-Lubricating Material. *Adv. Eng. Mater.* **2023**, *25*, 2201745. [CrossRef]
66. Ashby, M.; Cebon, D.; Silva, A. *Teaching Engineering Materials: The CES EduPack*; Engineering Department, Cambridge University: Cambridge, UK, 2007; pp. 1–13.
67. Myshkin, N.; Kovalev, A.; Spaltman, D.; Woydt, M. Contact mechanics and tribology of polymer composites. *J. Appl. Polym. Sci.* **2014**, *131*, 39870. [CrossRef]
68. Yamaguchi, T.; Sugawara, T.; Takahashi, M.; Shibata, K.; Moriyasu, K.; Nishiwaki, T.; Hokkirigawa, K. Dry sliding friction of ethylene vinyl acetate blocks: Effect of the porosity. *Tribol. Int.* **2017**, *116*, 264–271. [CrossRef]
69. Rathod, H.J.; Nagaraju, T.; Prashanth, K.; Ramamurty, U. Tribological properties of selective laser melted Al12Si alloy. *Tribol. Int.* **2019**, *137*, 94–101. [CrossRef]
70. Salam, A.; Xie, G.; Guo, D.; Xu, W. Fabrication and tribological behavior of self-lubricating composite impregnated with synthesized inorganic hollow fullerene-like MoS_2. *Compos. Part B Eng.* **2020**, *200*, 108284. [CrossRef]
71. Tsukerman, S.A. *Powder Metallurgy*; Elsevier: Amsterdam, The Netherlands, 2013.
72. Kumar, V. Porous metal bearings—A critical review. *Wear* **1980**, *63*, 271–287. [CrossRef]
73. Guo, J.; Du, H.; Zhang, G.; Cao, Y.; Shi, J.; Cao, W. Fabrication and tribological behavior of Fe-Cu-Ni-Sn-Graphite porous oil-bearing self-lubricating composite layer for maintenance-free sliding components. *Mater. Res. Express* **2020**, *8*, 015801. [CrossRef]
74. Fatih Tunay, R.; Kurbanoğlu, C. Investigation of the tribological properties of Cu-based porous bearings. *Ind. Lubr. Tribol.* **2012**, *64*, 111–119. [CrossRef]
75. Susilowati, S.E.; Fudholi, A.; Sumardiyanto, D. Mechanical and microstructural characteristics of Cu–Sn–Zn/Gr metal matrix composites processed by powder metallurgy for bearing materials. *Results Eng.* **2022**, *14*, 100377. [CrossRef]
76. Martin, F.; García, C.; Blanco, Y. Influence of residual porosity on the dry and lubricated sliding wear of a powder metallurgy austenitic stainless steel. *Wear* **2015**, *328*, 1–7. [CrossRef]
77. Rowthu, S.; Hoffmann, P. Perfluoropolyether-Impregnated Mesoporous Alumina Composites Overcome the Dewetting–Tribological Properties Trade-Off. *ACS Appl. Mater. Interfaces* **2018**, *10*, 10560–10570. [CrossRef] [PubMed]
78. Kasar, A.K.; Rahman, M.H.; D'Souza, B.; Menezes, P.L. Tribological performance of ionic liquid impregnated porous aluminum borate ceramic. *Tribol. Int.* **2023**, *180*, 108219. [CrossRef]
79. Zhang, Z.; Breidt, C.; Chang, L.; Friedrich, K. Wear of PEEK composites related to their mechanical performances. *Tribol. Int.* **2004**, *37*, 271–277. [CrossRef]
80. Tevet, O.; Von-Huth, P.; Popovitz-Biro, R.; Rosentsveig, R.; Wagner, H.D.; Tenne, R. Friction mechanism of individual multilayered nanoparticles. *Proc. Natl. Acad. Sci. USA* **2011**, *108*, 19901–19906. [CrossRef]
81. Branco, A.; Silva, R.; Jorge, H.; Santos, T.; Lorenz, K.; Polido, M.; Colaço, R.; Serro, A.; Figueiredo-Pina, C. Tribological performance of the pair human teeth vs 3D printed zirconia: An in vitro chewing simulation study. *J. Mech. Behav. Biomed. Mater.* **2020**, *110*, 103900. [CrossRef]

materials

MDPI

Article

Construction and Application of a Quantitative Perforation Erosion Model Based on Field Experiments

Bo Wang [1], Huan Li [1], Enyu Zhang [1], Jinglong Ma [1], Zichen Shang [1] and Xiongfei Liu [2,*]

[1] Petroleum Institute, China University of Petroleum-Beijing at Karamay, Karamay 834000, China; wangbo@cupk.edu.cn (B.W.); lihuancupk@163.com (H.L.); 17599901216@163.com (E.Z.); 2024216884@st.cupk.edu.cn (J.M.); zc13663280176@163.com (Z.S.)

[2] Research Institute of Unconventional Oil and Gas Science and Technology, China University of Petroleum-Beijing, Beijing 102249, China

* Correspondence: lxfei@cup.edu.cn

Abstract: Perforation erosion is one of the critical factors influencing the effectiveness of hydraulic fracturing and the productivity of oil and gas wells. This study developed a mathematical model for perforation erosion based on the field experimental data and theoretical analysis. This model comprehensively considers the effects of the rate of change in perforation diameter and the flow coefficient. Through field experiments, the values of the perforation diameter correlation coefficient (α) and the flow coefficient correlation coefficient (β) were determined. The wear behavior of perforations under high-pressure sand-carrying fluid conditions was thoroughly investigated, and the primary factors influencing perforation erosion were systematically analyzed. The results indicate that perforation erosion under high-pressure sand-carrying fluid conditions undergoes two distinct stages: the roundness erosion stage, characterized by a sharp pressure drop (greater than 30%) and the diameter erosion stage, marked by a gradual pressure decline (less than 5%), ultimately forming a trumpet-shaped perforation channel. The study further revealed that larger proppants cause significantly severe erosion than smaller proppants, resulting in 18.19% greater perforation diameter enlargement. In comparison tests, ceramic proppants produced 16.87% more diameter expansion than quartz sand under identical erosion conditions. Innovatively, this study proposes a "limited entry and temporary plugging" synergistic composite process. The timing of temporary plugging and the selection criteria for diverter size were clarified and optimized by determining the critical perforation friction for limited-entry failure based on inter-cluster stress differences. Field applications demonstrate that the optimized approach reduces erosion rates by 35–50%, improves fracture uniformity to over 80%, and increases single-well productivity by 18–25%. This research provides a quantitative basis and practical guidance for optimizing fracturing operation parameters, offering significant insights for enhancing the efficiency and productivity of hydraulic fracturing in oil and gas wells.

Keywords: unconventional oil and gas; sand fracturing; perforation hole erosion; fracture propagation

1. Introduction

The tight conglomerate reservoirs in the Mahu Sag of the Junggar Basin represent a critical replacement area for China's onshore deep oil and gas resources. However, their development faces multiple challenges, including significant burial depth (>4500 m), firm stress heterogeneity (variation coefficient >0.65), and complex natural fracture systems [1–3].

The multi-cluster fracturing technology in horizontal wells, achieved through optimized cluster spacing (5–10 m) and cluster density (6–12 clusters per stage), has demonstrated significant advantages in achieving volumetric stimulation of reservoirs [4,5]. Nevertheless, with the large-scale application of "three-high" fracturing techniques (high pumping pressure >90 MPa, high flow rate >12 m^3/min, and high sand ratio >30%), perforation holes undergo morphological distortion under the continuous erosion of high-pressure sand-carrying fluids. This leads to abnormal near-wellbore friction fluctuations, severely restricting the uniform propagation of fractures [6,7].

Current research on perforation erosion primarily follows three directions:

(1) Numerical Simulation: Liu Yanshu et al. [8] revealed the synergistic effect of flow rate and sand concentration on erosion rates using the discrete element method, but the characterization of multi-physical field coupling remains challenging. Huang et al. [9] developed a multi-field coupled fracturing model integrating stress–seepage–fracture–erosion mechanisms using the continuous-discontinuous element method (CDEM). Through benchmark case studies, they investigated the impact of perforation erosion on multi-fracture propagation. They analyzed the influence of flow rate and perforation parameters on perforation erosion and fracture extension. The results demonstrate that perforation erosion non-uniformly enlarges perforation diameter and increases the flow coefficient, leading to redistributed flow among multiple fractures and exacerbating uneven fracture propagation. Hu et al. [10] established a 3D CFD wellbore model for single- and dual-cluster fractures. By incorporating erosion effects, they studied proppant transport behavior in slickwater fracturing. The findings reveal that proppant distribution is significantly affected by perforation erosion location, erosion angle, erosion severity, and fluid injection velocity.

(2) Experimental Studies: The high-speed two-phase flow erosion system developed by Li Zhen's team [11] can simulate the bridging process of temporary diverters. However, the centimeter-scale perforations and static stress conditions result in scale effect errors exceeding 40%. Wu et al. [12] developed a surface-perforation erosion simulation system to quantitatively investigate the erosion extent of perforations during sand-carrying fracturing. They studied the erosive effects of sand-laden fluids on individual perforations under varying operational conditions, clarifying the changes in perforation geometry and post-erosion perforation friction in real-world fracturing scenarios. Wang Bo et al. [13] constructed a high-pressure perforation plugging simulation system and prepared five perforation models with distinct geometries. Their research focused on the influence of temporary plugging ball diameter, bridging particle concentration, filler particle concentration, pumping rate, and perforation shape on the effectiveness of perforation sealing.

(3) Field Monitoring Techniques: Zang Chuanzhen et al. [14] employed distributed optical fiber monitoring to observe that perforation erosion causes over 30% of perforation clusters to lose fluid intake capacity, yet quantitative description methods are lacking. The limitations of these approaches are evident: numerical models oversimplify boundary conditions, physical experiments struggle to replicate actual scale effects, and field monitoring lacks dynamic process analysis [15,16]. Wheaton et al. [17] monitored multi-cluster fracturing operations in the Eagle Ford Shale using DAS (distributed acoustic sensing) and DTS (distributed temperature sensing) technologies. The results revealed that heel-side perforations dominated fluid intake. Under the exact total perforation count, increasing the number of perforation clusters led to a gradual rise in the proportion of ineffective clusters. Additionally, due to stress shadow effects, minimal cluster spacing severely inhibited the initiation of specific perforation clusters. Cramer et al. [18] employed DAS, treatment pressure analysis, and perforation imaging to monitor fluid and proppant distribution during limited-entry fracturing. They qualitatively evaluated fluid intake per cluster by comparing perforation diameters before and after fracturing. The results indicated that

wellbore pressure dropped after proppant addition, with some clusters showing no fluid intake, while heel-side perforations exhibited more pronounced erosion. Roberts et al. [19] utilized downhole video imaging (e.g., Eagle Eye) to collect detailed perforation size data from over 6000 clusters and 600 stages. By analyzing erosion patterns, they identified the impacts of stage length, cluster count, cluster spacing, perforation density, charge type, and stage number on proppant distribution. The study confirmed that heel-side perforations experienced more severe erosion than toe-side perforations, demonstrating non-uniform proppant distribution along the stage, with heel clusters receiving a higher proppant volume.

This study innovatively constructs a tripartite research methodology integrating field experiments, theoretical modeling, and engineering control. A field-scale perforation erosion simulation device was developed, enabling dynamic erosion observation of full-scale perforations (Φ8–12 mm) under realistic in situ stress conditions. An improved dynamic prediction model for perforation erosion based on the Long model was established, incorporating a dynamic friction correction factor to overcome the limitations of traditional models in characterizing inter-cluster interference effects. A temporary plugging parameter optimization method was proposed, using "inter-cluster stress difference" as the core criterion. Providing a new technical pathway for the efficient development of Mahu reservoirs.

2. Field-Scale Perforation Erosion Experiments

During hydraulic fracturing operations in the Mahu conglomerate reservoirs, the pumping rate, proppant concentration, and injection pressure reach up to 14 m^3/min, 420 kg/m^3, and 90 MPa, respectively. As the sand-laden fluid passes through the perforations, erosion of the perforation holes is inevitable. This erosion compromises the effectiveness of limited-entry fracturing. It increases the difficulty and uncertainty of temporary perforation plugging, leading to uneven propagation of multiple fracture clusters within a single stage of horizontal well fracturing. Current laboratory studies often employ low pumping rates, low pressures, and small proppant volumes, significantly different from actual field fracturing conditions. To address this gap, this study utilizes a field-scale, high-intensity perforation erosion simulation system, incorporating fracturing trucks and actual downhole casings, achieving perforation flow velocities of up to 200 m/s and injection pressures of up to 105 MPa. This research delves into the dynamic characteristics and influencing mechanisms of perforation erosion by replicating actual field conditions.

2.1. Design of the Field-Scale High-Intensity Perforation Erosion System

The field-scale perforation erosion experiments aim to investigate the impact of sand-carrying fluids on perforation erosion under actual fracturing conditions (high rate, high pressure, and high sand ratio). As illustrated in Figures 1 and 2, the system comprises fracturing trucks, liquid storage trucks, sand transport trucks, mixing trucks, instrumentation trucks, high-pressure pipelines, and perforated casings. The fracturing truck assembly enables maximum fluid velocities of 200 m/s and injection pressures of up to 100 MPa. Liquid storage trucks continuously supply fracturing fluid, each capable of storing 30 m^3 of fluid, ensuring operational flexibility and on-demand fluid replenishment. Sand transport trucks are primarily used for proppant storage and supply, with each tank capable of holding 20 m^3 of proppant. Mixing trucks are responsible for blending, agitating, and transporting the proppant. The instrumentation truck provides real-time control and monitoring of pressure, flow rate, fluid density, and hydraulic power, allowing for dynamic adjustment of pumping parameters to ensure experimental safety. The high-pressure manifold serves as the main injection channel for the fracturing fluid, with a maximum pressure rating of

110.0 MPa. A specially designed fluid storage tank is also used for fracturing fluid storage and flowback.

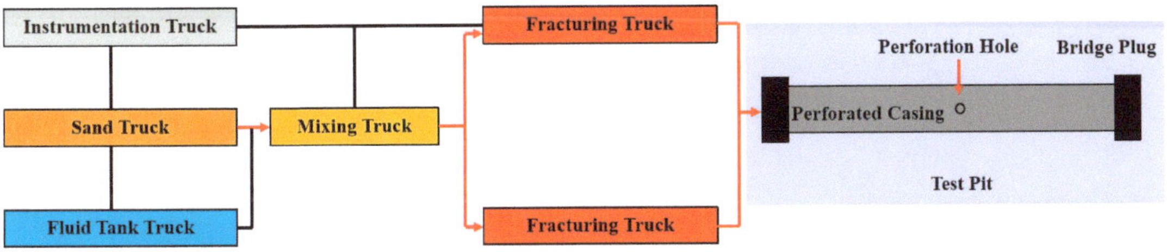

Figure 1. Schematic diagram of the perforation erosion evaluation system.

Figure 2. Field-scale perforation erosion evaluation system.

A 1 m long casing section, cut from the fracturing casing used in the Mahu conglomerate reservoirs, is prepared for the experiments. The TP125V steel casing has a pressure rating of up to 90.70 MPa, with an outer diameter of 139.7 mm and a wall thickness of 10.54 mm. One end the casing is threaded to connect to the high-pressure manifold, while the other is sealed by welding. A 10 mm diameter perforation hole is drilled in the middle of the casing section using a diamond drill bit.

2.2. Experimental Methods and Procedures

The perforated casing was fixed in the experimental pool, connected to high-pressure pipelines, and prepared with fracturing fluid and proppant. Pressure, sealing, and safety tests were conducted before proceeding with the experiments according to the design outlined in Table 1.

Experiment 1: High-viscosity fracturing fluid was used to study the effect of fluid viscosity on the erosion coefficient.

Experiment 4: 20/40 mesh quartz sand was employed to investigate the influence of proppant size on the erosion coefficient.

Experiment 2: 40/70 mesh ceramic proppant was used to examine the impact of proppant type on the erosion coefficient.

Experiment 3: Different sand concentrations were tested to analyze the effect of proppant concentration on perforation erosion.

Table 1. Experimental Plan.

Experiment	Flow Rate (m^3/min)	Proppant Concentration (kg/m^3)	Proppant Size and Type	Erosion Coefficient (α)	Erosion Coefficient (β)
1	0.9 (High Viscosity)	90	40/70 Mesh Quartz Sand	1.17×10^{-13}	9.01×10^{-11}
2	0.9	90	40/70 Mesh Ceramic	1.34×10^{-13}	5.66×10^{-11}
3	0.9	90 160 240	40/70 Mesh Quartz Sand	3.96×10^{-13} 4.51×10^{-13} 3.80×10^{-13}	1.43×10^{-10} 1.60×10^{-10} 1.68×10^{-10}
4	0.9	160	20/40 Mesh Quartz Sand	3.10×10^{-13}	1.16×10^{-10}

3. Perforation Erosion Model

3.1. Construction of the Perforation Erosion Prediction Model

When high-pressure fluid flows through the perforations in oil and gas wells, the surrounding material is subjected to fluid impact forces, leading to gradual erosion and wear. This increases perforation size, adversely affecting well productivity, flow characteristics, and wellbore integrity. The Long model [20] is a mathematical model that describes and predicts perforation erosion caused by the flow of liquids or gases through perforations, typically employed in well-completion operations. The model focuses on the changes in perforation size, flow resistance, and material loss at the perforation edges due to the impact and shear forces exerted by high-pressure, high-velocity fluids.

The core of the Long model lies in its mathematical formulation, which describes the geometric changes in perforations under fluid erosion, particularly the enlargement of the perforation diameter and the variation in the flow coefficient. The model incorporates factors such as fluid velocity, particle concentration, fluid density, and the wear resistance of the perforation material. The key equations of the Long model are as follows:

1. Perforation diameter change equation [20]:

$$\frac{dD}{dt} = \alpha C v^2 \tag{1}$$

where D is the perforation diameter (m), t is time (s), the parameter α is the perforation diameter erosion coefficient, which characterizes the influence of proppant kinetic energy, represented by the product of sand concentration and the square of flow velocity, on the growth rate of perforation diameter. Its unit is ($m^2 \cdot s/kg$). C is the sand concentration (kg/m^3), and v is the fluid velocity (m/s).

2. Flow coefficient change equation [20]:

$$\frac{dC_d}{dt} = \beta C^{v^2} \left(1 - \frac{C_d}{C_d^{max}}\right) \tag{2}$$

where C_d is the flow coefficient (dimensionless), the parameter β is the flow coefficient erosion coefficient, reflecting the effect of kinetic energy on the smoothing of perforation edges, that is, its contribution to the growth rate of the flow coefficient C_d. Its unit is ($m \cdot s/kg$). C_{max} is the maximum flow coefficient, C is the sand concentration (kg/m^3), and v is the fluid velocity (m/s).

3. Fluid Velocity Expression:

$$v = \frac{4q}{n\pi D^2} \tag{3}$$

where v is the fluid velocity (m/s), Q is the flow rate (m^3/s), n is the number of perforations (dimensionless), and D is the perforation diameter (m).

4. Pressure drop equation:

$$P_r = \frac{8\rho}{\pi^2 C_d^2 D^4}\left(\frac{q}{n}\right)^2 \tag{4}$$

where P_r is the pressure drop (Pa), ρ is the fluid density (kg/m^3), Q is the flow rate (m^3/s), n is the number of perforations (dimensionless), C_d is the flow coefficient (dimensionless), and D is the perforation diameter (m).

3.2. Calculation of α and β Based on Experimental Data

Solid particles impact the perforation edges at high speeds when high-velocity fluid flows through perforations. Due to the complex flow regime, different regions of the perforation edges experience varying forces, leading to localized mechanical wear. As particles continuously impact the edges, the material gradually erodes, causing the perforation edges to become irregular and uneven. This phenomenon, known as roundness erosion, results in the perforation shape deviating from its original circular form, becoming asymmetric or rough. Additionally, diameter erosion occurs as the overall perforation diameter increases due to prolonged particle impact and continuous wear on the perforation walls. Over time, the perforation diameter significantly enlarges.

In summary, roundness erosion primarily affects the shape of the perforation edges, reducing their smoothness and causing irregularity. In contrast, diameter erosion refers to the overall increase in perforation size due to erosion. The changes in perforation shape and size directly influence the flow characteristics of fluids passing through the perforations. As shown in Figure 3, the roundness erosion stage is characterized by a sharp pressure drop (>30%), while the diameter erosion stage exhibits a gradual pressure decline (<5%). The process involves roundness loss (40–55% reduction in roundness) and diameter expansion (1.8–2.5 times the original diameter). A trumpet-shaped perforation with an inner diameter more than twice the outer diameter is formed. Specifically, perforation expansion reduces pressure loss and improves fluid flow efficiency, while irregular perforation shapes may cause localized flow disturbances, increase flow resistance, and mitigate fracturing fluid performance.

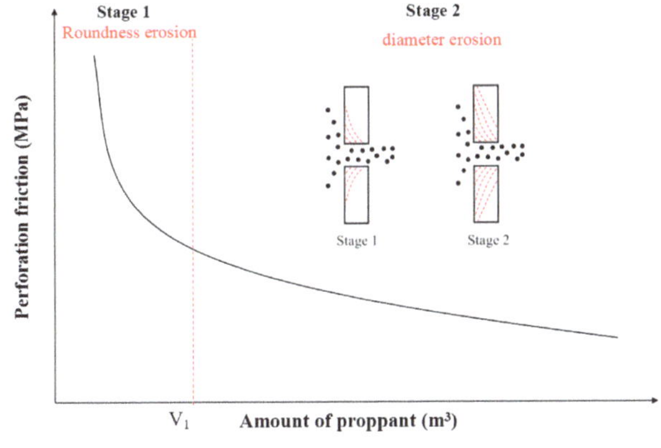

Figure 3. Roundness erosion stage and diameter erosion stage during sand-laden erosion [21].

Equation (1) characterizes diameter erosion, where the perforation diameter changes as proppant particles are continuously injected. Equation (1) is only applicable to the "diameter erosion stage"; during the "roundness erosion stage", the perforation evolution should be characterized by the change in the flow coefficient C_d. Equation (2) characterizes roundness erosion, where the perforation shape changes significantly as proppant particles erode the edges, leading to variations in the flow coefficient. To solve for α and β, Equation (3) is first substituted into Equation (1), and the resulting expression is integrated to obtain:

$$D^5 = \frac{80\alpha Cq^2}{n^2\pi^2}t + D_0^5 \tag{5}$$

Substituting Equation (5) into Equation (4), the solution reveals that roundness erosion occurs first, followed by diameter erosion. Therefore, during diameter erosion, the flow coefficient reaches its maximum value:

$$P_r^{-1.25} = \alpha\left(\frac{n^2\pi^2}{8\rho q^2}\right)(C_d^{max})^{2.5}\frac{80Cq^2}{n^2\pi^2}t + \left(\frac{n^2\pi^2}{8\rho q^2}\right)^{1.25}(C_d^{max})^{2.5}D_0^5 \tag{6}$$

As shown in Table 1, the experimental parameters for Test 3 include a flow rate of 0.9 m^3/min, 40/70 mesh quartz sand, and a sand concentration of 90 kg/m^3. By dividing the treatment curve of Test 3 (as shown in Figure 4) into the roundness erosion stage and the diameter erosion stage, the diameter erosion coefficient α was determined. Figures 5 and 6 correspond to two different stages of the erosion process. Due to the shift in dominant mechanisms, we adopted different forms of pressure, such as the vertical axis in each figure. Figure 5 corresponds to the "diameter erosion stage", where C_d has nearly reached its maximum value, and the expansion of the orifice diameter D becomes the dominant factor. From the model, pressure follows $P_r \propto 1/D^4$, and combining this with the geometric relationship $D^5 \propto t$, it follows that $P_r{}^{-1.25} \propto t$. Therefore, we use $P_r{}^{-1.25}$ as the vertical axis to linearize the relation for regression and extract parameter α.

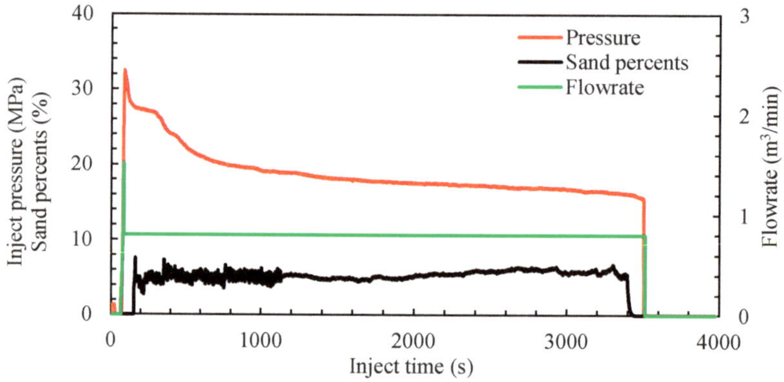

Figure 4. Test 3 operational curve.

Figure 6 analyzes the "roundness erosion stage", during which the orifice diameter remains nearly constant. At this stage, the primary change arises from increased discharge coefficient C_d. According to the model, pressure satisfies the relationship $P_r \propto 1/C_d{}^2$. Therefore, the original pressure P is used as the vertical axis. By taking the time derivative of Equation (4), the regression of parameter β is performed. Considering that both the discharge coefficient C_d and the perforation diameter D change during the erosion process, Equations (1) and (2) are also incorporated, yielding:

$$\frac{dP_r}{dt} = -\frac{16q^2Cv^2\rho}{n^2\pi^2C_d^2D^4}\left[\beta\left(\frac{1}{C_d} - \frac{1}{C_d^{max}}\right) + \frac{2\alpha}{D}\right] \tag{7}$$

when $t = 0$, $C_d = C_{d0}$ and $D = D_0$, leading to:

$$\frac{\mathrm{d}P_r}{\mathrm{d}t} = -P_{r0}\frac{32Cq^2}{n^2\pi^2D_0^4}\left[\beta\left(\frac{1}{C_{d0}} - \frac{1}{C_d^{max}}\right) + \frac{2\alpha}{D}\right] \tag{8}$$

As shown in Figure 6, the flow coefficient correlation coefficient β is calculated using the above equations.

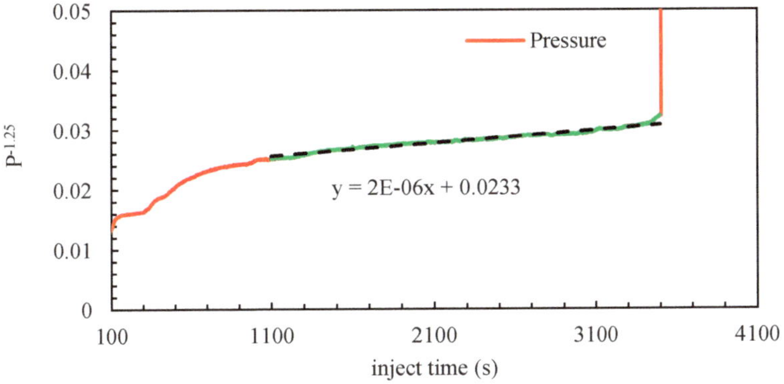

Figure 5. The diameter erosion stage $P^{-1.25}$-t curve.

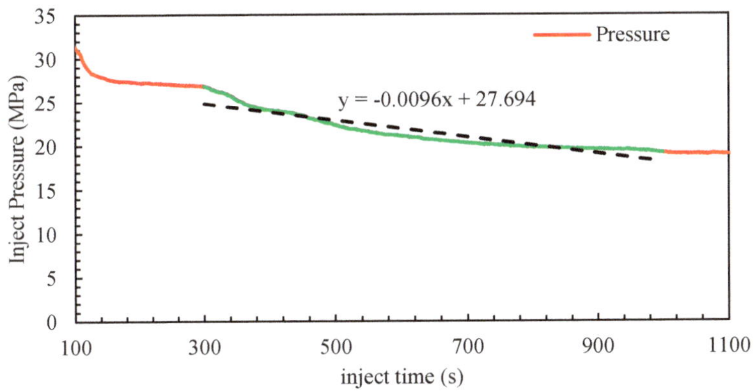

Figure 6. The roundness erosion stage P-t curve.

4. Quantitative Characteristics of Perforation Erosion

4.1. Perforation Erosion Under Different Fluid Viscosities

While considering proppant suspension characteristics, the influence of fracturing fluid viscosity on perforation erosion exhibits significant variations. A comparative study between Experiment 1 and Experiment 3 was conducted to analyze the erosion patterns under different fluid viscosity conditions. Based on the perforation erosion model, this research evaluated the dynamic evolution of perforation diameter, flow coefficient, and pressure drop during erosion (as illustrated in Figures 7–9). High-viscosity fracturing fluids (e.g., crosslinked gels) demonstrate enhanced proppant transport capacity, effectively reducing particle settling and mitigating localized impact erosion along perforation edges. Furthermore, the lower flow velocity of high-viscosity fluids under equivalent pumping rates diminishes shear stress on perforation walls, alleviating material wear. Experimental data reveal that when fluid viscosity decreases from high (60 mPa·s) to low (10 mPa·s), calculated based on comparative analysis of model results, the perforation diameter increases by 18.84%, the flow coefficient rises by 7.32%, and the perforation pressure drops by 21.99%. In contrast, low-viscosity fluids (e.g., slickwater) exhibit inferior suspension performance, leading to proppant settling and accumulation near the perforation base. This results in elevated local particle concentration and intensified impact erosion. The higher flow velocity of low-viscosity fluids also amplifies turbulence and shear forces, directly accelerating the mechanical wear of perforation edges. Experimental observations confirm more severe erosion intensity and the potential formation of trumpet-shaped perforations. To address these issues, field applications should prioritize high-viscosity fracturing fluids (≥50 cP) to optimize the balance between proppant transport and erosion control. In summary, fluid viscosity governs erosion behavior through dual mechanisms: proppant distribution and hydrodynamic shear. High-viscosity fluids suppress erosion via uniform particle suspension and reduced shear, whereas low-viscosity fluids exacerbate erosion through particle settling and enhanced shear stress.

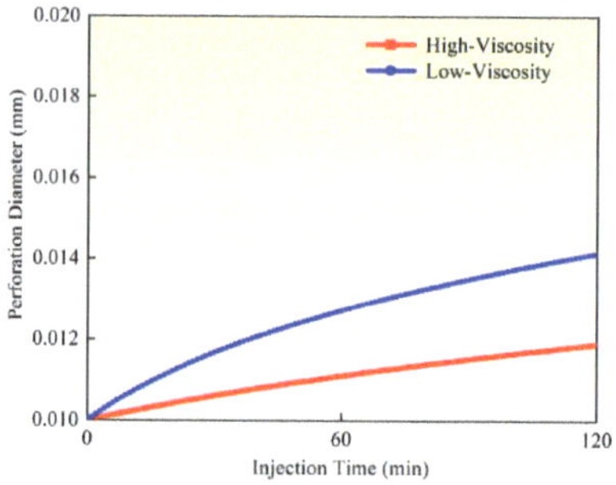

Figure 7. Hole diameter dynamic change.

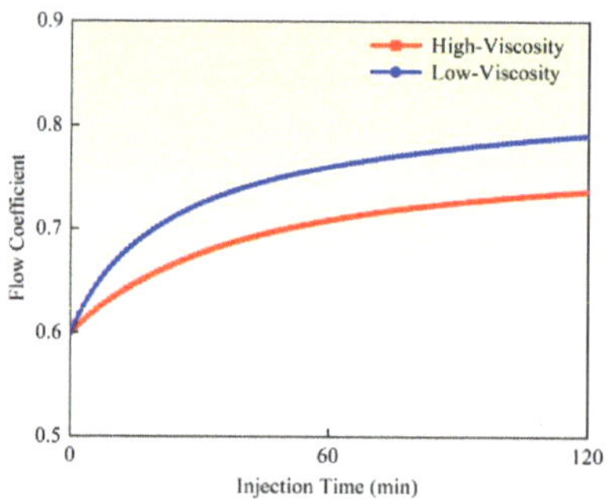

Figure 8. Flow coefficient dynamic changes.

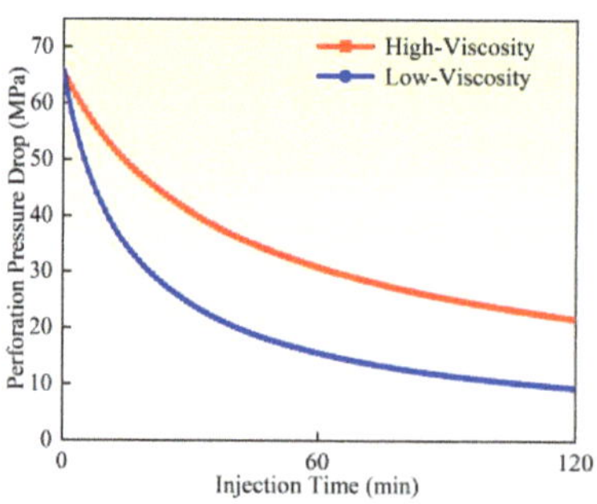

Figure 9. Hole pressure drop dynamic changes.

4.2. Perforation Erosion Under Different Proppant Types

The influence of proppant type on perforation erosion exhibits significant variations, as demonstrated by experimental studies (Experiment 1 vs. Experiment 2) and field observation data. Ceramic proppants show markedly higher erosion levels than quartz sand (as illustrated in Figures 10–12). Calculated based on comparative analysis of model results, with a 120 min pumping operation resulting in 16.87% and 6.91% increases in perforation diameter and flow coefficient, respectively, along with a 19.04% reduction in perforation pressure drop. Ceramic proppants' superior compressive and shear strength leads to more severe material erosion under identical pumping conditions. In contrast, quartz sand's lower mechanical strength causes particle fragmentation upon contact with casing surfaces, paradoxically reducing its cumulative erosive effect. Field-scale tests reveal that ceramic proppants form more pronounced trumpet-shaped perforations with inner diameters exceeding twice the original size, while quartz sand produces relatively uniform diameter enlargements. This divergence stems from ceramic particles maintaining structural integrity during high-velocity impacts, enabling sustained erosive action, whereas quartz sand dissipates erosive energy through particle breakage. Based on these findings, quartz sand proppants are recommended for field applications to mitigate the adverse effects of perforation erosion on pipeline materials and operational environments.

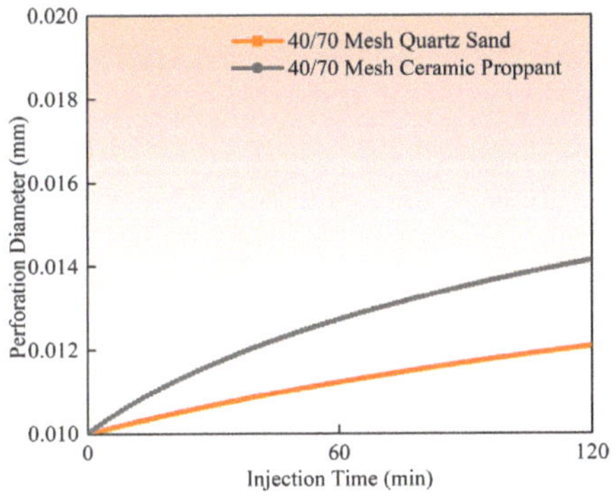

Figure 10. Hole diameter dynamic change.

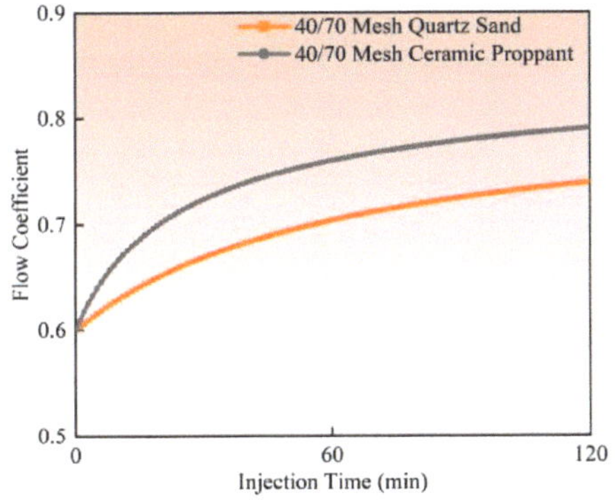

Figure 11. Flow coefficient dynamic changes.

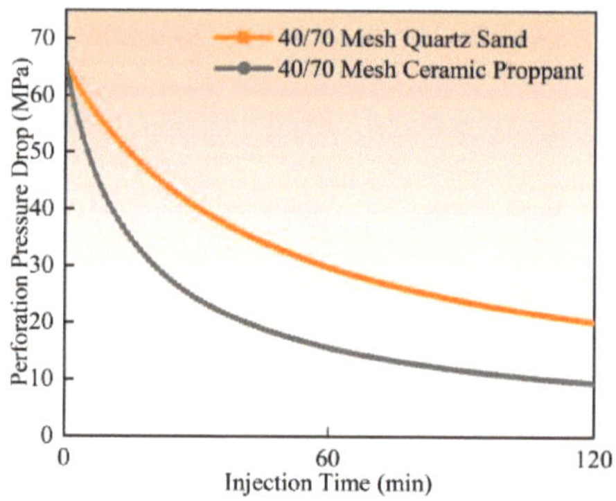

Figure 12. Hole pressure drop dynamic changes.

4.3. *Perforation Erosion Under Different Proppant Sizes*

Experimental studies and field observations (Experiments 2 and 3) demonstrate that under identical pumping conditions (flow rate 0.9 m^3/min, proppant concentration 90 kg/m^3), larger-sized proppants (20/40 mesh) induce more severe perforation erosion than smaller-sized proppants (40/70 mesh) (as illustrated in Figures 13–15). Calculated based on comparative analysis of model results, after 120 min of pumping, the larger proppants caused an 18.19% increase in perforation diameter, a 5.47% improvement in flow coefficient, and an 8.34% reduction in perforation pressure drop. This phenomenon primarily results from the greater contact area between larger proppants and perforation surfaces, combined with their higher kinetic energy per particle and more concentrated impact forces. Full-scale field tests further confirm that 20/40 mesh proppants create more substantial erosion on perforation surfaces, leading to trumpet-shaped deformation of the perforations. The underlying mechanical mechanism lies in the superior momentum conservation of larger particles during high-velocity flow, which generates more intense point impacts on perforation edges. In contrast, although smaller particles are more numerous, their impact energy is more dispersed, and their contact area with surfaces is comparatively reduced. The engineering practice commonly adopts a "small-to-large" proppant injection strategy; finer proppants (40/70 mesh) are preferentially injected to penetrate deeper into fracture networks. In comparison, coarser proppants (20/40 mesh) are subsequently placed near the fracture entrance to establish high-conductivity proppant packs. Field applications should prioritize smaller mesh proppants to mitigate perforation erosion while enhancing proppant placement efficiency.

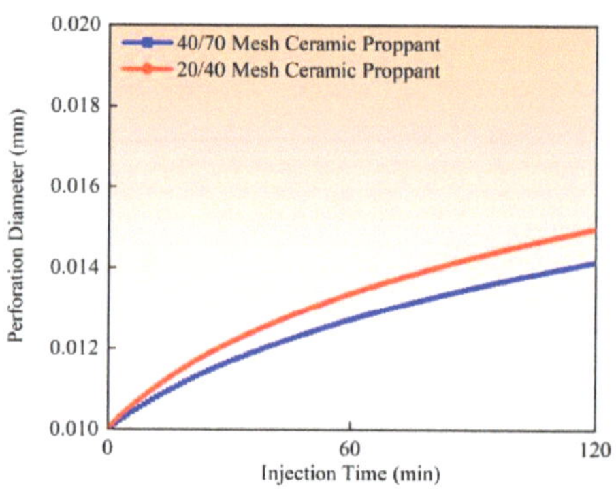

Figure 13. Hole diameter dynamic change.

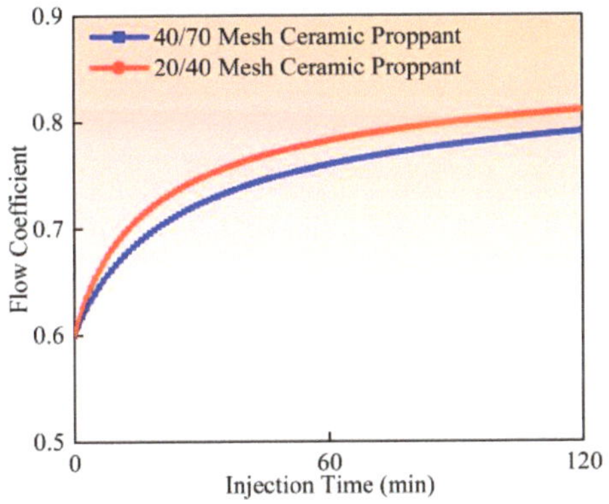

Figure 14. Flow coefficient dynamic changes.

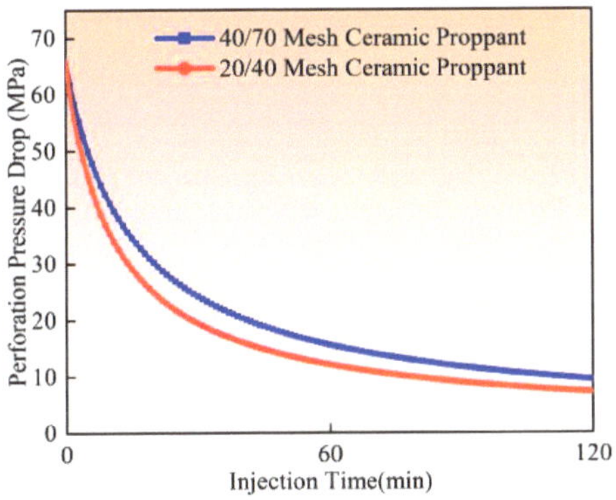

Figure 15. Hole pressure drop dynamic changes.

4.4. Perforation Erosion Under Different Sand Concentrations

Experimental studies and field observations (Experiment 3) demonstrate that under identical pumping conditions, perforation erosion intensifies significantly as proppant concentration increases from 90 kg/m^3 to 240 kg/m^3 (Figures 16–18). Calculated based on comparative analysis of model results, after 120 min of pumping, the perforation diameter and flow coefficient increased by 18.19% and 5.47%, respectively, while the perforation pressure drop decreased by 8.34%. Under high proppant concentrations, the perforation surfaces exhibit more severe localized erosion characteristics, forming erosion pits of varying depths and distinct trumpet-shaped deformation, with the inner diameter expanding up to twice its original size. The underlying mechanism reveals that as proppant concentration increases, the number of particles impacting the perforation wall per unit of time grows linearly, while enhanced inter-particle interactions leading to substantially more significant cumulative erosion effects. Based on these findings, field operations should employ optimized proppant concentration gradient designs (e.g., using lower concentrations initially and gradually increasing during middle-to-late stages) to balance erosion effects with proppant placement requirements while simultaneously implementing real-time pressure monitoring to prevent screen outs.

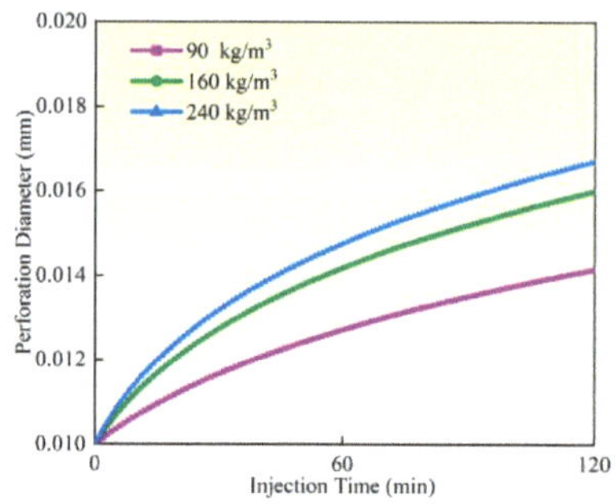

Figure 16. Hole diameter dynamic change.

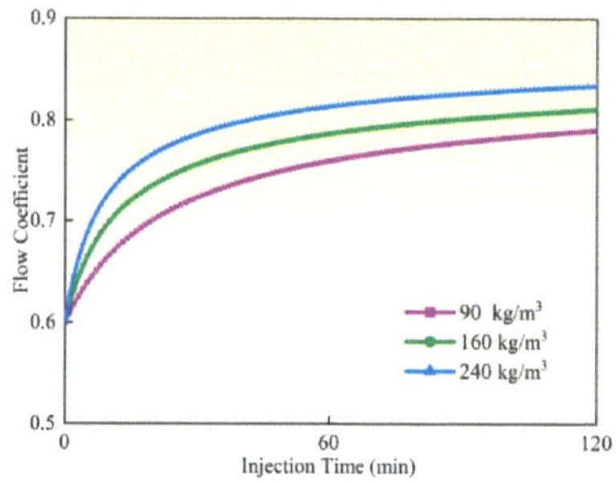

Figure 17. Flow coefficient dynamic changes.

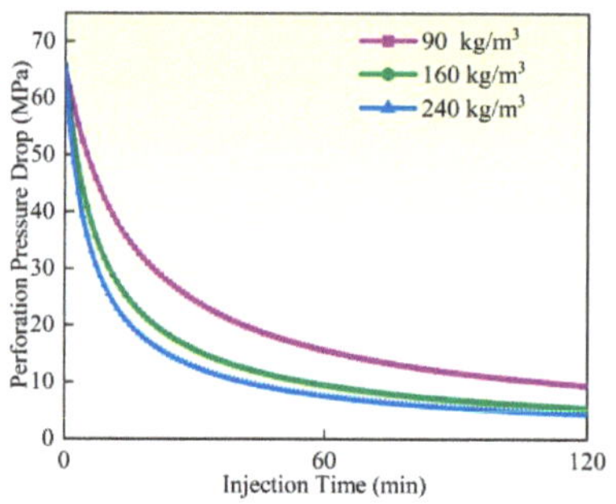

Figure 18. Hole pressure drop dynamic changes.

5. Application Case Study of Perforation Erosion Model

5.1. Application Methodology of the Perforation Erosion Model

Limited entry and temporary plugging fracturing techniques are commonly employed to promote simultaneous initiation and propagation of multiple fracture clusters in horizontal wells [22–24]. The core principle of limited-entry fracturing is to increase the perforation friction within the horizontal wellbore, thereby restricting the fluid intake capacity of each cluster and ensuring simultaneous fracture initiation and propagation [25]. However, field-distributed fiber optic monitoring [26] and downhole camera observations reveal that fracture propagation is often uneven, and the continuous erosion of perforations by sand-laden fluids leads to the failure of limited-entry techniques. This underscores the significant impact of perforation erosion on the effectiveness of limited-entry fracturing.

This study constructs a perforation erosion model that accounts for the effects of erosion on fracture propagation. A "limited entry + temporary plugging" composite process is proposed to achieve balanced reservoir stimulation. As the number of perforations decreases, perforation friction increases, overcoming inter-cluster stress differences and promoting uniform fluid intake and fracture propagation. However, during fracturing operations, the injection of large volumes of quartz sand causes perforation erosion, leading to the deformation and enlargement of the perforations. This reduces perforation friction, rendering the limited-entry technique ineffective. To address this issue, a temporary plugging process is introduced, utilizing degradable fiber-based diverters to seal dominant perforations, creating localized blockages dynamically. This redistributes fluid flow within the wellbore, forcing fluid into unsealed perforations and promoting the extension of underdeveloped fractures. The diverters, made of degradable fiber materials, exhibit high bridging capability, rapidly forming effective seals and naturally dissolving post-operation to avoid wellbore or fracture blockages.

5.2. Field Application and Analysis

Well A, a typical horizontal well in the Mahu oilfield, was selected for this study. Located in the Xia 72 fault block of the Mahu 131 well area, the target formation is the Triassic Baikouquan Formation ($T_1b_3{}^2$). The well has a vertical depth of 2.721.76 m, a measured depth of 4850 m, and a horizontal section length of 1940 m. The well employs a dense-cluster design for volumetric fracturing, utilizing plug-and-perf staged fracturing for large-scale stimulation, with consideration given to perforation erosion effects. Temporary plugging was applied during the fracturing of stages 7–15 and 17, using fiber-based diverters to seal perforations. Table 2 shows that three different perforation densities per cluster (3, 5, and 8 perforations per cluster) were tested for limited-entry fracturing.

Table 2. Perforation Scheme for Well A.

Perforations per Cluster	Clusters per Stage	Perforations per Stage	Fracturing Stages
3	6	18	Stages 7, 10, 13
5	6	30	Stages 8, 11, 14
8	6	48	Stages 9, 12, 15

A comparative test was conducted on Well A, stages 7–15 (9 fracturing stages). Each stage was designed with 6 clusters, and each cluster was created with a proppant volume of 25 m^3, resulting in an average proppant intensity of 1.2 m^3/m. To eliminate spatial randomness, stages 7, 10, and 13 were designed with three perforations per cluster, stages 8, 11, and 14 with five perforations per cluster, and stages 9, 12, and 15 with eight perforations per cluster. The perforation erosion parameters, calculated from field tests in the Mahu area, are $\alpha = 9.85 \times 10^{-14}$ and $\beta = 9.48 \times 10^{-10}$ and were used to construct the perforation erosion model.

Stage 7: Designed with a pumping rate of 10 m^3/min, 18 perforations per stage, and 160 kg/m^3 sand concentrations. The inter-cluster stress difference was 4.85 MPa. Based on the perforation erosion model, the dynamic erosion changes are shown in Figure 19. When perforation friction could no longer balance the inter-cluster stress difference, temporary plugging was initiated at 101 min using 13–15 mm fiber-based diverters.

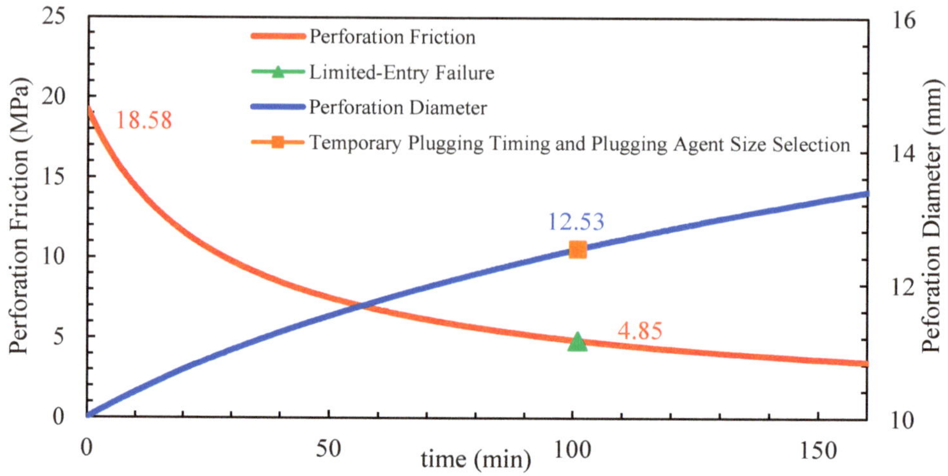

Figure 19. Single-stage 18 perforation holes erosion dynamics prediction chart.

Stage 8: Designed with a pumping rate of 10 m^3/min, 30 perforations per stage, and 160 kg/m^3 sand concentrations. The inter-cluster stress difference was 3.34 MPa. Based on the perforation erosion model, the dynamic erosion changes are shown in Figure 20. Temporary plugging was initiated at 33 min using 13–15 mm fiber-based diverters.

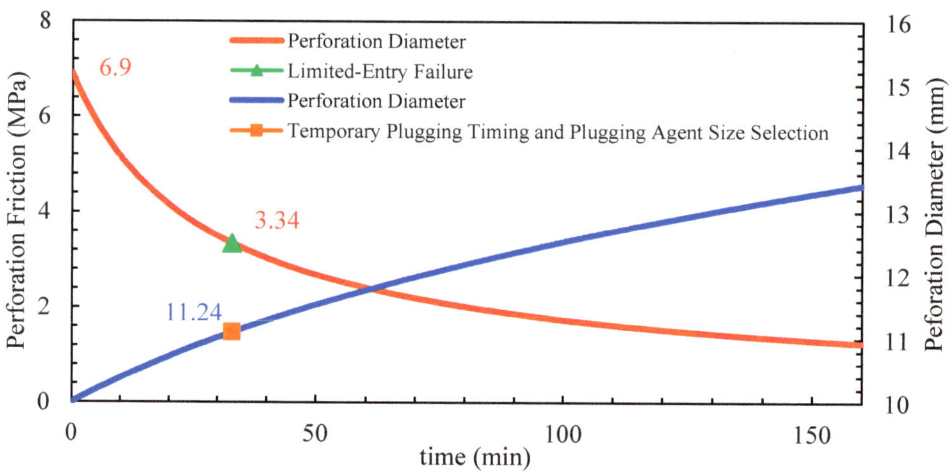

Figure 20. Single-stage 30 perforation holes erosion dynamics prediction chart.

Stage 9: Designed with a pumping rate of 10 m^3/min, 48 perforations per stage, and 160 kg/m^3 sand concentrations. The inter-cluster stress difference was 3.8 MPa. Based on the perforation erosion model, the dynamic erosion changes are shown in Figure 21. With 48 perforations, the initial perforation friction was 3.17 MPa, which is insufficient for limited-entry effectiveness. Pre-stage plugging was applied before proppant injection to redirect fluid flow and promote fracture extension.

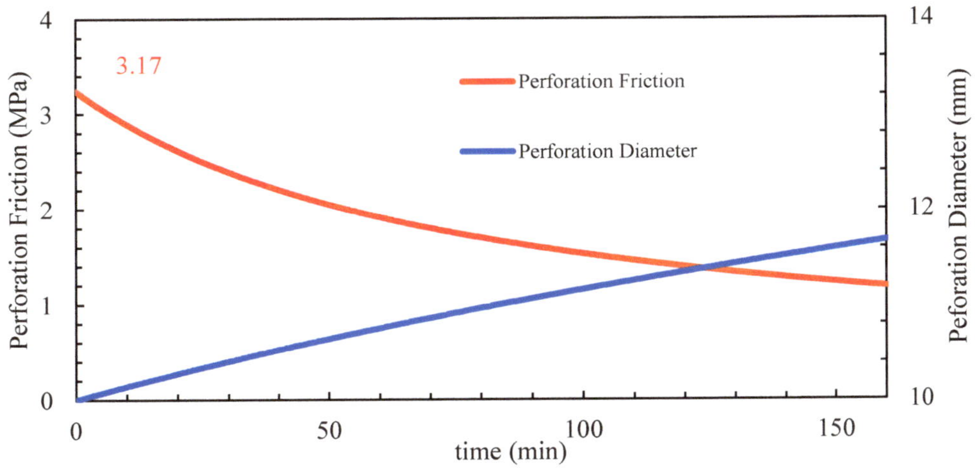

Figure 21. Single-stage 48 perforation holes erosion dynamics prediction chart.

The limited-entry technique alone is insufficient to address the challenges of low porosity and permeability in the Baikouquan Formation of the Mahu Sag, as well as uneven fracture propagation under heterogeneous formation conditions and large-scale sand fracturing. The integration of temporary plugging optimizes fluid distribution, mitigates perforation erosion, and improves fracture uniformity. This study introduces a "limited entry + temporary plugging" composite process supported by a dynamic perforation erosion model, achieving effective stimulation in Well A. The erosion rate was reduced by 35–50%, fracture uniformity improved to over 80%, and single-well productivity increased by 18–25%. The well achieved a peak daily oil production of 40.55 m^3/d and a cumulative oil production of 2083.11 m^3 over 165 days, demonstrating significant reservoir stimulation and enhanced development outcomes.

6. Conclusions

Addressing the severe perforation erosion in horizontal well staged fracturing under "high pressure, high rate, and high sand ratio" conditions in the Mahu reservoirs, this study establishes a quantitative perforation erosion model through field experiments, yielding the following key conclusions:

(1) Field experiments indicate that pumping rate, proppant size, concentration, and fluid viscosity are the primary controlling factors of erosion. Ceramic proppants and low-viscosity fluids exacerbate erosion. High-viscosity sand-carrying fluids ($\geq$50 mPa·s) and high-mesh quartz sand (40/70 mesh) are recommended to mitigate erosion risks.

(2) The improved Long model reveals a two-stage erosion process: a sharp pressure drop (>30%) during the roundness erosion stage, followed by a gradual pressure decline (<5%) during the diameter erosion stage, ultimately forming a trumpet-shaped perforation with an inner diameter exceeding twice the outer diameter.

(3) A "limited entry + temporary plugging" dynamic control composite process is proposed. By determining the critical perforation friction for limited-entry failure based on inter-cluster stress differences, the timing and size selection criteria for temporary diverters are optimized, achieving balanced fracture propagation. Field applications resulted in productivity increases of 18–25%.

Author Contributions: Conceptualization, H.L.; Methodology, H.L.; Validation, H.L.; Formal analysis, Z.S.; Investigation, J.M.; Data curation, E.Z.; Writing—original draft, B.W. and H.L.; Writing—

review & editing, B.W. and X.L.; Supervision, X.L. All authors have read and agreed to the published version of the manuscript.

Funding: The authors would like to acknowledge the financial support from the National Natural Science Foundation of China (No. 52374057 and U23B2084), the "Tianshan Talent" Training Program (2023TSYCCX0004), and the autonomous region key research and development project (No. 2024B01013-1).

Institutional Review Board Statement: Not applicable.

Informed Consent Statement: Not applicable.

Data Availability Statement: The original contributions presented in the study are included in the article, further inquiries can be directed to the corresponding author.

Conflicts of Interest: The authors declare that they have no known competing financial interests or personal relationships that could have appeared to influence the work reported in this paper.

References

1. Zhuo, R.; Ma, X.; Li, J.; Ren, L.; Cheng, N.; Ma, J.; Zhang, S. Influence of limited-entry fracturing on perforation erosion in horizontal wells. *Drill. Prod. Technol.* **2023**, *46*, 77–82.
2. Li, G.; Lei, Z.; Dong, W.; Wang, H.; Zheng, X.; Tan, J. Progress, challenges, and prospects of unconventional oil and gas development in China. *China Pet. Explor.* **2022**, *27*, 1–11.
3. Zhao, J.; Ren, L.; Jiang, T.; Hu, D.; Wu, L.; Wu, J.; Yin, C.; Li, Y.; Hu, Y.; Lin, R.; et al. A decade of shale gas fracturing in China: Review and outlook. *Nat. Gas Ind.* **2021**, *41*, 121–142.
4. Wu, Q.; Xu, Y.; Zhang, S.; Wang, T.; Guan, B. Core theories and key optimization designs of volumetric fracturing technology for unconventional reservoirs. *Acta Pet. Sin.* **2014**, *35*, 706–714.
5. Zou, C.; Yang, Z.; Zhu, R.; Zhang, G.; Hou, L.; Wu, S.; Tao, S.; Yuan, X.; Dong, D.; Wang, Y.; et al. Progress in exploration, development, and theoretical technologies of unconventional oil and gas in China. *Acta Geol. Sin.* **2015**, *89*, 979–1007.
6. Zhang, J.; Lu, Z.; Shi, S. Large-scale field experimental study on casing perforation erosion during fracturing. *Drill. Prod. Technol.* **2023**, *46*, 85–91.
7. Liu, Y.; Qin, L.; Wang, Z.; Dou, Y. Numerical simulation of perforation erosion during casing fracturing. *Pet. Mach.* **2015**, *43*, 66–69.
8. Zhang, S.; Zeng, D.; Pan, Y.; Tian, G.; Yu, H.; Shi, S. Research progress on the erosion of fracturing strings during large-scale sand fracturing. *Xinjiang Oil Gas* **2022**, *18*, 65–72.
9. Huang, G.; Hu, X.; Zhou, F.; Li, X.; Dong, E.; Li, Z. A new multi-cluster fracturing simulation model coupled with perforation erosion: Based on the continuous–discontinuous method. *Rock Mech. Rock Eng.* **2023**, *56*, 3887–3901. [CrossRef]
10. Hu, X.; Li, X.; Zhou, F.; Bai, Y.; Chen, C.; Zhang, P. Simulation study on proppant transport in a horizontal wellbore considering perforation erosion. *Geoenergy Sci. Eng.* **2023**, *231*, 212282. [CrossRef]
11. Li, Z.; Li, Z.; Cheng, J.; Cui, L. Experimental study on erosion and hole enlargement by high-speed jet flow. *Shandong Chem. Ind.* **2022**, *51*, 1–4.
12. Wu, B.; Zhou, F.; Wang, M.; Lv, Z.; Li, M.; Wang, B.; Guo, X.; Zhang, J. Field-Scale Experimental Study on the Perforation Erosion in Horizontal Wellbore under Real Fracturing Conditions. *Processes* **2022**, *10*, 1058. [CrossRef]
13. Wang, B.; Gu, X.; Li, W.; Zhang, G.; Ge, H.; Zhou, F. Experimental study on plugging law of irregular perforations under strong erosion. *J. Xi'an Shiyou Univ. Nat. Sci. Ed.* **2023**, *38*, 46–53.
14. Zang, C.; Jiang, H.; Shi, S.; Li, J.; Zou, Y.; Zhang, S.; Tian, G.; Yang, P. Analysis of uniform fracture initiation in multi-cluster fracturing based on perforation imaging monitoring: A case study of the tight conglomerate in Mahu Sag, unggar Basin. *Pet. Explor. Dev.* **2022**, *49*, 394–402. [CrossRef]
15. Xu, G. Numerical simulation study on hydraulic jet erosion of coiled tubing. *Oil Field Equip.* **2016**, *45*, 12–15.
16. Guo, X.; Cheng, L.; Zhang, J.; Li, H.; Li, Y. Numerical simulation study of hole erosion under directional perforating conditions. *Xinjiang Oil Gas* **2022**, *18*, 56–61.
17. Eckdahl, R.; Vigrass, A.; Singh, A.; Liu, X.; Chrusch, L. Assessment of limited entry cluster distribution effectiveness and impactful variables using perforation erosion measurements. In Proceedings of the SPE Hydraulic Fracturing Technology Conference and Exhibition, The Woodlands, TX, USA, 4–6 February 2022.
18. Cramer, D.; Friehauf, K.; Roberts, G.; Whittaker, J. Integrating DAS, treatment pressure analysis and video-based perforation imaging to evaluate limited entry treatment effectiveness. In Proceedings of the SPE Hydraulic Fracturing Technology Conference and Exhibition, The Woodlands, TX, USA, 4–6 February 2019.

19. Roberts, G.; Whittaker, J.; McDonald, J.; Paxson, T. Proppant Distribution Observations from 20,000+ Perforation Erosion Measurements. In Proceedings of the SPE Hydraulic Fracturing Technology Conference and Exhibition, The Woodlands, TX, USA, 4–6 February 2020.
20. Long, G.; Liu, S.; Xu, G.; Wong, S.-W.; Chen, H.; Xiao, B. A perforation-erosion model for hydraulic-fracturing applications. *SPE Prod. Oper.* **2018**, *33*, 770–783. [CrossRef]
21. Wang, F.; Li, J.; Xiang, Z.; Chen, R.; Zhao, Y.; Mou, Y. Simulation and countermeasures of casing erosion at downhole decoupled sandblaster during fracturing. *Xinjiang Oil Gas* **2023**, *19*, 57–65.
22. Wang, Y.; Lu, Y.; Li, Y.; Wang, X.; Yan, X.; Zhang, Z. Progress and application of fracturing technology in unconventional reservoirs. *Acta Pet. Sin.* **2012**, *33* (Suppl. S1), 149–158.
23. Zhou, T.; Zhang, S.; Chen, M.; Huang, Z. Competitive propagation and control of multi-cluster fractures in horizontal wells. *Sci. Sin. Technol.* **2019**, *49*, 469–478. [CrossRef]
24. Zhou, F.; Su, H.; Liang, X.; Meng, L.; Yuan, L.; Li, X.; Liang, T. Integrated technology of efficient fracture network stimulation and enhanced oil recovery in tight oil reservoirs. *Pet. Explor. Dev.* **2019**, *46*, 1007–1014. [CrossRef]
25. Crump, J.B.; Conway, M.W. Effects of perforation-entry friction on bottom hole treating analysis. *J. Pet. Technol.* **1988**, *40*, 1041–1048. [CrossRef]
26. Stephen, R.; Thomas, L.; Tim, L.; Kacper, W.; Andrew, E.; Blake, H.; Michael, O. Acoustic imaging of perforation erosion in hydraulically fractured wells for optimizing cluster efficiency. In Proceedings of the SPE Hydraulic Fracturing Technology Conference and Exhibition, The Woodlands, TX, USA, 4–6 February 2020.

materials

MDPI

Article

Performance Study of Wearable Thermoelectric Cooler with Phase-Change Composite Heat Sink

Zhanglong Xia, Wei Cao, Xiaolong Sun, Qianfeng Ding, Zheng Zhu, Wenjie Zhou, Sijia Yan, Yue Hou * and Ziyu Wang *

The Institute of Technological Sciences, Wuhan University, Wuhan 430072, China; xiazhanglong@whu.edu.cn (Z.X.); wei_cao@whu.edu.cn (W.C.); xiaolongsun@whu.edu.cn (X.S.); qfding@whu.edu.cn (Q.D.); zhengzhu@whu.edu.cn (Z.Z.); wenjiezhou@whu.edu.cn (W.Z.); yansijia@whu.edu.cn (S.Y.)

* Correspondence: yuehou@whu.edu.cn (Y.H.); zywang@whu.edu.cn (Z.W.)

Abstract: Based on existing studies, we identified that the heat sinks used in wearable thermoelectric coolers (WTECs) are predominantly bulky, which limits their practicality and comfort. To address this issue, we propose the use of phase-change composite materials (PCCMs) due to their inherent flexibility and thermal properties. Through comprehensive theoretical analysis, numerical simulations, and experimental validation, we successfully optimized the design of a WTEC.

Keywords: wearable thermoelectric cooler; personal thermal management; phase-change composite material; simulation analysis

1. Introduction

Facing the reality of existing extremely hot weather, personal thermal management technology, which enhances the overall thermal comfort by cooling the specific area, is garnering increasing attention [1–5]. Within multiple cooling performance, thermoelectric devices hold significant potential due to their temperature-controllable cooling ability [6–10]. However, the devices' cooling durations are highly restricted by the poor heat dissipation on their hot ends [11–16]. Previous studies have shown that adding heat sinks or improving heat ventilation (or heat exchange) can both enhance cooling performance [17–22].

The sophisticated manufacturing technology employed by the thermoelectric device will also serve as a cornerstone for enhancing its operational flexibility and resilience [23,24]. Hou et al. introduced ultra-flexible fabric-based thermoelectric generators, incorporating conductive fabric electrodes and elastic fabric substrates [24]. Combined with serpentine-structured fabric electrodes, the device exhibits exceptional structural integrity and flexibility. At the same time, conductive polymers are highly suitable for use in wearable thermoelectric devices in low-temperature regions [25,26], as their flexibility and biocompatibility provide significant advantages over inorganic materials. To tackle the issue of low carrier mobility in conducting polymers, Gharahcheshmeh et al. investigated the nanostructures of poly(3,4-ethylenedioxythiophene) films fabricated via oxidative chemical vapor deposition at various deposition temperatures [26], and finally they achieved a substantial improvement in carrier mobility and conductivity. Indeed, efficient heat dissipation remains a critical consideration for both fabric-based devices and those utilizing conductive polymers.

He et al. investigated the performance of plate-fin heat sinks using rectangular and V-shaped fins under lateral and vertical convection conditions and in multiple arrangements.

Their results showed that rectangular fins arranged in double rows exhibited the highest heat dissipation capacity on the hot end of the thermoelectric cooler under vertical jet conditions [18]. Mark Baldry et al. developed an optimized natural convection heat sink through iterative numerical simulations for a small thermoelectric cooler (TEC) module, achieving a steady-state reference temperature 11.7 °C lower than a conventional design [27]. Liu et al. demonstrated the potential of integrating heat pipes into thermoelectric cooling systems for electronic device applications [28]. Y. Lyu et al. developed a battery thermal management system combining thermoelectric, forced air, and liquid cooling methods, and the experimental results showed good cooling effectiveness and reasonable power consumption [29]. However, the application of the aforementioned heat dissipation methods in wearable thermal management is not applicable due to their limitations, including bulky size and lack of flexibility.

To address the cooling issue and at the same time provide a wearable-compatible solution for TEC heat dissipation, phase-change material (PCM) becomes a perfect solution. In multiple application scenarios, thermal management based on PCM has been reported [30–39]. Xiang et al. prepared polyurethane rigid foam composites which are made via in situ foaming the polyurethane prepolymer blended with microencapsulated phase-change material [30]. Huang et al. prepared a novel anti-leakage and anti-vibration thermally induced flexible phase-change composite material which utilized Ethylene–Propylene–Diene Monomer to increase the crosslinking of styrene–butadienestyrene block copolymer so as to improve the adsorption of paraffin (PA) [31]. Rohan Bhuiya et al. utilized ANSYS 18.1 to analyze the impact of phase-change materials on heat dissipation, specifically investigating various shapes of heat sinks integrated with these materials on thermoelectric coolers. Their findings indicated that the use of phase-change materials significantly lowered the hot-side temperature of the thermoelectric cooler compared to scenarios without such materials, highlighting their potential in thermoelectric cooling system thermal management [32]. Wang et al. performed a numerical comparative study on heat sink modules transitioning between single-phase and solid–liquid phase, examining both solid and unconsolidated porous structures. Their findings clearly demonstrated that phase-transitioned materials can markedly decrease the cold surface temperature and enhance the transient coefficient of performance (COP) of TEC [33]. Since pure PCM is prone to leakage, the current research on thermoelectric devices with PCM is mainly focused on simulation.

At this stage, the phase-change composite materials (PCCMs) listed above are not flexible and they do not have a mature application when combined with TEC. Meanwhile, in this work, we developed anti-leakage PCCM to be used as a heat sink for flexible TECs. The base material, Epoxy Resin, can effectively encapsulate the PCM particles, thus avoiding leakage of liquid PCM. Due to the good flexibility of PCCM and the high heat capacity, the PCCM can be applied to flexible TEC as a heat dissipation layer for wearable use. Both COMSOL simulation (COMSOL Multiphysics 6.2) and experiments have found that higher PCM content and geometry (thickness) are more favorable to improve the cooling effect of wearable thermoelectric coolers (WTECs). Subsequent experiments demonstrated that the WTEC in this study could achieve a temperature drop of 2.5 °C on the human skin and sustain the cooling duration for up to 10 min.

In this work, to address the issue of bulky heat sinks in existing WTECs, we propose the use of flexible PCCMs for efficient heat dissipation. Through theoretical analysis, simulation studies, and experimental validation, we demonstrate that the integration of PCCM enables the WTEC to achieve a temperature reduction of 2.5 °C on human skin, offering a promising solution for enhanced thermal comfort.

2. Experimental Section

2.1. Materials and Reagents

The materials we use are all commercially available. The p-type $Bi_{0.5}Sb_{1.5}Te_3$ and n-type $Bi_2Te_{0.3}Sb_{2.7}$ thermoelectric legs with the size of $1.4 \times 1.4 \times 2.5$ mm^3 were purchased from Hubei Segre New Energy Technology, Company Ltd., Wuhan, China. Our thermoelectric leg material parameters are consistent with those used in another study [40]. The zT value of the p-type $Bi_{0.5}Sb_{1.5}Te_3$ thermoelectric legs is 0.927, while the zT value of the n-type $Bi_2Te_{0.3}Sb_{2.7}$ thermoelectric legs is 0.965, both measured at 303 K. The flexible printed circuit board (FPCB) was purchased from Jialichuang Technology Group, Company Ltd., Shenzhen, China. Epoxy Resin (Leini 105) was purchased from Huizhou Tinglan Technology Company, Huizhou, China, and PA (PCM-A-30) was purchased from Dongguan Zhangmutou Shengbang Plastic Raw Material Operation Department, Dongguan, China, as PCM.

2.2. Fabrication of PCCM and WTEC

The fabrication process of the PCCM is shown in Figure S2. Epoxy Resin A and B agents were added sequentially, mixed, and stirred at 50 °C for 10 min. PA was then added, mixed, and stirred at ratios of 5, 10, and 15 wt% for another 10 min. The mixture was then vacuumed for 5 min to remove any bubbles before being poured into a mold which measures 24×44 mm^2. The mixture was then cured at 50 °C for 8 h, after which the PCCM was obtained.

The preparation detail for the WTEC is demonstrated in Figure S3. The overall FPCB size is 24×44 mm^2. First, the FPCB was placed on an unheated heating plate and low-temperature solder paste [$Sn_{42}Bi_{58}$ (melting point is 138 °C)] was loaded onto the mask and swiftly printed onto the substrate electrode with a squeegee. Second, the p-type and n-type thermoelectric legs were deposited onto the bottom electrode of the FPCB. Next, the whole sample was heated up to 200 °C for solder bonding and the soldering time was 5 min. Subsequently, the top electrode was welded using the same technological process.

2.3. Testing Instruments and Methods

Pure Epoxy Resin and our PCCM underwent differential scanning calorimetry (DSC) experiments (NETZSCH, Munich, Germany) to assess their heat absorption and release capabilities. The DSC experiments were set to start at 18 °C and end at 55 °C, while the ramp-up and ramp-down rates were all 3 °C/min. All samples were tested in a nitrogen (N_2) atmosphere. The optical photos of Pure Epoxy Resin and PCCM were characterized by OLYMPUS (Center Valley, PA, USA). The scanning electron microscope (SEM) images of p-type and n-type thermoelectric legs were characterized by Zeiss SIGMA (Oberkochen, Germany). The Fourier transform infrared spectrometer (FT-IR) curves of pure PA, pure Epoxy Resin, and PCCM were characterized with a Thermo Fisher Scientific (Waltham, MA, USA) Nicolet iS20, and the scanning range was 400–4000 cm^{-1}. The bending and tensile properties were evaluated using a flexible electronic tester (Prtronic FT2000, Shanghai Mifang Electronic Technology, Company Ltd., Shanghai, China). The skin temperature before and after cooling was recorded using a thermal infrared imager (FLUKE Ti480 PRO, FLUKE, Norwich, UK). At the same time, a thermocouple was placed at the interface between the skin and the cold end of the WTEC to measure the temperature. The internal resistance of the device was measured by using an electrochemical analyzer (CHI650e, CH Instruments, Inc., Bee Cave, TX, USA). The data tested are the average values of at least three measurements.

3. Results and Discussion

3.1. Theoretical Model

Figure 1 illustrates the schematic diagram of the WTEC, which includes p-type and n-type thermoelectric legs, copper connecting electrodes, and a polyimide substrate. The fundamental principle governing TEC devices is the Peltier effect.

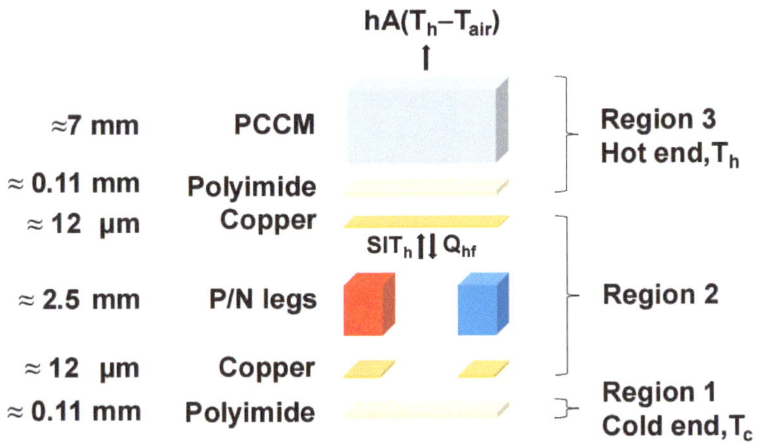

Figure 1. Schematic diagram of WTEC.

We are very concerned with the cooling efficiency, or COP. The COP can be calculated using the following equation [34,35]:

$$COP = \frac{T_c}{T_h - T_c} \frac{\sqrt{1+zT_m} - \frac{T_h}{T_c}}{1+\sqrt{1+zT_m}} \tag{1}$$

In this equation, T_c represents the temperature at the cold end, T_h represents the temperature at the hot end, z is the figure of merit of thermoelectric material, and T_m is the average temperature between the cold end and the hot end. In this work, the transient response of the TEC is our primary focus. For the TEC depicted in the diagram, the energy equation for the hot end is as follows [34]:

$$(M_{hpi}C_{hpi} + M_{PCCM}C_{PCCM})\frac{dT_h}{dt} = SIT_h - Q_{hf} - hA(T_h - T_{air}) \tag{2}$$

Here, M_{hpi} represents the mass of polyimide at the hot end, C_{hpi} denotes the specific heat capacity of polyimide at the hot end, M_{PCCM} is the mass of PCCM, C_{PCCM} is the specific heat capacity of PCCM, T_h represents the temperature at the hot end, S is the Seebeck coefficient of the thermoelectric legs, I is the applied current, Q_{hf} denotes the heat transfer rate between the hot end polyimide and the thermoelectric legs, h is the heat transfer coefficient at the hot surface of the thermoelectric device, A is total heat transfer surface area of the PCCM, and T_{air} represents the ambient temperature [41].

From Equation (2), it can be observed that increasing the mass M_{PCCM} and specific heat capacity [36] C_{PCCM} of the PCCM can significantly enhance the heat capacity at the hot end of TEC, thereby reducing the rate of hot end temperature recovery $\frac{T_h}{t}$. The most straightforward method to increase the mass M_{PCCM} of the PCCM is to adjust its geometric parameters. Notably, increasing the thickness not only enhances the heat capacity at the hot end but also increases the heat dissipation area, both of which contribute positively to controlling the temperature at the hot end. According to the apparent heat capacity method, increasing the content of PCM can significantly boost the specific heat capacity of the composite material [42]. This indicates that a reasonable enhancement of the thickness and PCM content of the PCCM is beneficial for the thermal management of the hot end of the device.

3.2. Simulation Module and Results

3.2.1. Phase-Change Module

In the subsequent experiments, we observed that the PCCM had excellent shape stability during the phase-change process, which means the appearance is almost unchanged during the phase-change process, so we simulated the solid–solid phase change as a solid–solid phase change, and at the same time, we investigated the phase change by the apparent heat capacity method [43]. The following equation controls the phase-change module [44]:

$$\rho_{pccm} c_{p,pccm} \frac{\partial T}{\partial t} = \nabla \left(k_{pccm} \nabla_T \right) \tag{3}$$

where ρ_{pccm} represents the equivalent density of the PCCM, $c_{p,pccm}$ represents the equivalent specific heat capacity, T represents the temperature, k_{pccm} represents the thermal conductivity of the PCCM, and t represents the time.

The equivalent thermal conductivity, equivalent density, and equivalent heat capacity of the PCCM are related to the phase-change fraction $\theta(T)$ and can be used to calculate the heat transfer in the material during the phase-change process. The values k_{pccm}, ρ_{pccm}, and $C_{p,pccm}$ of the PCCM are determined as follows during the phase-change process, respectively:

$$k_{pccm} = k_{s1} + (k_{s2} - k_{s1})\theta(T) \tag{4}$$

$$\rho_{pccm} = \frac{\rho_{s2} C_{p,s2} \theta(T) + \rho_{s1} C_{\rho,s1} (1 - \theta(T))}{C_{p,s2} \theta(T) + (1 - \theta(T)) C_{p,s1}} \tag{5}$$

$$C_{p,pccm} = C_{p,s1} + \left(C_{p,s2} - C_{p,s1} \right) \theta(T) + \Delta H_{composite} \cdot D(T) \tag{6}$$

k_{s1} and k_{s2} represent the thermal conductivity of the PCM in the PCCM before and after melting, respectively; ρ_{s1} and ρ_{s2} represent the density of the PCM in the PCCM before and after melting, respectively; $C_{\rho,s1}$ and $C_{p,s2}$ represent the constant-pressure heat capacity of the PCM in the PCCM before and after melting, respectively; $\Delta H_{composite}$ represents the latent heat of the PCCM value; $D(T)$ is a Gaussian function to account for the latent heat in the melting temperature range; and the phase transition fraction $\theta(T)$ is used to describe the phase transition of the PCCM, and according to the principle of phase transition, the value of $\theta(T)$ can also be expressed as the following equation.

$$\theta(T) = \begin{cases} 0, T < T_m - \frac{\Delta T}{2} \\ \frac{T - T_m + \Delta T/2}{\Delta T}, T_m - \frac{\Delta T}{2} < T < T_m + \frac{\Delta T}{2} \\ 1, T > T_m + \frac{\Delta T}{2} \end{cases} \tag{7}$$

T_m represents the melting temperature of the PCCM and ΔT represents the melting temperature range of the PCCM. In Section 2, we mention that we choose PA as the PCM. The PA we chose has a melting point of 30 °C, which meets the standard that it will not melt at room temperature. Since the PCCM is a shape-stabilized material, as confirmed by subsequent experiments, we can study the PCCM as a whole [44].

The formula for calculating the latent heat of the composite materials is shown below:

$$\Delta H_{composite} = w_{PA} \times \Delta H_{PA} \tag{8}$$

$\Delta H_{composite}$ is the latent heat value of the shape-stabilized PCCM [45–47], w_{PA} is the mass fraction of the phase-change material to the PCCM, and ΔH_{PA} is the latent heat value of the phase-change material.

In the simulations, the selected PA contents are 5 wt%, 10 wt%, 15 wt%, and 20 wt%, respectively.

3.2.2. Thermoelectric Module

The non-steady-state energy equation for a p/n-type semiconductor column can be described as follows:

$$\left(\rho_{p/n}C_{p/n}\right)\frac{\partial T}{\partial t} = \nabla\left(k_{p/n}\nabla T\right) + \frac{\vec{J}^2}{\sigma_{p/n}(T)} - \nabla S_{p/n}\vec{J}T_{p/n} \tag{9}$$

Here, $\rho_{p/n}$ denotes the equivalent density of the thermoelectric material, $C_{p/n}$ denotes the specific heat capacity of the thermoelectric material, T represents the temperature, t represents the time, $k_{p/n}$ denotes the thermal conductivity of the thermoelectric material, $\sigma_{p/n}(T)$ represents the electrical conductivity of the thermoelectric material, $T_{p/n}$ represents the temperature of the thermoelectric material, and $S_{p/n}$ denotes the Seebeck coefficient.

$$(\rho_{cop}C_{cop})\frac{\partial T}{\partial t} = \nabla(k_{cop}\nabla T) + \frac{\vec{J}^2}{\sigma_{cop}} \tag{10}$$

$$(\rho_{pi}C_{pi})\frac{\partial T}{\partial t} = \nabla(k_{pi}\nabla T) \tag{11}$$

For the copper bonded substrate, the subscript 'cop' denotes Cu, and the subscript 'pi' denotes polyimide. Here, ρ_{cop} denotes the equivalent density of the Cu electrode, C_{cop} denotes the specific heat capacity of the Cu electrode, σ_{cop} represents the electrical conductivity of the Cu electrode, T represents the temperature, t represents the time, k_{cop} denotes the thermal conductivity of the Cu electrode, ρ_{pi} denotes the equivalent density of the polyimide, C_{pi} denotes the specific heat capacity of the polyimide, and k_{pi} represents the thermal conductivity of the polyimide.

$$\vec{J} = \sigma\vec{E} \tag{12}$$

The electric current density vector corresponds to the electric field density vector and the electric resistivity. Here, $\vec{J}$ denotes the electric current density vector, $\vec{E}$ denotes the electric field density vector, and σ denotes the electric resistivity.

$$\vec{E} = -\nabla\varphi + S_{p/n}(T)\nabla T \tag{13}$$

The electric field density vector can be obtained by the electric potential and Seebeck motive force. Here, $\vec{E}$ denotes the electric field density vector, φ denotes the electric potential, $S_{p/n}(T)\nabla T$ represents the Seebeck motive force.

Because of the electrical current's continuity properties, the current flowing through the components follows this rule below:

$$\nabla\cdot\vec{J} = 0 \tag{14}$$

Here, $\vec{J}$ denotes the current density. After setting up the phase-change module and thermoelectric module, we mainly focus on the transient temperature changes and COP of the WTEC.

3.2.3. Assumed Conditions

To simplify this study, we make the following assumptions in the simulation:

1. Three-dimensional transient heat transfer is considered;
2. Negligible contact thermal resistance is assumed;
3. The surface of the PCCM is thermally insulated and isotropic;
4. The ambient temperature is 25 °C, corresponding to room temperature.

3.2.4. Simulation Results

Based on Equation (2), both the PA content and geometric dimensions (thickness) of PCCM significantly influence its thermal capacity. Consequently, we systematically investigated the impact of these parameters on the cooling performance of the TEC system.

Figure 2a illustrates the temperature distribution of the WTEC system following 400 s of electrical current application, revealing an inverse relationship between PA content and the hot end temperature. Figure 2b depicts the temporal evolution of the phase-change percentage in the PCCM. The complete phase-change durations for PCM mass fractions of 5, 10, 15, and 20 wt% were determined to be 297, 373, 605, and 871 s, respectively. These results demonstrate a direct correlation between PCM concentration and the material's thermal energy storage capacity. Figure 2c demonstrates the temperature variation profile on the cold side of the WTEC as a function of PA content. The cooling durations ($\Delta T > 2$ °C) are also summarized in Figure 2d. Specifically, the WTEC maintains cooling above 2 °C for 360, 460, 690, and 950 s when the PA content is 5, 10, 15, and 20 wt%, respectively. Following 400 s of current application, the cold surface temperatures were recorded at 23.3, 22.3, 21.1, and 20.6 °C for increasing PA contents, as illustrated in Figure 2e. The collective analysis of Figure 2c–e indicates a significant enhancement in cooling performance with elevated PA content, manifested through both extended cooling duration and reduced operational temperatures.

According to the research literature, the increase in PCM content can directly lead to the larger specific heat capacity of the thermal layer in the phase-transition temperature range [48]. Equation (2) reveals that the thermal capacity at the hot end of the thermoelectric device undergoes significant enhancement during the phase transition process. This phenomenon is clearly demonstrated in Figure 2f, which shows a marked reduction in the temperature rise rate at the hot end during the material's phase-transition period. Concurrently, the COP was systematically evaluated, as presented in Figure 2g. The results demonstrate a positive correlation between the COP value and increasing PA content. This phenomenon can be attributed to the enhanced heat absorption capacity of the PCCM with increased PA content, which promotes more efficient thermal energy transfer from the hot surface. Consequently, this mechanism effectively reduces the temperature gradient between the hot and cold surfaces.

In addition to the PA content, we also investigated the influence of the geometric dimensions (thickness) of the PCCM on the cooling performance of thermoelectric devices. Figure 3a presents the temperature distribution of the WTEC following 400 s of current application, demonstrating that increased PCCM thickness results in significantly reduced operational temperatures at the thermal surface. Figure 3b depicts the temporal evolution of phase transition percentage within the PCCM material system. The complete phase transition durations were measured to be 206, 376, 605, and 885 s for PCCM thicknesses of 3, 5, 7, and 9 mm, respectively, revealing a positive correlation between PCCM thickness and phase-change completion time. Figure 3c demonstrates a characteristic trend in cold surface temperature variation with increasing PCCM thickness: an initial decrease to a minimum threshold value, followed by a gradual temperature recovery. The experimental results reveal that increased PCCM thickness significantly reduces the cold surface temperature of the WTEC. The cooling durations (ΔT > 2 °C) are also summarized in Figure 3d. Specifically, the WTEC maintains cooling above 2 °C for 280, 480, 690, and 940 s when the PCCM thickness is 3, 5, 7, and 9 mm, respectively. As quantitatively demonstrated in Figure 3e, the cold surface temperatures were recorded at 24.4, 22.1, 21.1, and 20.9 °C, respectively, following 400 s of current application.

As the thickness of the PCCM increases, so does the mass M_{PCCM}. Consequently, the rate of temperature increase at the hot end $\frac{T_h}{t}$ decreases during the phase transition of the composite material, as shown in Figure 3f. Utilizing Equation (1) for calculation, Figure 3g illustrates the obtained instantaneous COP, which increases with the thickness of the PCCM during its operation. This phenomenon arises because the increased thickness of the PCCM enhances the overall heat capacity at the hot end and accelerates heat transfer from the hot surface to the PCCM, thereby reducing the temperature difference between the hot and cold surfaces.

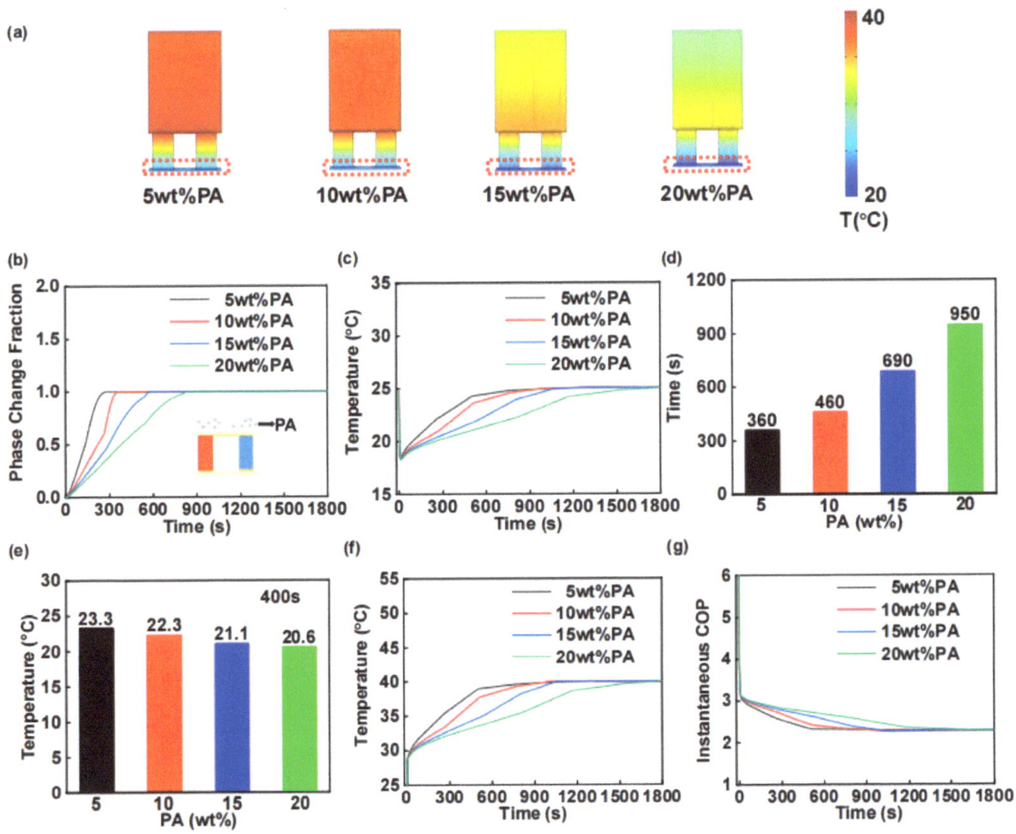

Figure 2. Simulation analysis of the effect of PA content on the cooling performance of WTEC, with PA content from 5 wt% to 20 wt% with 5 wt% interval. (**a**) Temperature cloud of WTEC after 400 s of on-current. (**b**) Phase-change fraction curve. (**c**) WTEC cold surface temperature curve. (**d**) Cooling duration of WTEC. (**e**) Cold surface temperature after 400 s of on-current. (**f**) WTEC hot surface temperature. (**g**) Instantaneous COP of WTEC.

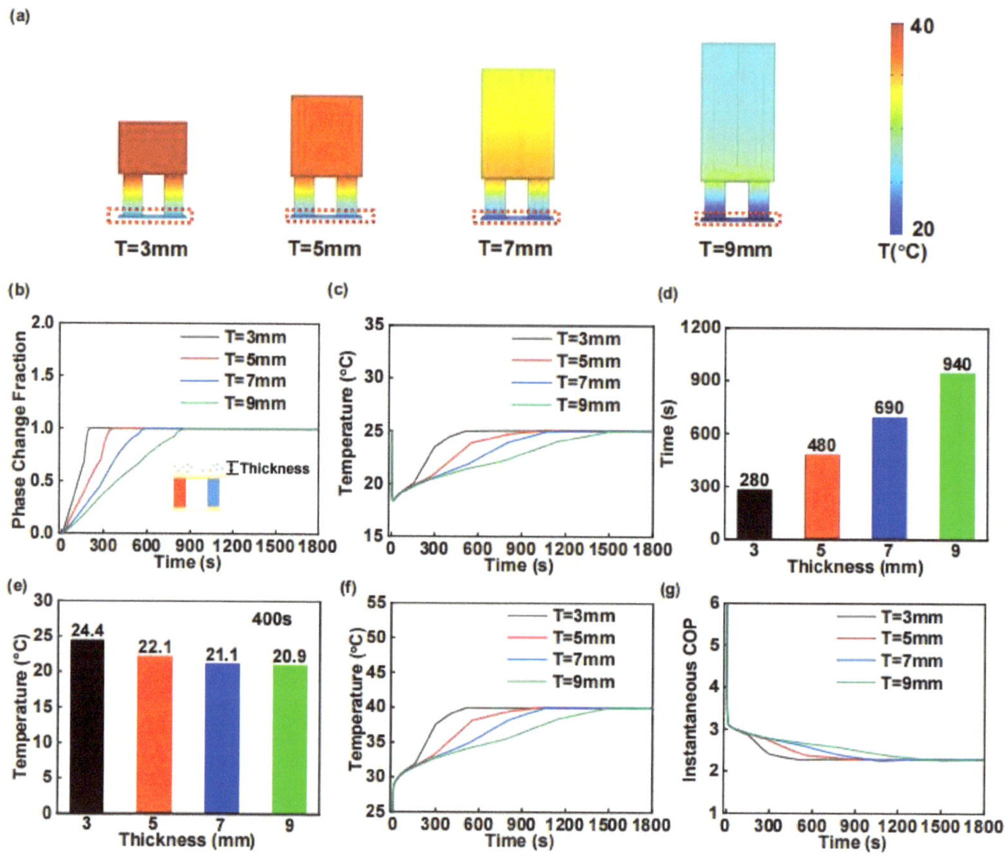

Figure 3. Simulation analysis of the effect of PCCM thickness on the cooling performance of WTEC, with thickness from 3 mm to 9 mm with 2 mm interval. (**a**) Temperature cloud of WTEC after 400 s of on-current. (**b**) Phase-change fraction curve. (**c**) Cold surface temperature curve. (**d**) Cooling duration of WTEC. (**e**) Cold surface temperature after 400 s of on-current. (**f**) Hot surface temperature. (**g**) Instantaneous COP of WTEC.

3.3. Experimental Results

3.3.1. Experimental Verification

Figure 4a–c reveal a significant positive correlation between cooling duration and PA content at a fixed device thickness of 7 mm. Quantitative analysis shows that increasing the PA content from 5 to 15 wt% results in a progressive extension of the cooling period from 450 to 1080 s, with a measured duration of 670 s at the intermediate content of 10 wt%. Furthermore, the minimum achievable temperature shows an inverse relationship with PA content, decreasing progressively with higher PA content.

Additionally, when the PA content is fixed at 15 wt%, increasing the PCCM thickness from 3 mm to 9 mm (in 2 mm increments) leads to longer cooling durations of 520, 850, 1080, and 1300 s, respectively, as shown in Figure 4d–f. These experimental results exhibit the same trends as the simulation outcomes.

It should be noted that a larger geometric thickness will deteriorate the flexibility of the heat dissipation layer. Therefore, considering both flexibility and heat dissipation capacity, we ultimately chose a geometric dimension of 7 mm. In summary, the final selected heat dissipation layer has a PA content of 15 wt% and a geometric thickness of 7 mm.

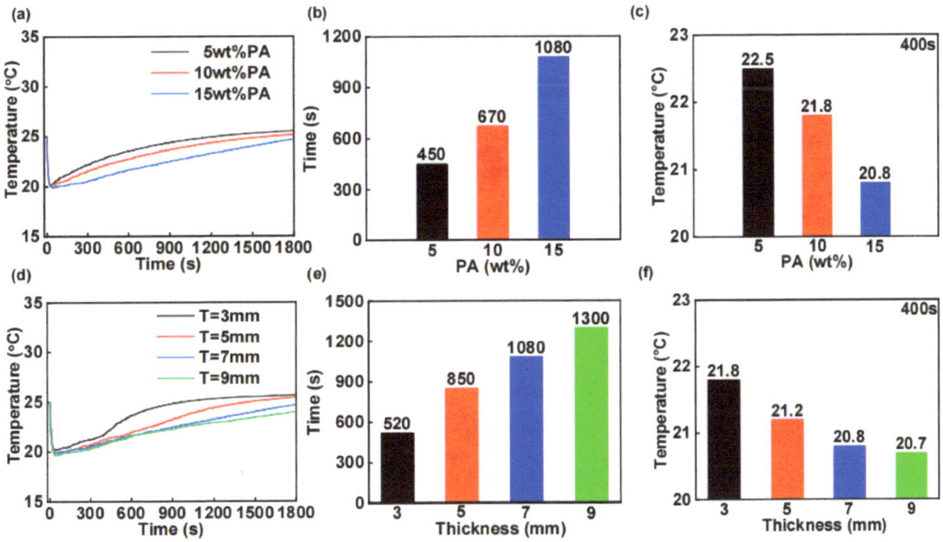

Figure 4. Experimental analysis of the effect of PA content on the cooling performance of WTEC is shown in (**a**–**c**). (**a**) Cold surface temperature of WTEC. (**b**) Cooling duration of WTEC. (**c**) Cold surface temperature after 400 s of on-current. Experimental analysis of the effect of PCCM thickness on the cooling performance of WTEC is shown in (**d**–**f**). (**d**) Cold surface temperature of WTEC. (**e**) Cooling duration of WTEC. (**f**) Cold surface temperature after 400 s of on-current.

To enhance thermoelectric response or ensure optimal adaptation of the entire device to the electrical load, we applied varying currents and monitored the resulting variation in the cold surface temperature of the WTEC over time [49]. The findings are presented in Figure S4. It is evident that the cold surface temperature of the WTEC reaches its minimum value at an applied current of 0.30 A. However, the temperature does not decrease further at 0.35 A, as the Joule heating generated by the device begins to exceed the heat absorbed by the Peltier effect [34]. Additionally, the rate of recovery of the cold surface temperature at 0.35 A is faster than that at 0.30 A. Based on these experimental observations, we selected a current of 0.30 A to achieve optimal electrical load.

3.3.2. WTEC and Materials Performance

For body temperature regulation applications, the flexibility and structural stability of the WTEC system represent critical parameters that significantly influence its practical implementation and operational reliability.

Figure 5a illustrates the WTEC's flexibility [50], demonstrating its ability to bend with a curvature radius of 40 mm, along with its internal resistance stability after undergoing 1000 repeated bending cycles. Figure 5b highlights the superior anti-leakage properties of the PCCM. The pure PA fully melted within 30 min at a temperature of 30 °C, while the PCCM exhibits no significant leakage, where 30 °C is the melting point of the PA. Figure 5c presents the cooling test results for the WTEC under conditions of no bending and 500 and 1000 bending cycles, indicating a consistent cooling performance across these conditions. Figure 5d contrasts the cold end temperatures of the thermoelectric cooler with and without the PCCM, highlighting that the phase-change heat sink effectively lowers the cold surface temperature of the thermoelectric cooler. The lowest temperature of the cold surface of the WTEC with phase-change heat sink is 2.3 °C lower than that without phase-change heat sink. Figure 5e exhibits the stable cooling performance of the

WTEC following three extended exposures to the melting point temperature and cooling, affirming the reusability of the WTEC.

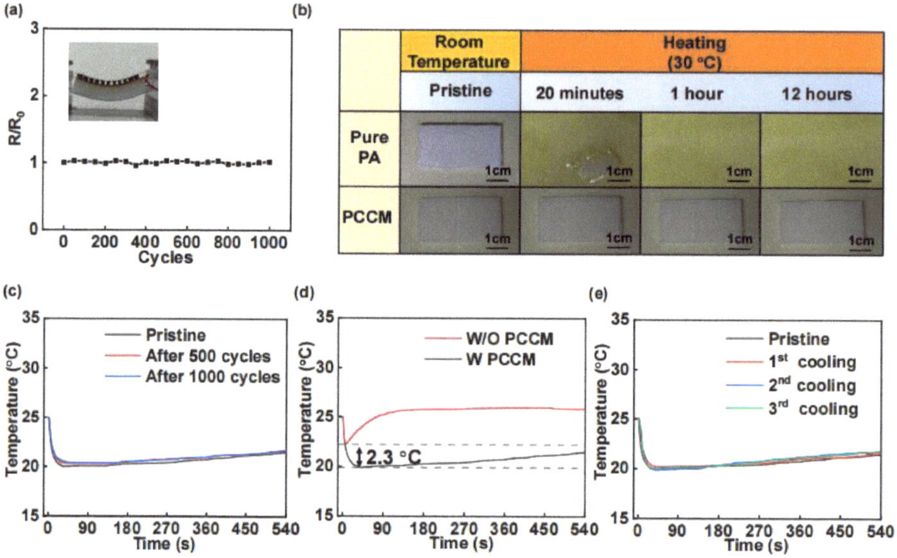

Figure 5. Stability testing of WTEC and PCCM. (**a**) The remarkable stability of resistance during 1000 times of bending of the WTEC. (**b**) Anti-leakage features of PCCM. (**c**) The stability of the cooling effect of the WTEC before bending, after 500 times of bending, and after 1000 times of bending. (**d**) The cold surface temperature of the thermoelectric cooler with and without PCCM. (**e**) The stability of the cooling performance of the WTEC after three prolonged exposures to the melting point temperature and then cooled.

Optical microscope images of pure Epoxy Resin and PCCM (15 wt%) are shown in Figure 6a, indicating that the PA particles are well encapsulated within the Epoxy Resin. Figure 6b displays the DSC curves [51] of pure Epoxy Resin and the PCCM. The range of heat absorption peaks for the PCCM demonstrates its effectiveness in thermal management within the desired temperature range. Figure 6c depicts an optical photograph of the PCCM, illustrating its good flexibility. FT-IR spectra of pure Epoxy Resin, pure PA, and the PCCM are presented in Figure 6d. The PA curve depicted that there were peaks at 2921 and 2850 cm^{-1},which were characteristic peaks [52]. The curves for both PCCM and Epoxy Resin exhibit a characteristic peak at 910 cm^{-1}, which is indicative of the Epoxy Resin. The FT-IR spectra of the PCCM closely resemble those of pure Epoxy Resin due to the low PA content. No new peaks appear in the FT-IR spectrum of the PCCM, indicating that the Epoxy Resin and PA are physically mixed without undergoing chemical reactions [53]. The SEM images of the p-type and n-type thermoelectric legs are presented in Figure 6e and Figure 6f, respectively. It is evident that the layer structure of the n-type material is significantly larger than that of the p-type material. Additionally, the p-type material exhibits lower density compared to the n-type material. These structural differences explain why the thermal conductivity of the n-type material is higher than that of the p-type material.

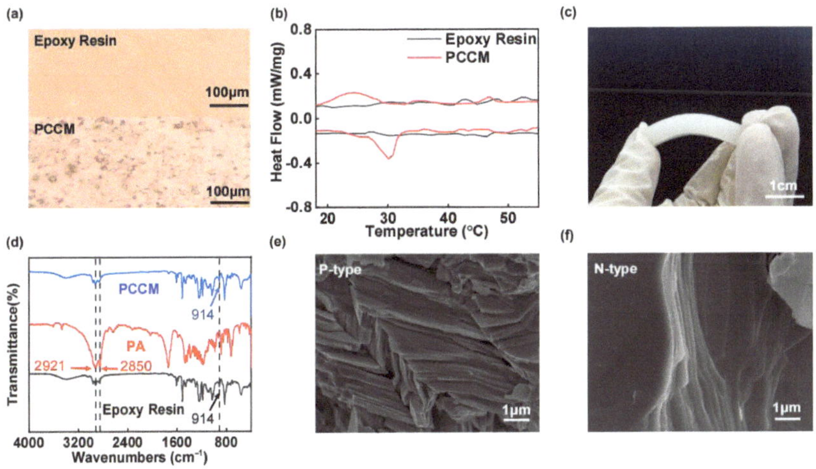

Figure 6. (**a**) Optical microscope photo of Epoxy Resin and PCCM. (**b**) DSC curve of Epoxy Resin and PCCM. (**c**) Optical photo of PCCM. (**d**) FT-IR curve of PCCM, PA, and Epoxy Resin. (**e**) SEM image of p-type thermoelectric leg. (**f**) SEM image of n-type thermoelectric leg.

3.3.3. Device Application

Human beings rely on their skin to perceive external temperatures. In hot environments, lowering the temperature of the skin's surface can effectively enhance human comfort. Of course, excessively low skin temperature can also cause discomfort.

Given that most of the human skin surface is curved [24], flexible cooling devices better meet the needs of wearable temperature control. Here, we also conducted tests and validation of the wearable cooling characteristics of the WTEC on human skin. Figure 7a illustrates the superior skin adherence of the WTEC due to its flexibility. The on-body temperature when wearing the WTEC was tested in Figure 7b, showing the largest temperature drop of 2.5 °C with a cooling duration of up to 10 min, and its cooling performance was further proved by the infrared images in Figure 7c,d, which were in accord with the WTEC off and on, respectively.

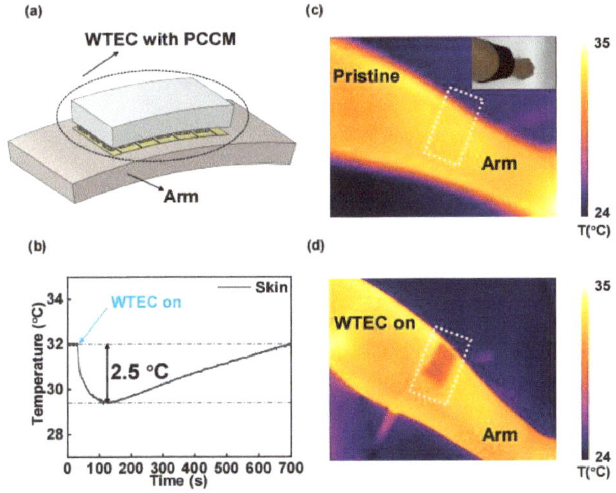

Figure 7. Cooling effect of human skin with WTEC. (**a**) Schematic of WTEC cooling on human arm. (**b**) Skin temperature change curve. Infrared images of arm (**c**) before cooling and (**d**) after cooling.

4. Conclusions

In this work, we successfully prepared PCCM with both flexibility and the property and it does not leak, which could serve as the heat sink layer for flexible TECs. Through theoretical analysis, simulation, and experimental tests, it was verified that both the PA content and geometry parameters, namely the thickness, could affect the heat capacity of the PCCM heat sink, thereby leading to the cooling performance optimization of the WTEC. The experimental results show that the higher the PA content in the PCCM, the better the cooling effect of the WTEC, and the longest cooling time can reach 1080 s; similarly, the larger the geometry (thickness) of the PCCM, the better the cooling effect of the WTEC, of which the longest cooling time can reach 1080 s when the PCCM is 7 mm. The PCCM was also proved to be anti-leakage even at the transition temperature of 30 °C for at least 12 h. Moreover, the cold surface temperature of the WTEC with PCCM can be further reduced by 2.3 °C compared to the WTEC without PCCM. The WTEC combined with the PCCM heat sink was also fabricated with a proper temperature drop of 2.5 °C and a cooling duration of up to 10 min on human skin. The study reveals an approach that may be used to address long-lasting body temperature regulation in the future.

Supplementary Materials: The following supporting information can be downloaded at: https://www.mdpi.com/article/10.3390/ma18071576/s1, Figure S1: Rendering of the FPCB; Figure S2: Fabrication process of phase-change composite material; Figure S3: Preparation process of thermoelectric cooler; Figure S4: Variation in the cold surface temperature of WTEC with time under different applied currents; Table S1: Comparison of this study with other studies about WTEC; Table S2: Thermal conductivity of WTEC and PCCM (303 K).

Author Contributions: Conceptualization, Z.X., Y.H. and Z.W.; methodology, software, validation, Z.X., W.C. and X.S.; formal analysis, investigation, Z.X., Q.D. and Z.Z.; resources, data curation, Z.X., W.Z. and S.Y.; writing—original draft preparation, visualization, Z.X.; writing—review and editing, supervision, Z.X. and Y.H.; project administration, Y.H. and Z.W. All authors have read and agreed to the published version of the manuscript.

Funding: This research received no external funding.

Institutional Review Board Statement: Not applicable.

Informed Consent Statement: Not applicable.

Data Availability Statement: The original contributions presented in this study are included in the article/Supplementary Materials. Further inquiries can be directed to the corresponding authors.

Acknowledgments: The authors thank Zhengyou Liu for providing the right to use COMSOL software.

Conflicts of Interest: The authors declare no conflicts of interest.

Abbreviations

The following abbreviations are used in this manuscript:

WTEC	Wearable thermoelectric cooler
TEC	Thermoelectric cooler
PCCM	Phase-change composite material
COP	Coefficient of performance
PCM	Phase-change material
PA	Paraffin
FPCB	Flexible printed circuit board
DSC	Differential scanning calorimetry
SEM	Scanning electron microscope
N_2	Nitrogen
FT-IR	Fourier transform infrared spectrometer

References

1. Hu, R.; Liu, Y.; Shin, S.; Huang, S.; Ren, X.; Shu, W.; Cheng, J.; Tao, G.; Xu, W.; Chen, R.; et al. Emerging Materials and Strategies for Personal Thermal Management. *Adv. Energy Mater.* **2020**, *10*, 1903921. [CrossRef]
2. Ma, Z.; Zhao, D.; She, C.; Yang, Y.; Yang, R. Personal Thermal Management Techniques for Thermal Comfort and Building Energy Saving. *Mater. Today Phys.* **2021**, *20*, 100465. [CrossRef]
3. Veselý, M.; Zeiler, W. Personalized Conditioning and Its Impact on Thermal Comfort and Energy Performance—A Review. *Renew. Sustain. Energy Rev.* **2014**, *34*, 401–408. [CrossRef]
4. Xue, S.; Huang, G.; Chen, Q.; Wang, X.; Fan, J.; Shou, D. Personal Thermal Management by Radiative Cooling and Heating. *Nano-Micro Lett.* **2024**, *16*, 153. [CrossRef]
5. Shoeibi, S.; Kargarsharifabad, H.; Sadi, M.; Arabkoohsar, A.; Mirjalily, S.A.A. A Review on Using Thermoelectric Cooling, Heating, and Electricity Generators in Solar Energy Applications. *Sustain. Energy Technol. Assess.* **2022**, *52*, 102105. [CrossRef]
6. Chen, W.-Y.; Shi, X.-L.; Zou, J.; Chen, Z.-G. Thermoelectric Coolers: Progress, Challenges, and Opportunities. *Small Methods* **2022**, *6*, 2101235. [CrossRef]
7. Zaferani, S.H.; Sams, M.W.; Ghomashchi, R.; Chen, Z.-G. Thermoelectric Coolers as Thermal Management Systems for Medical Applications: Design, Optimization, and Advancement. *Nano Energy* **2021**, *90*, 106572. [CrossRef]
8. Jiang, L.; Zhang, H.; Li, J.; Xia, P. Thermal Performance of a Cylindrical Battery Module Impregnated with PCM Composite Based on Thermoelectric Cooling. *Energy* **2019**, *188*, 116048. [CrossRef]
9. Huo, W.; Xia, Z.; Gao, Y.; Guo, R.; Huang, X. Flexible Thermoelectric Devices with Flexible Heatsinks of Phase-Change Materials and Stretchable Interconnectors of Semi-Liquid Metals. *ACS Appl. Mater. Interfaces* **2023**, *15*, 29330–29340. [CrossRef]
10. Chandel, R.; Chandel, S.S.; Prasad, D.; Dwivedi, R.P. Prospects of Sustainable Photovoltaic Powered Thermoelectric Cooling in Zero Energy Buildings: A Review. *Int. J. Energy Res.* **2022**, *46*, 19319–19340. [CrossRef]
11. Sun, X.; Ling, L.; Liao, S.; Chu, Y.; Fan, S.; Mo, Y. A Thermoelectric Cooler Coupled with a Gravity-Assisted Heat Pipe: An Analysis from Heat Pipe Perspective. *Energy Convers. Manag.* **2018**, *155*, 230–242. [CrossRef]
12. Hyland, M.; Hunter, H.; Liu, J.; Veety, E.; Vashaee, D. Wearable Thermoelectric Generators for Human Body Heat Harvesting. *Appl. Energy* **2016**, *182*, 518–524. [CrossRef]
13. Cai, Y.; Wang, Y.; Liu, D.; Zhao, F.-Y. Thermoelectric Cooling Technology Applied in the Field of Electronic Devices: Updated Review on the Parametric Investigations and Model Developments. *Appl. Therm. Eng.* **2019**, *148*, 238–255. [CrossRef]
14. Guclu, T.; Cuce, E. Thermoelectric Coolers (TECs): From Theory to Practice. *J. Electron. Mater.* **2019**, *48*, 211–230. [CrossRef]
15. Zhao, D.; Lu, X.; Fan, T.; Wu, Y.S.; Lou, L.; Wang, Q.; Fan, J.; Yang, R. Personal Thermal Management Using Portable Thermoelectrics for Potential Building Energy Saving. *Appl. Energy* **2018**, *218*, 282–291. [CrossRef]
16. Siddique, A.R.M.; Venkateshwar, K.; Mahmud, S.; Van Heyst, B. Performance Analysis of Bismuth-Antimony-Telluride-Selenium Alloy-Based Trapezoidal-Shaped Thermoelectric Pallet for a Cooling Application. *Energy Convers. Manag.* **2020**, *222*, 113245. [CrossRef]
17. Liu, X.; Zheng, F.; Fu, Q.; Song, G.; Xiong, Y. Thermoelectric Cooler with Embedded Teardrop-Shaped Milli-Channel Heat Sink for Electronics Cooling. *Appl. Therm. Eng.* **2024**, *244*, 122763. [CrossRef]
18. He, Z.; Yu, Q.; Ye, J.; Yan, F.; Li, Y. Optimization of Plate-Fin Heat Exchanger Performance for Heat Dissipation of Thermoelectric Cooler. *Case Stud. Therm. Eng.* **2024**, *53*, 103953. [CrossRef]
19. Wu, B.; Lin, Y.; Tian, Y.; Wei, W.; Xu, Y.; Hu, Y.; Li, J.; Li, K.; Hou, C.; Zhang, Q.; et al. Bioinspired Wearable Thermoelectric Device Constructed with Soft-Rigid Assembly for Personal Thermal Management. *Adv. Funct. Mater.* **2024**, *34*, 2402319. [CrossRef]
20. Park, H.; Lee, D.; Kim, D.; Cho, H.; Eom, Y.; Hwang, J.; Kim, H.; Kim, J.; Han, S.; Kim, W. High Power Output from Body Heat Harvesting Based on Flexible Thermoelectric System with Low Thermal Contact Resistance. *J. Phys. D Appl. Phys.* **2018**, *51*, 365501. [CrossRef]
21. Park, H.; Kim, D.; Eom, Y.; Wijethunge, D.; Hwang, J.; Kim, H.; Kim, W. Mat-like Flexible Thermoelectric System Based on Rigid Inorganic Bulk Materials. *J. Phys. D Appl. Phys.* **2017**, *50*, 494006. [CrossRef]
22. Wei, H.; Zhang, J.; Han, Y.; Xu, D. Soft-Covered Wearable Thermoelectric Device for Body Heat Harvesting and on-Skin Cooling. *Appl. Energy* **2022**, *326*, 119941. [CrossRef]
23. Lund, A.; Tian, Y.; Darabi, S.; Müller, C. A Polymer-Based Textile Thermoelectric Generator for Wearable Energy Harvesting. *J. Power Sources* **2020**, *480*, 228836. [CrossRef]
24. Hou, Y.; Yang, Y.; Wang, Z.; Li, Z.; Zhang, X.; Bethers, B.; Xiong, R.; Guo, H.; Yu, H. Whole Fabric-Assisted Thermoelectric Devices for Wearable Electronics. *Adv. Sci.* **2022**, *9*, 2103574. [CrossRef]
25. Prunet, G.; Pawula, F.; Fleury, G.; Cloutet, E.; Robinson, A.J.; Hadziioannou, G.; Pakdel, A. A Review on Conductive Polymers and Their Hybrids for Flexible and Wearable Thermoelectric Applications. *Mater. Today Phys.* **2021**, *18*, 100402. [CrossRef]
26. Heydari Gharahcheshmeh, M.; Dautel, B.; Chowdhury, K. Enhanced Carrier Mobility and Thermoelectric Performance by Nanostructure Engineering of PEDOT Thin Films Fabricated via the OCVD Method Using SbCl5 Oxidant. *Adv. Funct. Mater.* **2025**, *35*, 2418331. [CrossRef]

27. Shen, L.; Zhang, W.; Liu, G.; Tu, Z.; Lu, Q.; Chen, H.; Huang, Q. Performance Enhancement Investigation of Thermoelectric Cooler with Segmented Configuration. *Appl. Therm. Eng.* **2020**, *168*, 114852. [CrossRef]
28. Liu, D.; Cai, Y.; Zhao, F.-Y. Optimal Design of Thermoelectric Cooling System Integrated Heat Pipes for Electric Devices. *Energy* **2017**, *128*, 403–413. [CrossRef]
29. Lyu, Y.; Siddique, A.R.M.; Majid, S.H.; Biglarbegian, M.; Gadsden, S.A.; Mahmud, S. Electric Vehicle Battery Thermal Management System with Thermoelectric Cooling. *Energy Rep.* **2019**, *5*, 822–827. [CrossRef]
30. Xiang, H.; An, J.; Zeng, X.; Liu, X.; Li, Y.; Yang, C.; Xia, X. Preparation and Properties of Polyurethane Rigid Foam Materials Modified by Microencapsulated Phase Change Materials. *Polym. Compos.* **2019**, *41*, 1662–1672. [CrossRef]
31. Huang, Q.; Li, X.; Zhang, G.; Kan, Y.; Li, C.; Deng, J.; Wang, C. Flexible Composite Phase Change Material with Anti-Leakage and Anti-Vibration Properties for Battery Thermal Management. *Appl. Energy* **2022**, *309*, 118434. [CrossRef]
32. Bhuiya, R.; Shah, N.; Arora, D.; Krishna, N.V.; Manikandan, S.; Selvam, C.; Lamba, R. Thermal Management of Phase Change Material Integrated Thermoelectric Cooler with Different Heat Sink Geometries. *J. Energy Storage* **2022**, *51*, 104304. [CrossRef]
33. Alkan, C.; Sari, A. Fatty Acid/Poly(Methyl Methacrylate) (PMMA) Blends as Form-Stable Phase Change Materials for Latent Heat Thermal Energy Storage. *Sol. Energy* **2008**, *82*, 118–124. [CrossRef]
34. Cheng, C.-H.; Huang, S.-Y.; Cheng, T.-C. A Three-Dimensional Theoretical Model for Predicting Transient Thermal Behavior of Thermoelectric Coolers. *Int. J. Heat Mass Transf.* **2010**, *53*, 2001–2011. [CrossRef]
35. Alvarez-Rodriguez, M.; Alonso-Martinez, M.; Suarez-Ramon, I.; José García-Nieto, P. Numerical Model for Determining the Effective Heat Capacity of Macroencapsulated PCM for Building Applications. *Appl. Therm. Eng.* **2024**, *242*, 122478. [CrossRef]
36. Chern, B.-C.; Moon, T.J.; Howell, J.R.; Tan, W. New Experimental Data for Enthalpy of Reaction and Temperature- and Degree-of-Cure-Dependent Specific Heat and Thermal Conductivity of the Hercules 3501-6 Epoxy System. *J. Compos. Mater.* **2002**, *36*, 2061–2072. [CrossRef]
37. Said, Z.; Pandey, A.K.; Tiwari, A.K.; Kalidasan, B.; Jamil, F.; Thakur, A.K.; Tyagi, V.V.; Sarı, A.; Ali, H.M. Nano-Enhanced Phase Change Materials: Fundamentals and Applications. *Prog. Energy Combust. Sci.* **2024**, *104*, 101162. [CrossRef]
38. Aftab, W.; Usman, A.; Shi, J.; Yuan, K.; Qin, M.; Zou, R. Phase Change Material-Integrated Latent Heat Storage Systems for Sustainable Energy Solutions. *Energy Environ. Sci.* **2021**, *14*, 4268–4291. [CrossRef]
39. Talele, V.; Moralı, U.; Najafi Khaboshan, H.; Patil, M.S.; Panchal, S.; Fraser, R.; Fowler, M. Improving Battery Safety by Utilizing Composite Phase Change Material to Delay the Occurrence of Thermal Runaway Event. *Int. Commun. Heat Mass Transf.* **2024**, *155*, 107527. [CrossRef]
40. Li, C.; Jin, J.; Cao, W.; Sun, X.; Ding, Q.; Hou, Y.; Wang, Z. Finite Element Analysis and Design of a Flexible Thermoelectric Generator with a Rhombus Gap Structure. *ACS Appl. Mater. Interfaces* **2024**, *16*, 23406–23415. [CrossRef]
41. Wang, C.; Mobedi, M.; Yang, X.; Shen, Y.; Zhao, H.; Chen, H.; Zhang, T.; Zheng, X. A Comparison Study of Heat Dissipation Module between the Consolidated and Unconsolidated Porous Structures for Thermoelectric Cooler. *Appl. Therm. Eng.* **2024**, *247*, 123085. [CrossRef]
42. Choi, J.; Dun, C.; Forsythe, C.; Gordon, M.P.; Urban, J.J. Lightweight Wearable Thermoelectric Cooler with Rationally Designed Flexible Heatsink Consisting of Phase-Change Material/Graphite/Silicone Elastomer. *J. Mater. Chem. A* **2021**, *9*, 15696–15703. [CrossRef]
43. Amberkar, T.; Mahanwar, P. Thermal Energy Management in Buildings and Constructions with Phase Change Material-Epoxy Composites: A Review. *Energy Sources Part A Recovery Util. Environ. Eff.* **2023**, *45*, 727–761. [CrossRef]
44. Baldry, M.; Timchenko, V.; Menictas, C. Optimal Design of a Natural Convection Heat Sink for Small Thermoelectric Cooling Modules. *Appl. Therm. Eng.* **2019**, *160*, 114062. [CrossRef]
45. Cao, X.; Zhang, R.; Zhang, N.; Chen, L.; Li, X. Flexible Composite Phase Change Material with Improved Hydrophobicity and Thermal Conductivity Characters for Thermal Management. *J. Energy Storage* **2023**, *72*, 108641. [CrossRef]
46. Cai, Y.; Hong, B.-H.; Wu, W.-X.; Wang, W.-W.; Zhao, F.-Y. Active Cooling Performance of a PCM-Based Thermoelectric Device: Dynamic Characteristics and Parametric Investigations. *Energy* **2022**, *254*, 124356. [CrossRef]
47. Cong, B.; Kong, Y.; Ye, Y.; Liu, R.; Du, X.; Yu, L.; Jia, S.; Qu, Z.; Jiao, B. A Combined Solution of Thermoelectric Coolers and Microchannels for Multi-Chip Heat Dissipation with Precise Temperature Uniformity Control. *Appl. Therm. Eng.* **2023**, *219*, 119370. [CrossRef]
48. Ling, Z.; Chen, J.; Xu, T.; Fang, X.; Gao, X.; Zhang, Z. Thermal Conductivity of an Organic Phase Change Material/Expanded Graphite Composite across the Phase Change Temperature Range and a Novel Thermal Conductivity Model. *Energy Convers. Manag.* **2015**, *102*, 202–208. [CrossRef]
49. Prete, D.; Dimaggio, E.; Demontis, V.; Zannier, V.; Rodriguez-Douton, M.J.; Guazzelli, L.; Beltram, F.; Sorba, L.; Pennelli, G.; Rossella, F. Electrostatic Control of the Thermoelectric Figure of Merit in Ion-Gated Nanotransistors. *Adv. Funct. Mater.* **2021**, *31*, 2104175. [CrossRef]

50. Zhu, P.; Luo, X.; Lin, X.; Qiu, Z.; Chen, R.; Wang, X.; Wang, Y.; Deng, Y.; Mao, Y. A Self-Healable, Recyclable, and Flexible Thermoelectric Device for Wearable Energy Harvesting and Personal Thermal Management. *Energy Convers. Manag.* **2023**, *285*, 117017. [CrossRef]
51. Ling, Z.; Li, S.; Cai, C.; Lin, S.; Fang, X.; Zhang, Z. Battery Thermal Management Based on Multiscale Encapsulated Inorganic Phase Change Material of High Stability. *Appl. Therm. Eng.* **2021**, *193*, 117002. [CrossRef]
52. Zhao, X.; Lei, K.; Wang, S.; Wang, B.; Huang, L.; Zou, D. A Shape-Memory, Room-Temperature Flexible Phase Change Material Based on PA/TPEE/EG for Battery Thermal Management. *Chem. Eng. J.* **2023**, *463*, 142514. [CrossRef]
53. Li, M.; Guo, Q.; Nutt, S. Carbon Nanotube/Paraffin/Montmorillonite Composite Phase Change Material for Thermal Energy Storage. *Sol. Energy* **2017**, *146*, 1–7. [CrossRef] [PubMed]

materials

MDPI

Article

A New 3D Mathematical Model for Simulating Nanofluid Flooding in a Porous Medium for Enhanced Oil Recovery

Abdullah Al-Yaari [1,2,*], Dennis Ling Chuan Ching [1], Hamzah Sakidin [1], Mohana Sundaram Muthuvalu [1], Mudasar Zafar [1], Abdurrashid Haruna [1,3], Zulkifli Merican Aljunid Merican [1] and Abdus Samad Azad [1]

[1] Department of Fundamental and Applied Sciences, Universiti Teknologi PETRONAS, Seri Iskandar 32610, Malaysia; dennis.ling@utp.edu.my (D.L.C.C.); hamzah.sakidin@utp.edu.my (H.S.); mohana.muthuvalu@utp.edu.my (M.S.M.); mudasar_20000296@utp.edu.my (M.Z.); abdurrashid_21002269@utp.edu.my (A.H.); zulkifli.aljunid@utp.edu.my (Z.M.A.M.); abdus_19000955@utp.edu.my (A.S.A.)

[2] Department of Mathematics, Faculty of Applied Science, Thamar University, Dhamar 00967, Yemen

[3] Department of Chemistry, Ahmadu Bello University, Zaria 810107, Nigeria

* Correspondence: abdullah_20001447@utp.edu.my; Tel.: +60-183103931

Abstract: Two-phase Darcy's law is a well-known mathematical model used in the petrochemical industry. It predicts the fluid flow in reservoirs and can be used to optimize oil production using recent technology. Indeed, various models have been proposed for predicting oil recovery using injected nanofluids (NFs). Among them, numerical modeling is attracting the attention of scientists and engineers owing to its ability to modify the thermophysical properties of NFs such as density, viscosity, and thermal conductivity. Herein, a new model for simulating NF injection into a 3D porous media for enhanced oil recovery (EOR) is investigated. This model has been developed for its ability to predict oil recovery across a wide range of temperatures and volume fractions (VFs). For the first time, the model can examine the changes and effects of thermophysical properties on the EOR process based on empirical correlations depending on two variables, VF and inlet temperature. The governing equations obtained from Darcy's law, mass conservation, concentration, and energy equations were numerically evaluated using a time-dependent finite-element method. The findings indicated that optimizing the temperature and VF could significantly improve the thermophysical properties of the EOR process. We observed that increasing the inlet temperature (353.15 K) and volume fraction (4%) resulted in better oil displacement, improved sweep efficiency, and enhanced mobility of the NF. The oil recovery decreased when the VF (>4%) and temperature exceeded 353.15 K. Remarkably, the optimal VF and inlet temperature for changing the thermophysical properties increased the oil production by 30%.

Keywords: mathematical model; EOR; nanofluid flooding; thermophysical properties; porous medium

1. Introduction

Enhanced oil recovery methods are typically implemented to increase petroleum extraction from depleted oil reservoirs [1–3]. One of the methods used for EOR is nanofluid (NF) flooding, which involves the injection of a mixture of base fluid, such as water, and nanoparticles (NPs) into a reservoir to enhance oil mobility and recovery [4,5]. Nanofluids have attracted interest because of their distinctive characteristics and their ability to modify density, viscosity, and thermal conductivity with temperature and volume-fraction variations.

Over the past decade, numerous investigations have been carried out on the movement of NPs through PM [6,7]. These studies have explored how nanotechnology can be utilized in the oil industry to enhance oil recovery (OR). It has already been demonstrated that nanotechnology can be useful in the oil and gas sector, with NPs being suggested for various applications in the upstream oil industry such as NF flooding to help recover oil more effectively. There is a significant need to increase hydrocarbon energy production to meet growing energy demand. This problem can only be met by extracting as much as possible from the active reservoir, where a significant proportion of the oil remains, using EOR methods [8].

The initial phase of oil extraction, referred to as primary recovery, relies on the natural pressure within the reservoir to force oil to the surface. Over time, this pressure diminishes, resulting in a decline in oil production. Fewer than 30% of the reservoir's oil is usually recovered during the first recovery. Following this, a secondary recovery phase is initiated, which often entails flooding the reservoir with water to increase OR by up to 50%. All extraction techniques utilized after primary and secondary recoveries are collectively known as EOR. According to the petroleum sector, the recovery factor in primary and secondary recovery methods ranges from 25% to 50%. Therefore, to meet the growing annual demand for oil, a tertiary recovery method such as chemical EOR is being implemented globally to obtain greater oil production from existing reservoirs [9].

Many recent experimental investigations have been conducted to evaluate the potential of NF flooding to increase OR. With the addition of NPs to water, oil recovery from porous media (PM) has significantly increased compared to conventional water flooding. These experiments commonly involve examining changes in properties such as viscosity, density, interfacial tension, wettability, and geometries to assess the effectiveness and efficiency of EOR. The focus of research on NFs in EOR has been on how various NPs influence these properties and the amount of oil that can be recovered from the PM. SiO_2, TiO_2, CuO, and SiO_2 NPs are among the chemicals utilized in these experiments. To develop EOR techniques using NF flooding, it is necessary to understand the mechanism of NP interaction. Recent experimental research has shed light on how NFs, oil, and PM interact, which is a crucial aspect of the EOR process. More research is required to fully understand the EOR process via NF flooding [10].

Most NF flooding studies have been conducted experimentally. Still, researchers suggest modeling can be a more effective way to improve the EOR process and bring it closer to industrialization [11,12]. Models can supplement experiments and provide a detailed understanding of the EOR process. The addition of NPs to the base fluid for EOR has been researched numerically over the last decade by simulating oil displacement using mathematical models [13,14]. Using mathematical modeling to simulate oil displacement by NFs requires a computational grid to describe the structure of the PM and significant computational resources. Numerical studies can solve complex problems with fewer resources and accurately calculate modeling results. Two models, the mixing model and the immiscible fluid model, have been utilized to explain the multiphase flow of the base fluid and oil [15–17]. A model has been developed to simulate the NF using the concentration equation to control the movement of NPs through PM.

Several researchers have developed different approaches for developing a mathematical model to predict oil production after NFs are injected into oil reservoirs. Liu and Civan [12,18] proposed a prediction model for the oil recovery factor (ORF) by employing a two-phase flow of oil and water in sandstone. When NPs were added to the base fluid, the reservoir's wettability and permeability changed. These changes were measured mathematically while considering pore-clogging and porosity alterations. The impacts of NP and pore surface relative permeabilities, capillary pressure, and wettability on two-phase flow were investigated. However, the drawback of their model is that it neglects the fluid density and viscosity changes caused by adding NPs to the fluid.

The deposition of NPs inside the pores of a reservoir can result in modifications to its permeability and porosity. Ju and Fan [19] enhanced a mathematical model that includes deriving equations for calculating changes in the reservoir's porosity and permeability following the injection of NFs. This model could predict the OR after NF injection and evaluate the reservoir's permeability and porosity. It was one of the first models developed to determine NP transportation in PM and has been frequently referenced by researchers in subsequent studies. This model was employed by Sepehri et al. [20] in a carbonate system, and its accuracy was confirmed by comparing it to experimental data. The results indicated that the model could predict a 10% enhancement in the ORF compared to the conventional water flooding method due to changes in wettability.

Ju and Fan [21] created a mathematical model that shows the movement of NPs in a two-phase flow through one-dimensional PM by adjusting the net rate of NP loss in their previous model. The model's assumptions include (1) isothermal conditions with one-dimensional geometry; (2) incompressible PM and fluids; (3) oil and NFs flow through PM based on Darcy's law, without considering gravity force; (4) NPs are divided into n-sized intervals; and (5) despite the addition of NPs, the viscosity and density of the fluids remain unchanged. NPs exist in either the oil or water phases based on their properties. Hydrophobic NPs are found in the oil phase, whereas hydrophilic NPs are found in the water phase. The transport of NFs through PM is facilitated by Brownian diffusion due to the wettability characteristics of the NFs. The authors improved the model to enhance surface deposition prediction and found that adding polysilicon NP to the base fluid improved OR [21].

El-Amin and colleagues modified the model developed by Ju and Fan by including a 2D displacement mechanism for EOR using non-Newtonian fluids. The study focused on the flow behavior of two-phase immiscible and incompressible fluids within a homogenous permeable material governed by Darcy's law and the mass conservation equations for each phase. The authors developed a mathematical model to describe the absorption of NPs in a water–oil flow. The model incorporated capillary pressure and mixed relative permeability correlations for mixed-wet systems. The authors injected NFs into the oil reservoir to alter the wettability of the medium, causing it to change from oil-wet to water-wet. As a result, the rock was reclassified as a mixed-wet rock. During their research, they examined how NPs affected the properties of fluids and materials and conducted numerical studies to determine whether the OR process could be improved. However, the base fluid's viscosity and density variations were not considered in their model [22].

Feng et al. extended the model developed by El-Amin et al. to investigate the impact of NP retention on oil displacement efficiency in a two-phase displacement and NP transport simulation study. The authors found that NF flooding following water flooding enhanced OR, and higher injection rates improved flooding efficiency. However, the study did not examine capillary pressure or changes in wettability [23]. Sepehri et al. [20] used the model developed by Darcy to examine how NPs affect wettability and cavitation generation in rock. In contrast, Yu et al. [24] discovered that water salinity significantly affects NP movement. Rahmani et al. [25] proposed a model to study the movement of superparamagnetic NPs within formations, aiming to assist in rock sounding using tracer diagnostics.

The loss of NPs in PM must be computed to predict their flow. Some researchers have investigated and computed the NP loss term using mechanical and empirical models. Gruesbeck and Collins [26] estimated NP deposition using an injection velocity-based expression to predict NP loss. Cullen et al. [27] proposed quantifying NP entrapment and modified the model to investigate how water saturation, NP concentration, porosity, and permeability affect the transport of NPs in PM. The improved model's nonlinear governing equations were solved using the cell-centered finite-difference method.

Zhang et al. [28] developed a model to calculate NP concentrations at various time periods and core lengths. The concentration of NPs determined the amount of energy absorbed by the NF during injection into the core. Studies such as this one could provide insights into the optimal ways to combine electromagnetic heating and NP concentration. A limitation of the model was that it did not consider fluid saturation.

An et al. [29] developed a model that simulated the transport of NPs in reservoirs, considering both micro- and macro-scale phenomena. The model incorporated factors such as Darcy's law, Brownian diffusion, gas diffusion, and capillary effects. The authors examined how oil and water flowed together in a two-phase system, where the water phase contained NPs and the oil phase contained methane. Although Darcy's law governed the water flow, the oil flow was affected differently due to the relative permeability of the nanoscale pores. Since the NPs' sizes were between 30 and 80 nm, the authors assumed that the flow was influenced by Brownian diffusion and convection. The authors made the following assumptions to examine NP mobility in shale reservoirs: NPs could only move in water; there was no NP concentration in the oil phase; and there was no interaction between NPs and the reservoir matrix.

Abdolfattah et al. [30] conducted a study on the mechanism of the wettability change in oil reservoirs during EOR using NPs based on the equations proposed by Liu and Sivan [18]. They developed a mathematical model to investigate how NPs adsorb in the reservoir, affecting wettability and changing the relative permeability of the two phases. In addition, Abdelfatah et al. [31] proposed a model combining Darcy's equation with the convection-diffusion equation to analyze the movement, interaction, and behavior of NPs in the fluid, taking into account factors such as permeability, injection rate, formation volume, and NP size.

The research by Khalilinezhad et al. [32] explored the influence of adding NPs to fluids on their flow behavior in petroleum reservoirs. They employed a numerical method to validate their rheological findings and confirmed that NPs are adsorbed on sandstone surfaces. Their results revealed that NPs can reduce sandstone retention and adsorption. In a separate study, Gharibshahi et al. examined the influence of wettability on oil removal in a miscible model. They discussed the impact of the shape and distribution of PM in a two-dimensional microscopic model on the EOR process. They also discussed the roles of various factors in the oil-displacement process, including NP type, fluid velocity, particle size, and temperature of the injected fluid.

Furthermore, Salama et al. [33] developed a mathematical model to study the movement of NPs in PM. They used the filtration theory to explain the relationship between the mass deposition rate and the governing balance equation, resulting in the migration of fines and NP deposition. Esfe and Esfandeh [34] devised a model to explore how NFs can enhance OR in petroleum reservoirs. Their research investigated how the flow rate of NFs, the VF of nanoparticles, and the porosity of the PM affect EOR. The model was developed by adjusting the TPs of the injected NFs. The study concluded that increasing the flow rate of NFs positively impacted the EOR process, resulting in the recovery of more trapped oil. Additionally, an increase in the VF led to a change in the TPs of the base fluid.

In addition, some scientists found that for EOR, a consistent flow rate was more efficient in a low-porosity medium than in a high-porosity medium. Another study by Esfe et al. [35] created a new flooding model to simulate the flow of NFs in a petroleum reservoir filled with oil using two-phase Darcy's law and mass transfer equations. The research examined the effects of mechanical parameters, such as fluid viscosity and density, on ORF. They also investigated the role of capillary pressure and permeability in the two-phase Darcy equations and found them to be critical parameters in the EOR process.

In their study, Al-Yaari et al. [36]. conducted a numerical simulation of NF flooding in PM, considering changes in thermophysical properties. Their results indicated that adding NPs at a VF ranging from 1% to 5% to a base fluid significantly increased ORF by over 25%. Furthermore, an increase in the inlet temperature resulted in EOR due to alterations in the viscosity and density of oil. The primary mechanism behind the EOR observed in the study was an increase in the relative permeability of the NF coupled with a simultaneous reduction in the relative permeability of the oil resulting from the presence of NPs.

Esfe et al. [37] developed a simulation model to replicate the EOR process by utilizing the two-phase Darcy equations. The model employed Raafat's permeability model and Brooks–Corey's capillary pressure model. The authors investigated the impact of the NF inlet temperature and VF on the OR rate. Although previous numerical studies have demonstrated the importance of modeling the oil-displacement process using NF injection to improve OR, only a few studies have addressed three-dimensional modeling. These models can predict how effective NFs will be in replacing oil using parameters such as NP mass fraction and size, displacement fluid flow rate, oil viscosity, and core permeability.

The wetting properties of the medium have been the primary focus of previous models. Still, the change in TPs of the NF, such as density and viscosity, is also an important factor affecting the EOR process [34]. Although several models have been proposed to predict the ORF, no generalized model has been able to account for the density and viscosity of NFs across a wide range of temperatures and VFs.

Al-Yaari et al. [38] conducted a study to optimize the use of NPs in enhancing oil recovery in a 3D PM. The authors examined the effects of different types of NPs, their VFs, and the inlet temperature on oil recovery efficiency. The results showed that SiO_2 NPs exhibited the best ORF compared to other NPs. Interestingly, SiO_2 also had the lowest density and the highest thermal capacity among the examined NPs. The authors found that the optimum volume fraction of SiO_2 was 4%, which increased the ORF. However, the ORF decreased when the volume fraction exceeded 4%. Their findings also revealed that the ORF improved as the viscosity and density of the oil decreased, which was achieved by increasing the injected inlet temperature. Specifically, the highest ORF of 37% was obtained when SiO_2 was used at a volume fraction of 4% and an inlet temperature of 353.15 K. These findings suggest that using SiO_2 NPs at an optimal volume fraction and inlet temperature can significantly improve oil recovery efficiency in a porous medium. This highlights the importance of considering the thermophysical properties of nanoparticles when selecting an optimal nanofluid for enhanced oil recovery applications.

This research suggests a three-dimensional mathematical model that can predict the ORF by simulating the EOR process. In the past, models have focused on how nanoparticles change the wettability of the PM surface and the interfacial tension between the fluids. However, the fluid's thermophysical properties change due to the nanoparticles, which affects the EOR process, and this aspect has not been previously modeled. The proposed model simultaneously considers the changes in TPs with varying temperatures and VFs using novel empirical correlations for the first time. It is observed that these properties play a critical role in evaluating the efficacy of NF flooding through the oil reservoirs for EOR. These correlations are used to compute the relative permeability formulas for water and oil. Additionally, we investigate how the VF and IT impact the nanofluid TPs and improve EOR performance.

2. Mathematical Modeling

Simulating the EOR process requires computational fluid dynamics (CFD) in oil reservoir engineering. This tool allows engineers to model fluid behavior in porous media, leading to process optimization and a better understanding of fluid interactions [39]. CFD examines different methods of EOR such as waterflooding, gas injection, chemical injection, and NF flooding [16]. Engineers can optimize the simulation of injection rates to evaluate their effectiveness. CFD considers factors such as NF properties, fluid compressibility, and fluid–rock interactions, which contribute to the analysis of EOR.

A study utilized COMSOL Multiphysics software version 4 and the finite-element method to simulate NF flooding and assess its impact on the increase in the ORF [40]. Parameters such as pressure, velocity, saturation, and temperature were analyzed in the research [41–43]. The equations for continuity, Darcy, energy, and concentration were solved. The empirical correlations determined thermal conductivity, density, and viscosity. Additionally, we examined the effects of NP deposition on porosity and permeability [44].

2.1. Governing Equations

To simulate NF flooding in PM for EOR, the mathematical equations usually consist of the two-phase Darcy's law and equations for energy and concentration. The two-phase Darcy's law accounts for the fluid's pressure gradient and viscous forces. The concentration and energy equations account for the concentration of NPs inside water, which refers to the number of NPs present in a given volume of water. The concentration of NPs can vary widely depending on the preparation method, including the size and stability of the NPs and the type of stabilizing agent used. The stability of NPs in water can be influenced by temperature. The equation for energy takes into consideration the transfer of heat that occurs between the fluids and the reservoir's PM.

The two-phase Darcy's law is a well-known model that is often used to explain the movement of two fluids that cannot be mixed, such as oil and NF, in a porous material. In particular, when NF flooding is used to enhance OR, the two-phase Darcy's law can be used to describe how oil and NF flow through the porous rock structure simultaneously. According to this law, there is a direct relationship between the pressure gradient and the velocity of the fluid, and a constant of proportionality known as permeability governs this relationship. The symbol K_m represents this constant, indicating how fluids can move through the PM.

The two-phase Darcy's law considers the relative permeabilities of NFs, $\kappa_{r,nf}$, and water, $\kappa_{r,w}$, which can be affected by the presence of NPs. By incorporating additional models, such as those related to heat and mass transfer, the complete behavior of the NF flooding process can be simulated by combining the two-phase Darcy's law with these other models. The specific forms of these equations depend on factors such as fluid properties, porosity, permeability, and temperature, among others. The solutions of these equations can provide insights into the flow and transport of NFs in PM, which can be used to optimize EOR processes.

In this section, the two-phase Darcy equations are used to calculate the pressure and velocity distribution of oil and NF. The saturation equation is used to measure the saturation of NF and oil phases and estimate other parameters associated with the NF flooding simulation process. The following equations model the NF flooding process inside the PM.

2.1.1. Transport Equations

Darcy's law and the continuity equations govern the flow of NF and oil in a PM. The PM has a porosity ε and an absolute permeability (K_m). Darcy's law describes the flow in the PM for each phase, with relative permeabilities $\kappa_{r,i}$ where i represents nf and o. The subscripts nf and o indicate the NF and oil fluids, respectively. The velocities $\boldsymbol{u}_i$ are as calculated as follows:

$$\boldsymbol{u}_{nf} = -\frac{K_m \kappa_{r,nf}}{\mu_{nf}}\left(\frac{\partial}{\partial x}\boldsymbol{i} + \frac{\partial}{\partial y}\boldsymbol{j} + \frac{\partial}{\partial z}\boldsymbol{k}\right)\left(P_{nf} - \rho_{nf} g z\right), \tag{1}$$

$$\boldsymbol{u}_{o} = -\frac{K_m \kappa_{r,o}}{\mu_{o}}\left(\frac{\partial}{\partial x}\boldsymbol{i} + \frac{\partial}{\partial y}\boldsymbol{j} + \frac{\partial}{\partial z}\boldsymbol{k}\right)(P_{o} - \rho_{o} g z), \tag{2}$$

where μ_{nf}, μ_o, ρ_{nf}, ρ_o, P_{nf}, and P_o are the viscosity, density, and pressure for NF and oil, respectively. The gravity (g) can be neglected due to the NF flooding horizontally.

Fluids are assumed to be incompressible. As a result, the two continuity equations and the saturation condition for S_{nf} and S_o can be expressed as follows:

$$S_{nf} + S_o = 1, \tag{3}$$

$$\varepsilon\rho_{nf}\frac{\partial S_{nf}}{\partial t} + \left(\frac{\partial}{\partial x}\boldsymbol{i} + \frac{\partial}{\partial y}\boldsymbol{j} + \frac{\partial}{\partial z}\boldsymbol{k}\right)\cdot\left(\rho_{nf}\boldsymbol{u}_{nf}\right) = 0, \tag{4}$$

$$\varepsilon\rho_{o}\frac{\partial S_{o}}{\partial t} + \left(\frac{\partial}{\partial x}\boldsymbol{i} + \frac{\partial}{\partial y}\boldsymbol{j} + \frac{\partial}{\partial z}\boldsymbol{k}\right)\cdot(\rho_{o}\boldsymbol{u}_{o}) = 0. \tag{5}$$

By adding Equation (4) to Equation (5), and then using Equation (3), Equation (6) can be expressed as follows:

$$\left(\frac{\partial}{\partial x}\boldsymbol{i} + \frac{\partial}{\partial y}\boldsymbol{j} + \frac{\partial}{\partial z}\boldsymbol{k}\right)\cdot\left(\boldsymbol{u}_{nf} + \boldsymbol{u}_{o}\right) = 0, \tag{6}$$

Then,

$$u_{nf} + u_o = u = \text{Constant m/s}, \tag{7}$$

where u_{nf} and u_o are the NF and oil velocities, respectively.

By adding Equation (1) to Equation (2), and utilizing the fact that $S_{nf} + S_o = 1$, Equation (7) can be applied, and Equation (8) can be written as

$$u = -K_m\left(\frac{\kappa_{r,nf}}{\mu_{nf}} + \frac{\kappa_{r,o}}{\mu_o}\right)\left(\frac{\partial}{\partial x}i + \frac{\partial}{\partial y}j + \frac{\partial}{\partial z}k\right)(P_{nf} + P_o). \tag{8}$$

When the NF is injected into the PM, it causes a change in the wettability of the medium from oil-wet to NF-wet. This change requires the inclusion of a capillary pressure correlation that considers the alteration in wettability through empirical relationships.

$$p_c = -B_C \times \log(Se), \tag{9}$$

where B_C and Se are the effective capillary pressure and saturation parameters, respectively.

$$P_c = P_{nf} - P_o, \tag{10}$$

where P_o is a constant, which can be set to 0 without loss of generality. Equation (8) can be reduced to

$$u = -K_m\left(\frac{\kappa_{r,nf}}{\mu_{nf}} + \frac{\kappa_{r,o}}{\mu_o}\right)\left(\frac{\partial}{\partial x}i + \frac{\partial}{\partial y}j + \frac{\partial}{\partial z}k\right)P_c. \tag{11}$$

Equation (11) is solved for $\nabla p_{nf} = \nabla p_c$, and the resulting expression is substituted into Equation (1), leading to the derivation of Equation (12):

$$u_{nf} = \left(\frac{\kappa_{r,nf}}{\mu_{nf}}u\right) \Big/ \left(\frac{\kappa_{r,nf}}{\mu_{nf}} + \frac{\kappa_{r,o}}{\mu_o}\right), \tag{12}$$

By substituting the value of u_{nf} from Equation (12) into Equation (4), Equation (13) can be derived, which is given as follows:

$$\nabla \cdot \left(\rho_{nf}\left(\left(\frac{\kappa_{r,nf}}{\mu_{nf}}u\right) \Big/ \left(\frac{\kappa_{r,nf}}{\mu_{nf}} + \frac{\kappa_{r,o}}{\mu_o}\right)\right)\right) = -\varepsilon\rho_{nf}\frac{\partial S_{nf}}{\partial t}. \tag{13}$$

In other words, the NF-oil flow (Equation (16)) can be obtained using Darcy's law (Equations (1) and (2)) and the continuity equations (Equations (4) and (5))

$$\frac{\partial \varepsilon \rho_{nf} S_{nf}}{\partial t} + \nabla \cdot (\rho_{nf} S_{nf} u) = \nabla \cdot \left(D_c \nabla \rho_{nf} S_{nf}\right), \tag{14}$$

$$D_C = \frac{\kappa_{r,nf}}{\mu_{nf}} \mathrm{K}\left(S_{nf} - 1\right)\frac{\partial P_c}{\partial S_{nf}}. \tag{15}$$

where D_c is the capillary diffusion coefficient parameter. As mentioned earlier, the capillary pressure relationship in Equation (10) (P_c) is utilized in this modeling.

To solve Equation (13), the effective density and viscosity of the NF must be considered since it is a liquid composed of water and NPs. Over the years, several models have been proposed to predict oil production based on empirical correlations of the density and viscosity with various VFs. The proposed model is applicable over a wide range of temperatures and VFs and can be used to determine the density and viscosity of NF as a function of the VF and temperature. As a result, Equations (16) and (17) are utilized to calculate the density and viscosity as a function of the temperature and VF [45]. A novel model based on the density, viscosity, and TC of the NF is proposed to simulate the EOR process:

$$\rho_{nf} = \rho_{\mathrm{w}}[1.032 + (0.03774 * \varnothing) + (-0.0004753 * T + (-0.00007523 * T * \varnothing)] \quad (16)$$

$$\mu_{nf} = \mu_{\mathrm{w}} * \left[1.116 + (0.3923 * \varnothing) + \left(\frac{-2.183}{T}\right) + \left(\frac{-3.849 * \varnothing}{T}\right)\right] \quad (17)$$

The results obtained using the model are compared to those obtained using the Maxwell [46] and Murshed models [47]. The Maxwell model [46] is used to calculate the TC of the particles.

$$k_{eff} = k_w \frac{k_{np} + 2k_w - 2\varnothing(k_w - k_{np})}{k_{np} + 2k_w + \varnothing(k_w - k_{np})} \quad (18)$$

where k_{np}, k_w, and $\varnothing$ represent the TC of the NPs, water, and VF, respectively.

The TC of NFs in the Murshed model [47] consists of two parts: static and dynamic

$$k_{eff-nf} = k_{\mathrm{st}} + k_{dy} \quad (19)$$

where

$$\begin{aligned} k_{eff-nf} &= \left\{k_{\mathrm{w}} \frac{\phi_{\mathrm{np}}\omega(k_{\mathrm{np}} - \omega k_{\mathrm{w}})\left[2\gamma_1^3 - \gamma^3 + 1\right] + (k_{\mathrm{np}} + 2\omega k_{\mathrm{w}})\gamma_1^3\left[\phi_{\mathrm{np}}\gamma^3(\omega - 1) + 1\right]}{\gamma_1^3(k_{\mathrm{np}} + 2\omega k_{\mathrm{w}}) - (k_{\mathrm{np}} - \omega k_{\mathrm{w}})\phi_{\mathrm{np}}\left[\gamma_1^3 + \gamma^3 - 1\right]}\right\} \\ &+ \left\{\phi_{\mathrm{np}}^2\gamma^6 k_{\mathrm{w}}\left(3\Lambda^2 + \frac{9\Lambda^3}{16}\frac{k_{\mathrm{cp}} + 2k_{\mathrm{w}}}{2k_{\mathrm{cp}} + 3k_{\mathrm{w}}} + \frac{3\Lambda^4}{2^6}\right)\right\} \\ &+ \left\{\frac{1}{2}\rho_{\mathrm{cp}}c_{\mathrm{np-cp}}d_{\mathrm{s}}\left[\sqrt{\frac{3K_{\mathrm{B}}T(1 - 1.5\gamma^3\phi_{\mathrm{p}})}{2\pi\rho_{\mathrm{cp}}\gamma^3 r_{\mathrm{np}}^3}} + \frac{G_{\mathrm{T}}}{6\pi\eta\gamma r_{\mathrm{np}}d_{\mathrm{s}}}\right]\right\} \end{aligned} \quad (20)$$

where $\gamma = 1 + \frac{t}{r_p}$ and $\gamma_1 = 1 + \frac{t}{2r_p}$, and a complex particle TC is defined by:

$$k_{\mathrm{cp}} = k_{\mathrm{lr}} \frac{2(k_{\mathrm{np}} - k_{\mathrm{lr}}) + \gamma^3(k_{\mathrm{np}} + 2k_{\mathrm{lr}})}{(k_{\mathrm{lr}} - k_{\mathrm{np}}) + \gamma^3(k_{\mathrm{np}} + 2k_{\mathrm{lr}})} \quad (21)$$

The TC of the nanolayer can be expressed as $k_{\mathrm{lr}} = \omega k_{\mathrm{f}}$, where ω is an empirical parameter that is influenced by the order and shape of the molecules present in the interface of the NPs, as well as their nature and surface chemistry (where $\omega > 1$). The value of Λ is defined as:

$$\Lambda = \frac{k_{\mathrm{cp}} - k_{\mathrm{f}}}{k_{\mathrm{cp}} + 2k_{\mathrm{f}}} \quad (22)$$

where G_T represents the total energy that exists between two NPs, which is mathematically defined as:

$$\begin{aligned} G_{\mathrm{T}} = -\frac{A}{6}&\left[\frac{2r_{\mathrm{cp}}^2}{d_{\mathrm{s}}^2 + 4r_{\mathrm{cp}}d_{\mathrm{s}}} + \frac{2r_{\mathrm{cp}}^2}{d_{\mathrm{s}}^2 + 4r_{\mathrm{cp}}d_{\mathrm{s}} + 4r_{\mathrm{cp}}^2}\right. \\ &\left. + \ln\left(\frac{d_{\mathrm{s}}^2 + 4r_{\mathrm{cp}}d_{\mathrm{s}}}{d_{\mathrm{s}}^2 + 4r_{\mathrm{cp}}d_{\mathrm{s}} + 4r_{\mathrm{cp}}^2}\right)\right] + 2\pi\varepsilon_{\mathrm{r}}\varepsilon_0 r_{\mathrm{cp}}\zeta^2 \exp(-\kappa d_{\mathrm{s}}) \end{aligned} \quad (23)$$

The Murshed model is defined in detail in [47].

The RP incorporates the necessary saturation (S_e) for determining the ORF, which can be calculated as

$$S_e = \frac{S_{nf} - S_{rnf}}{1 - S_{rnf} - S_{ro}}, \quad (24)$$

where (S_{rnf}) and (S_{ro}) represent the residual saturation of the NF and oil, respectively.

2.1.2. RPs of NF and Oil

When NPs are retained, they cause the wettability of the PM to shift from oil-wet to NF-wet, which leads to an expansion of the RP correlations to accommodate the mixed-wet conditions. Equations (25) and (26) are used to represent the RPs of both the NF and oil [48].

$$k_{rnf} = 3S_e + \frac{2}{\lambda}, \tag{25}$$

$$k_{ro} = \left(1 - S_e^2\right) \times \left(1 - S_e^{1+\frac{2}{\lambda}}\right), \tag{26}$$

where λ represents the index used to describe the size of the pores.

2.1.3. Nanoparticle Transport in PM

The transport of NPs within the PM throughout each phase of the flow is determined using the following equation [49]:

$$\boldsymbol{u}\frac{\partial C_i}{\partial t} + \varepsilon S_w \frac{\partial C_i}{\partial t} - \frac{\partial}{\partial x}\left(\varepsilon S_w D_c \frac{\partial C_i}{\partial x}\right) + R_i = 0, \tag{27}$$

$$D_c = \frac{k_{rw}}{\mu_w} k(S_w - 1)\frac{\partial P_c}{\partial S_w}, \tag{28}$$

where the concentration, diffusion, and NP rate loss in the PM are denoted by C_i, D_c, and R_i, respectively.

The term R_i can be divided into two terms:

- The number of NPs on the surface.
- The number of NPs located in the throat of the PM.

$$R_i = \frac{\partial \sigma}{\partial t} + \frac{\partial \sigma^*}{\partial t}, \tag{29}$$

where σ^* and σ represent the numbers of NPs. Specifically, σ^* refers to the number of NPs confined to the PM's throat (presumably some particle filter), and σ refers to the number of NPs deposited on the PM's surface per unit volume.

$$\frac{\partial \sigma}{\partial t} = k_d \boldsymbol{u} C, \tag{30}$$

$$\frac{\partial \sigma^*}{\partial t} = k_p \boldsymbol{u} C, \tag{31}$$

where k_d and k_p represent constants related to the deposition and entrainment of NPs, respectively. The vector $\boldsymbol{u}$ represents the required phase velocity of the NPs, which is 0.0025 m/s.

When NPs are deposited on the surface of a pore, it can cause the pore's opening to become obstructed, resulting in a decrease in the PM's overall ability to allow fluids to pass through it, as well as a reduction in the amount of space within the PM. The equations given below can be used to determine these values:

$$\varepsilon = \varepsilon_0 - (\sigma + \sigma^*), \tag{32}$$

where

$$K = K_0\left[(1 - F)\alpha_f + \frac{F\varepsilon}{\varepsilon_0}\right]^n, \tag{33}$$

In the given context, ε_0, K_0, K, ε, n, α_f, F, α_f, n, and F are used to represent specific physical properties or parameters. ε_0 denotes the permeability, K_0 represents the porosity,

K represents the fluid seepage constant, and ε represents the percentage of the original cross-sectional area that is available for fluid flow. The value of ε can range from 0 to 1.

The variable n is an empirical factor that can be any value between 1 and 3. Additionally, α_f and F denote the percentage of the initial cross-sectional area available for the flow of the NF. Although the value of α_f is constant, the value of F can vary.

The TPs of the oil and water are shown in Table 1. The properties of the NPs (SiO_2) and the PM are provided in Table 2, which are required for the computation of the model.

Table 1. Properties of the oil.

Property	Oil [50]	Water [51]	Unit
Density (ρ_o)	880	990	kg/m^3
Viscosity (μ_o)	4.5×10^{-4}	0.001	Pa.s
Specific heat capacity (C_o)	1670	4200	J/(kg·K)
TC coefficient (k_o)	0.13	0.6	W/(m·K)

Table 2. Properties of the NPs and porous media.

Property	SiO_2 [52]	Reservoir [53]	Unit
Density (ρ)	3970	2714	kg/m^3
Viscosity (μ)	0.0011	-	Pa·s
Specific heat capacity (C)	765	851	J/(kg·K)
TC coefficient (k)	40	2.2	W/(m·K)
Diameter of NPs	5	3	nm
Molecular weight	101.96	-	W/(m·K)

2.1.4. Heat-Transfer Equation

The equations for heat transfer in particulate materials (PM) are employed to investigate the heat transfer state in the PM. These equations utilize the effective TPs of the PM, consisting of the nanofluid and oil phases.

$$\left(\rho_{tot}C_p\right)_{\text{eff}}\frac{\partial T}{\partial t}+\rho_{tot}C_p\boldsymbol{u}\cdot\nabla T+\nabla.q=Q \tag{34}$$

$$q=-k_{\text{eff}}\,\nabla T \tag{35}$$

$$\left(\rho C_p\right)_{\text{eff}}=\theta_p\rho_p C_{p,p}+\left(1-\theta_p\right)\rho C_p \tag{36}$$

$$k_{\text{eff}}=\theta_p k_s+\left(1-\theta_p\right)k_{tot} \tag{37}$$

where θ_{p} is the VF of the porous rock.

2.1.5. Boundary and Initial Conditions

The velocity and inlet temperature of the NF injected into the PM through the inlet side are defined as follows:

Inlet Velocity Condition:

$$\boldsymbol{n}\cdot\rho_o\boldsymbol{u}\text{nf0}=0.01\text{ m/s}, \tag{38}$$

where $\boldsymbol{n}$ represents the unit normal vector, ρ_o is the density of the NF, and $\boldsymbol{u}$nf0 denotes the initial velocity of the NF.

Inlet Temperature Condition:

$$T=T_{\text{in}} \tag{39}$$

where T is the temperature and T_{in} is the specified inlet temperature.

At the beginning of the flooding process ($t = 0$) inside the PM, the NF saturation and the pressure at the outlet are defined as follows:

Outlet Pressure Condition:

$$p = p_{\text{nf}} = 0, \text{Pa} \tag{40}$$

where p represents the pressure and p_{nf} denotes the NF pressure at the outlet.

NF Saturation Condition:

$$S_{\text{nf0}} = 0 \quad \text{at} \quad t = 0 \tag{41}$$

where S_{nf0} represents the initial NF saturation.

The conditions at the cylinder outlet side are defined as follows:

Outlet Velocity Condition:

$$-\boldsymbol{n} \cdot \rho_o \boldsymbol{u}o = \rho_o \boldsymbol{u}o0 \tag{42}$$

where $\boldsymbol{u}o$ represents the velocity at the outlet, and $\boldsymbol{u}o0$ denotes the specified initial outlet velocity.

Temperature Gradient Condition:

$$-\boldsymbol{n} \cdot k_{\text{eff}} \nabla T = 0 \tag{43}$$

where k_{eff} is the effective thermal conductivity, and ∇T represents the temperature gradient.

The initial conditions for the simulation are defined as follows:

Initial NF Saturation:

$$S_{\text{nf}} = 0 \tag{44}$$

Initial Wetting Phase Saturation:

$$S_o = 1 \tag{45}$$

Initial Pressure:

$$P_o = 0 \tag{46}$$

Initial Temperature:

$$T_{\text{init}} = 293.15, \text{K} \tag{47}$$

Surface Tension:

$$\sigma = \sigma^* = 0 \tag{48}$$

In addition, the initial concentration of the nanoparticles (NPs) in the PM is assumed to be zero.

Initial Concentration of the NPs:

$$C_{\text{init}} = 0 \quad \text{at} \quad t = 0 \tag{49}$$

These conditions comprehensively describe the initial and boundary conditions for the problem under consideration.

2.2. Geometry

This research involves the simulation of the impact of an NF on the ORF. Previous studies have utilized different NPs mixed with water as NFs to enhance the ORF by injecting them into the PM to force out the oil flow, thereby increasing oil production. Table 3 presents the properties of the PM used in this study, which is a cylinder-shaped oil reservoir modeled using COMSOL Multiphysics software and based on the properties of an actual reservoir. The geometry and properties used in the current study are depicted in Figures 1a and 1b, respectively, and the boundary conditions for the problem are shown in Table 4. The NF layer closest to the surface remains stationary as it moves through the PM, and the triangular mesh in the CMS software version 6 is suitable for meshing the NF flow in the PM, as shown in Figure 2.

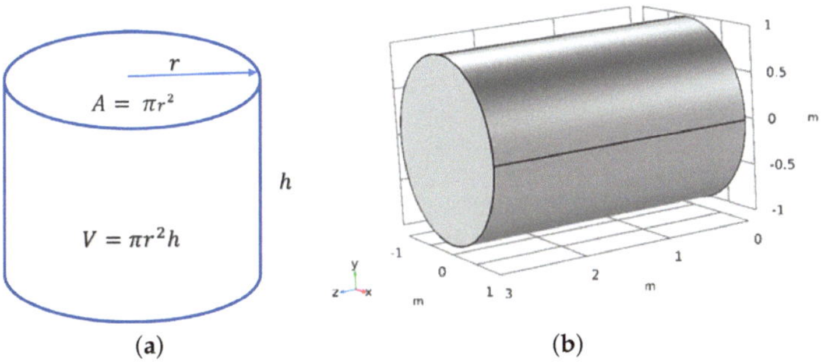

Figure 1. (**a**) Cylinder properties, and (**b**) model geometry.

Table 3. PM properties.

Property	Value	Unit
Cross-section area (A)	$A = \pi^2 h$	m^2
Volume (V)	$V = \pi^2 h$	m^3
Height (h)	3	m
Radius (r)	1	m
Porosity (ε)	0.27	%
Permeability (K)	1×10^{-9}	m^2
Initial NF saturation	0	%

Table 4. Boundary conditions.

The cylinder inlet surface	$S_o = 0$	$\mathbf{n}.\ \rho_{nf}\mathbf{u}_{nf} = 0.01\ \mathrm{m/s}$	$T = T_{in}$
The rectangular prism outlet surface	$p_o = 0\ \ \mathrm{Pa}$	$-\boldsymbol{n} \cdot \rho_o \boldsymbol{u}_o = \rho_o \boldsymbol{u}_{o0}$	$-\boldsymbol{n} \cdot k_{\mathrm{eff}} \nabla T = 0$
All surfaces except the inlet and outlet		$-\boldsymbol{n} \cdot \rho_o \boldsymbol{u}_o = 0$	$-\boldsymbol{n} \cdot k_{\mathrm{eff}} \nabla T = 0$
Initial condition	$S_o = 1$	$p_{nf} = 0\ \mathrm{Pa}$	$T_{init} = 293.15\ \mathrm{K}$

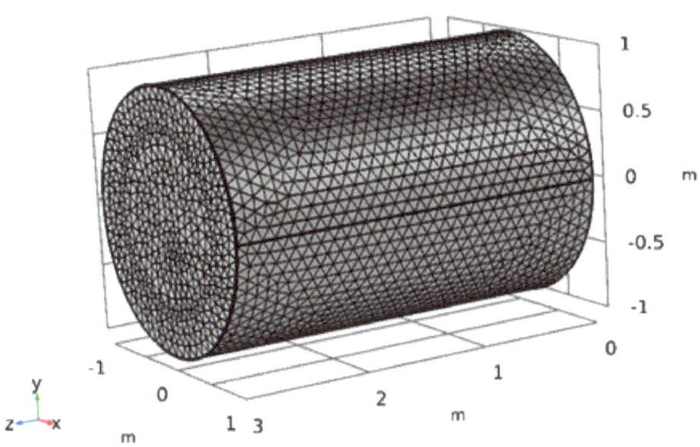

Figure 2. Mesh geometry.

2.3. Grid Independence

It was essential to consider the following steps in selecting the optimal mesh for EOR modeling to achieve accurate and reliable results. The first step was to determine the spatial resolution requirements. This involved evaluating the physical and chemical processes involved and determining the minimum size of the elements required to capture the relevant features of the system. This is important because the accuracy of the simulation results depends on the mesh's spatial resolution.

Next, a balance needed to be struck between accuracy and computational efficiency. The mesh needed to have elements of sufficient size to provide accurate results. However, it was also essential to consider the computational resources available and the cost-effectiveness of the simulation. The goal was to select a mesh that met the accuracy requirements while being computationally efficient.

It was also essential to consider the grid's adaptability. The mesh needed to be able to adapt to changes in the reservoir's physical properties over time such as variations in fluid saturation and pressure. This ensured that the simulation results remained accurate and relevant over time. Then, the mesh was validated, which involved comparing the simulation results with actual field data and refining the mesh to improve accuracy. By validating the mesh, it is possible to ensure that the simulation results are accurate and reliable.

Finally, six meshes with different grid elements (30,000, 40,000, 50,000, 60,000, and 70,000) were developed to examine grid independence. The model was run six times using water, and the parameters and properties of the water and oil, which are shown in Table 1, were fixed to compare the six ORF values and determine the optimal grid.

During the simulation, the NF was introduced into the reservoir through the inlet at a velocity of 0.01 m/s, and the oil pressure at the outlet was maintained at zero. The reservoir was configured so that the oil and NF saturations were 1 and 0, respectively. In addition, the remaining amount of oil left in the reservoir, known as the residual oil saturation, was only 0.2%. Table 5 presents the mesh with varying numbers of elements and the ORF value at a specific point (2 m height, 0.5 m radius, 600 s). In other words, the percentage of oil recovered from a particular area was calculated using a measurement point located 1 m away from the area. Measurements were taken at intervals of 0.5 m in two different directions, and the data were collected for 200 s. Various grid sizes were used for obtaining the measurements. The results showed that the ORF values remained primarily unchanged after mesh number 4, as increasing the number of elements led to longer computation times. Therefore, mesh number 4 was selected as the best option and used for all simulations in the study. In addition, the CMS generated a geometry mesh plot demonstrating the quality of the mesh elements. The mesh elements were considered of good quality if their color was green with a value of 1, indicating the highest quality. Figure 3a reveals that the meshing quality of the current study's geometry is good, as all elements are green. Furthermore, Figure 3b depicts the convergence of the governing equations at a time interval of 400 s. The error was less than 10^{-5}, which is depicted in Figure 3c.

Table 5. Oil recovery factor at specific points (1 m, 0.5 m, 0.5 m, and 200 s) with different grids.

No.	No. of Elements	ORF
1	30,000	0.572
2	40,000	0.623
3	50,000	0.630
4	60,000	0.6445
5	70,000	0.6449
6	80,000	0.6451

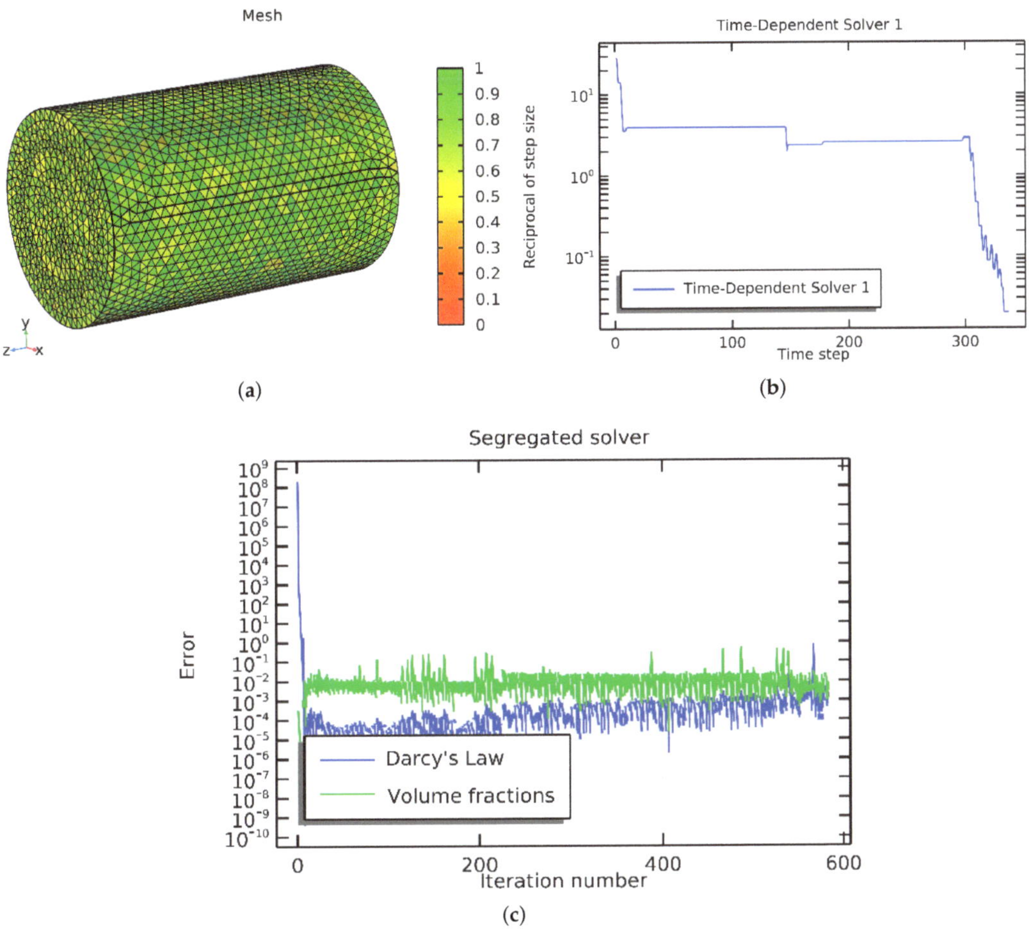

Figure 3. (**a**) Inspection of the mesh, (**b**) convergence plot, and (**c**) error of less than 1×10^{-5}.

Assumptions

- The NF (NP-water) is homogeneous, meaning it is considered one phase. The thermophysical parameters of the NF are computed using the experimental models.
- The PM is homogeneous, contributing to lower simulation processing costs.
- The NF and PM have incompressible flows.
- The EOR process is modeled using the two-phase Darcy's law.
- The flow and PM are considered to be isothermal.

2.4. Verification

In this study, we developed a model to simulate NF flooding in a three-dimensional PM. To validate our model, we compared our numerical findings with the laboratory results obtained by Minakov et al. [54], who conducted an experiment using NF (water-SiO_2). The characteristics of the PM are listed in Table 6. Our validation results, depicted in Figure 4, demonstrated a close agreement between our simulation and experimental data, with a relative error of only 2%.

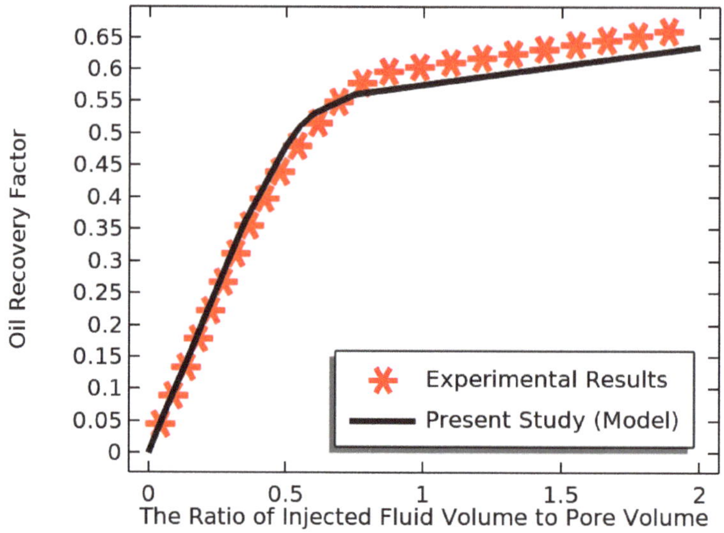

Figure 4. An examination of how data from the numerical analysis compared to data collected through practical experiments.

Table 6. PM specifications.

Property	Value
Size (m^3)	4.836×10^{-15}
Porosity (%)	0.25
Absolute permeability (m^2)	5.2×10^{-20}

3. Results and Discussion

Different mesh elements were used to solve the model geometry, and the variation in the ORF was compared. Mesh number 4, which had ideal elements, was selected for all modeling stages. After achieving good agreement between the numerical and experimental data, we analyzed the effects of various parameters. Specifically, we examined how the VF range of 1–5% at different inlet temperatures (293.15–500 K) affected the nanofluid's thermophysical properties for EOR.

3.1. The Effects of Nanofluid's TPs on EOR

During NF flooding at varying ITs, we analyzed how the addition of five different VFs to water affected the nanofluid's TPs. The VF of the NPs is a significant factor that can impact the TC of the NF. It is a common understanding that a higher VF results in an increased TC due to NPs acting as conductors. As the concentration of NPs in the fluid increases, the overall conductivity of the fluid improves.

The impact of the inlet temperature on the TC of the NF is a multifaceted issue that depends on various factors, including the type of NPs and base fluid, the size and form of the NPs, and the temperature range. Remarkably, the NF's thermal conductivity increases as the IT increases. Nonetheless, in some scenarios, the TC of the NF may decrease as the temperature rises, a phenomenon referred to as TC depression. The NPs then aggregate at higher temperatures, reducing their ability to effectively conduct heat.

We investigated how VFs and inlet temperatures can improve thermal conductivity and enhance oil recovery in the petroleum industry. Various factors, including the VF and IT, influenced the conductivity of the NF. Figure 5 shows that the increase in the VF and IT significantly increased the thermal conductivity. Furthermore, Figures 5 and 6 show that the addition of VFs improved the thermal conductivity compared to pure water. As the VF and IT increased, the thermal conductivity of the NF increased, leading to an increase in oil recovery. However, the gradual increment in thermal conductivity can be attributed to the tendency of NPs to agglomerate, which reduced the effectiveness of the NF on EOR. The IT also affected the conductivity of the NF. At higher ITs, the thermal conductivity of the NF increased due to increased Brownian motion, which helped break up agglomerates and improve particle dispersion. Therefore, the effect of VF and IT on NFs is crucial to enhancing oil recovery. However, very high temperatures and VFs caused a decrease in thermal conductivity, resulting in a decrease in oil recovery. Nanofluids are helpful in the EOR process as they have tailored thermal properties. One advantage of nanofluids is that they can improve oil extraction efficiency by increasing the thermal conductivity and heat transfer rate of the fluid that heats the reservoir. Adding SiO_2 to the base fluid (water) can significantly improve the TC, as SiO_2 has a high TC. The enhanced heat transfer rate reduces the energy required to heat the reservoir, resulting in cost savings and improved OR rates. When nanoparticles are added to water, they can change the density and thickness of the fluid. This can improve sweeping efficiency, increase oil recovery, and reduce the production of non-formation fluids. To determine the effectiveness of using NPs for enhanced oil recovery, it is essential to measure the VF, which is the proportion of NPs in an NF to the entire volume of the NF. A higher VF usually means better performance for EOR but it can also lead to increased viscosity, which can cause issues with injection and flow. The NF's inlet temperature is crucial to its performance in EOR. The NF's density, viscosity, and thermal conductivity affect its ability to transfer heat, which impacts the inlet temperature. By increasing the IT, the viscosity of the NF can be reduced, improving its flow properties and making it more effective for sweeping the reservoir.

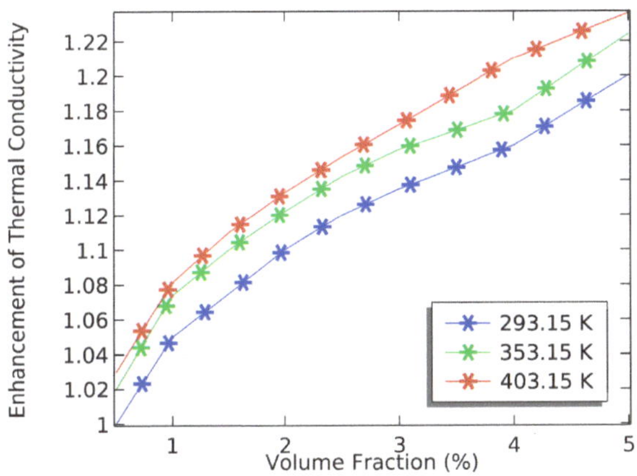

Figure 5. The effect of VF at different inlet temperatures on the nanofluid TC.

To optimize the performance of NFs in EOR, it is crucial to simultaneously consider their VF and IT. Our research delved into the impact of these factors on the NF's performance and identified an optimal VF and IT that can boost oil production. Increasing the VF of the base fluid resulted in an increase in the NF's density and viscosity, as shown in Figure 7a,b. Conversely, increasing the IT reduced the density and viscosity of the NF, as demonstrated in the same figures. However, Figure 8a,b illustrate that an increase in the VF led to an upsurge in the density and viscosity. Our study found that a VF of 0.4% and an IT of 353.15 K were the most effective in enhancing the TC of the NF and improving its performance in EOR.

We studied how the IT and VF affect the density and viscosity of the fluid used in EOR. We looked at different factors such as the properties of the oil and the reservoir. Adding

NPs increased the viscosity and density of the fluid since it improved the interaction between the fluid and the rock reservoir. However, the effects of the VF on the density and viscosity varied depending on the size, shape, and surface charge of the NPs used. We studied how temperatures and VFs can affect the density and viscosity of the NF used in EOR. Our results showed that when the VF increased, the fluid's viscosity also increased, which was more noticeable at higher temperatures. As a result, although higher temperatures led to lower densities, adding VFs increased the viscosity. There have been advancements in the field that involved improving the IT of the fluid used for EOR, leading to a decrease in the fluid's density due to thermal expansion and an increase in viscosity due to reduced intermolecular forces at higher temperatures. As a result, the improved IT led to increased fluid mobility and a remarkable ability to recover oil from the reservoir. Furthermore, adding NPs to the EOR fluid affected the density and viscosity of the water, ultimately increasing the fluid's viscosity and allowing for more efficient oil displacement from the reservoir.

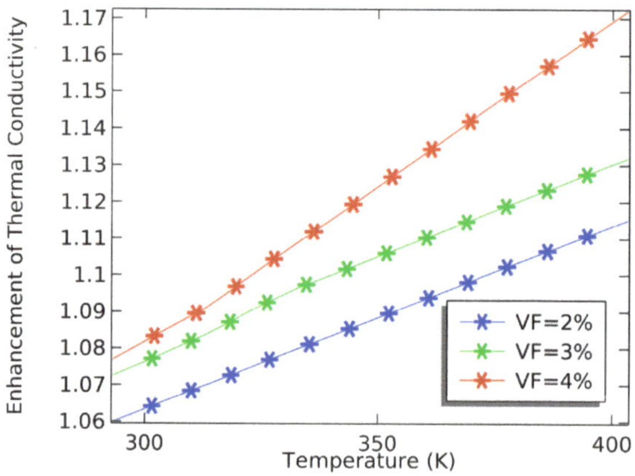

Figure 6. The effect of inlet temperature at different VFs on the nanofluid's TC.

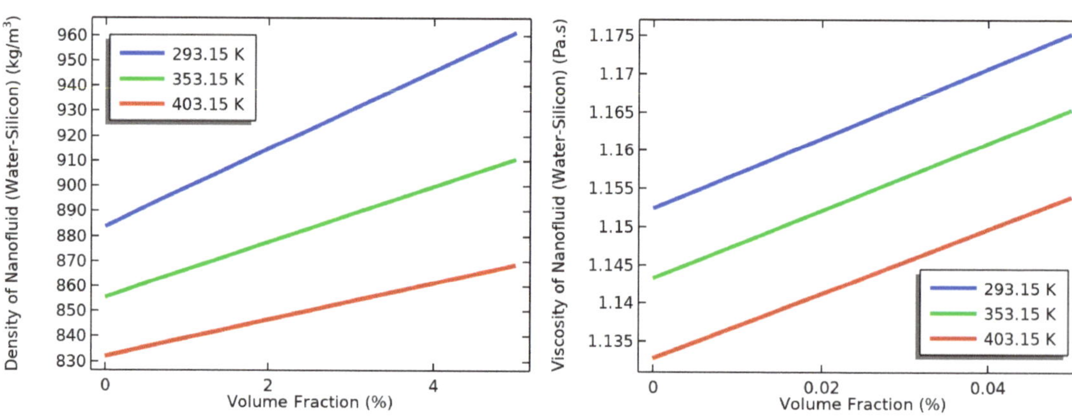

(**a**) (**b**)

Figure 7. (**a**) The effect of VF at different inlet temperatures on density, and (**b**) viscosity.

The effects of the VF and IT on the density and viscosity of the NF were examined. As shown in Figure 7a, we discovered that the density of the NF increased with the increase in the VF, irrespective of the IT value. The lowest density was observed at 403.15 K, whereas the highest was observed at 293.15 K. Additionally, Figure 7b shows that viscosity increased with the increase in the VF. The highest viscosity was observed at an IT of 293.15 K, whereas the lowest was observed at an IT of 403.15 K. Therefore, we can conclude that both the VF and IT impact the TPs of nanofluids.

The effect of the IT on the density and viscosity of the NF was analyzed. We utilized water-containing NPs with different volume fractions and determined the density of the NF using the empirical correlation provided in Equation (16). Figure 8a displays that as the IT increased, the density of the NF also decreased. We found that the NF density was lowest at a VF of 2% compared to other VFs, whereas the highest density was observed at a VF of 5%. Moreover, we investigated the impact of the IT on the viscosity of the NF with varying VFs. To calculate the viscosity, we used Equation (17). Figure 8b depicts that the NF viscosity decreased as the IT increased. A VF of 2% had the lowest viscosity compared to the other VFs, whereas a VF of 5% had the highest viscosity compared to the other VFs. We used empirical correlations of the TC, density, and viscosity to study the changes in the properties of the oil and NF. We also examined the continuity equations of both fluids. By adding SiO_2 to the NF, we were able to increase the viscosity of the fluid. This improved the fluid flow in the reservoir and helped enhance the OR rates.

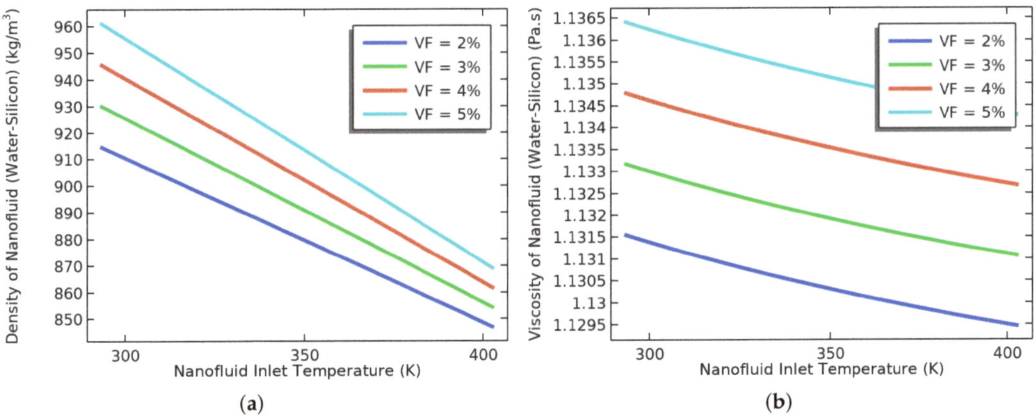

Figure 8. (**a**) The effect of the inlet temperature at different VFs on density, and (**b**) viscosity.

3.2. How Pressure and Velocity Are Distributed within the PM While Flooding with Nanofluid

To increase oil recovery, NF flooding injection into the oil reservoir for 10 min replaces the original oil. Various factors, including the properties of the base fluid and NPs, the size and shape of the pores in the PM, and the speed of NF injection, affect the pressure and velocity of the NF mixture. The model proposed in this study calculates the pressure using the continuity equation and the velocity using the Darcy equation while considering the initial and boundary conditions.

In enhanced oil recovery, the movement of nanofluid through porous media is smooth and steady, which is known as laminar flow. This occurs because the nanofluid has a low Reynolds number, meaning that the flow is mainly based on viscosity rather than inertia. As a result, the flow pattern is stable and organized due to the relatively weak inertial forces compared to the viscous forces.

The visual representation in Figure 9a shows the distribution of the nanofluid velocity in the porous medium during the oil phase, demonstrating how the nanofluid flowed evenly and continuously through the interconnected pores of the medium, leading to the effective displacement of the oil. An important factor to consider is the stability of nanofluid pressure over time. Because of its slow movement through the porous medium, there were minimal pressure fluctuations. This can be seen in Figure 9b, which shows a consistent pressure profile over time. The controlled movement of the nanofluid helped in maintaining a relatively constant pressure, which is crucial for an efficient and effective EOR process.

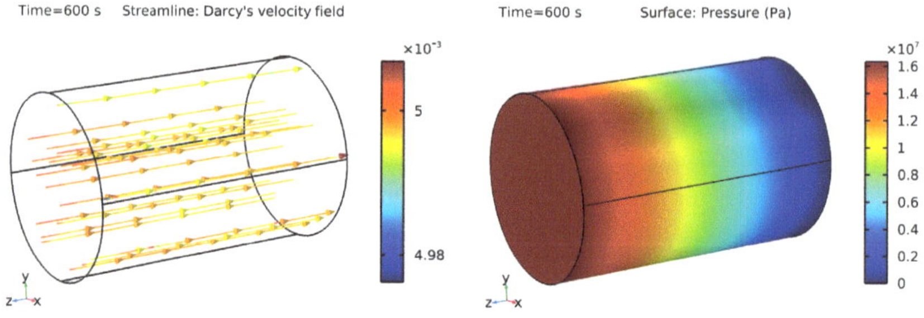

(a) (b)

Figure 9. (**a**) Pressure at 600 s. (**b**) The nanofluid's Darcy velocity at 600 s inside the porous medium.

3.3. NF and Water Saturation in PM

Water saturation refers to water in a porous material's empty spaces or pores. When water flows through the PM, it fills some of the empty spaces while the remaining spaces are filled with oil. The behavior of a nanofluid in the PM is influenced by the properties of the nanoparticles, volume fraction, and characteristics of the PM. Adding NPs to water affects its properties, including increasing its thermal conductivity and reducing its viscosity, which can lead to changes in the fluid flow behavior in the PM.

Figure 10a,b show the 3D layout of the water and oil phases during the 300 s and 600 s flooding processes. The color red represents water, whereas the color blue represents oil saturation. Figure 10b shows that the displacement of oil by water was more significant at 600 s than at 300 s, as seen in Figure 10a. Additionally, in Figures 10a,b and 11a,b, it can be seen that the saturation of the NF was higher at 300 s and 600 s compared to the saturation of water at the same time intervals. The NF saturation in the PM was higher because of the NPs present in the base fluid, which helped the fluid fill the pores more effectively. This resulted in more NF being retained in the PM, leading to a higher saturation. Additionally, the NPs occupied a more significant portion of the pore space on the PM's surface, increasing adsorption and further boosting saturation. However, it is essential to note that the saturation level depends on various factors such as the type of PM, the VF of the NPs, the IT, and the fluid itself [55,56].

The saturation of water containing SiO_2 with a VF of 4% was more significant, which occupied some of the void spaces and reduced the practical pore volume available for water. Additionally, the surface properties of the SiO_2 NPs affected their interaction with the pore surfaces of the PM, further decreasing the effective pore volume available for water. This contributed to a higher water saturation in the SiO_2 than in the PM.

The saturation level of the NF comprising SiO_2 with a VF of 4% and an inlet temperature of 353.15 K was higher than the water saturation in the PM for various reasons. Firstly, NFs have different TPs than water, meaning that the addition of NPs to water can increase its TC, thereby enhancing its heat transfer performance. As a result, the NF cools down more rapidly than pure water, resulting in a higher saturation level. Secondly, the nanoparticles in the nanofluid can alter the surface properties of the PM. The NPs adsorb onto the pore surfaces of the medium, changing their wettability and resulting in changes in the fluid flow and saturation level. In some cases, the NPs can also coat the PM's surface, reducing the effective pore size and increasing capillary forces, which leads to a higher saturation level. Thirdly, the saturation level of the nanofluid can also be influenced by its temperature. The thermal expansion of the NF at higher elevated temperatures increases its volume, leading to a higher saturation level.

By combining these factors, the saturation level in the NF containing silicon with a VF of 4% and an inlet temperature of 353.15 K was higher than the water saturation in the PM. Notably, the saturation level depends on the nanofluid's specific properties, the characteristics of the porous medium, and the conditions of their interaction.

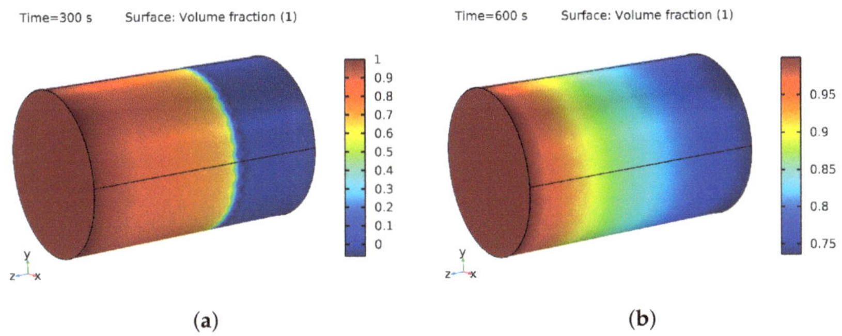

Figure 10. (**a**) Saturation of water at 300 s and (**b**) 600 s.

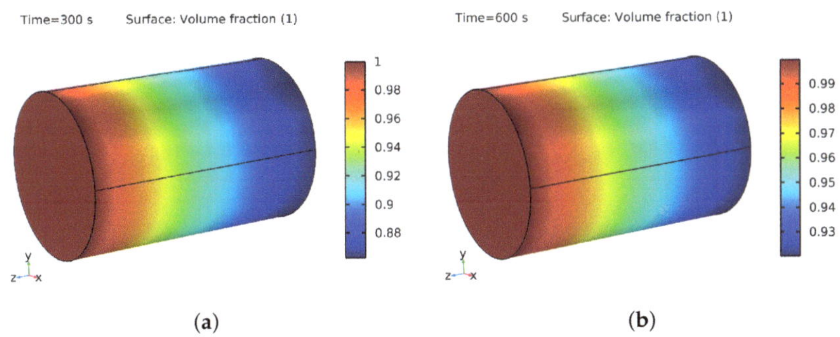

Figure 11. (**a**) Saturation of NF at 300 s and (**b**) 600 s.

3.4. The Effect of VF at Different Inlet Fluid Temperatures on ORF

The number of NPs in water can impact the effectiveness of EOR methods. When SiO_2 was added, the fluid's viscosity increased, which helped move oil through the PM, resulting in a higher recovery factor. Increasing the volume fraction made the fluid more viscous, which improved the ORF, as shown in Figure 12. However, when the volume fraction was too high, the fluid became too thick to be effectively injected into the rock, thereby decreasing the ORF. Figure 12 shows that the ORF decreased when the volume fraction exceeded 4%.

As seen in Figure 12, using an NF in EOR resulted in increased ORF when the fluid temperature at the inlet was raised from 293.15 K to 353.15 K. Several factors contributed to this improvement, including the increased fluid viscosity, reduced oil mobility ratio, and changed wettability of the reservoir rock. One mechanism that played a significant role was thermo-viscous displacement, which was influenced by the temperature. As the IT of the nanofluid increased, its viscosity increased, creating more resistance to oil flow and leading to more efficient oil displacement from the reservoir rock. This led to a higher ORF. Moreover, the higher temperature reduced the oil's viscosity, making it more mobile and thermally activated in the petroleum matrix.

It is important to consider how the IT affects the effectiveness of NPs in oil recovery. Figure 12 shows that higher ITs can increase OR rates as viscosity decreases and oil mobility increases. However, the optimal temperature may vary depending on the specific properties of the oil and NPs. To fully utilize NPs for OR to their fullest potential, it is crucial to carefully examine the impact of the VF on the ORF, which encompasses both the VF and its influence. When the temperature was at 293.15 K, the oil recovery factor increased as the silicon oxide content increased from 1% to 5%, as seen in Figure 12. This is because lower temperatures tend to increase viscosity and resistance to flow. Adding silicon oxide decreased the interfacial tension between the oil and the rock, allowing for more effortless flow and improving the recovery rate. At 353.15 K (80 °C), the higher temperature reduced the oil's viscosity. Adding silicon oxide also positively reduced the interfacial tension and enhanced OR efficiency, resulting in a higher recovery rate, as shown in Figure 12.

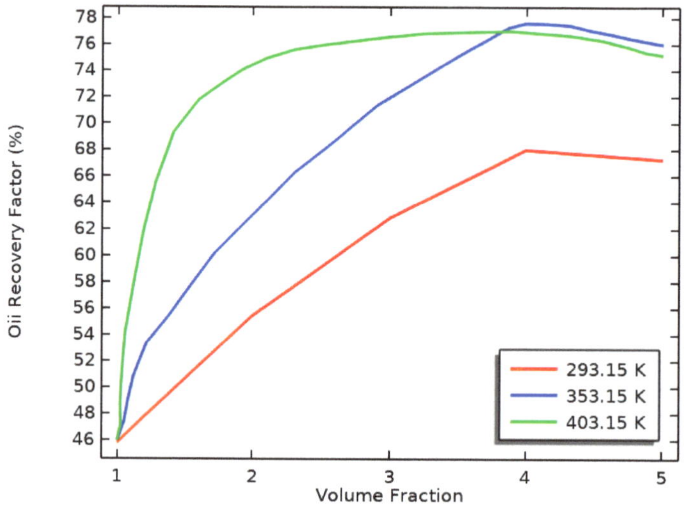

Figure 12. The effect of VF on the ORF at different NF inlet temperatures.

3.5. The Effect of NF Inlet Temperature on the ORF at Different VFs

Increasing the inlet temperature of the NF boosted its thermal conductivity, resulting in better heat transfer performance, as shown in Figure 6. Increasing the inlet temperature for water-silicon oxide NPs with a volume fraction of 4% or 5% likely positively affected their thermal conductivity. At a VF of 4% and a temperature of 353.15 K, the OR increased by 30%, as indicated in Figure 13, before decreasing.

When nanoparticles are well dispersed and do not form agglomerates, increasing their volume fraction in a nanofluid can increase thermal conductivity. However, when the VF of silicon oxide nanoparticles was constant and the inlet temperature of the NF was increased from 0 to 403.15 K, the effect on thermal conductivity was less noticeable than when the VF was increased to 4%. In addition, increasing the IT from 0 to 403.15 K at a VF of 4% produced more oil recovery compared to a VF of 5%.

The temperature of a specific fluid had a notable impact on the ability to extract oil from the reservoir. As the temperature increased, the oil became less viscous, improving its ability to move outside the reservoir, as demonstrated in Figure 13. However, if the temperature became too high, the nanofluid became less viscous, decreasing its ability to push the oil outside and flow through the reservoir. Therefore, the ideal temperature range for this particular fluid is between 293.15 and 353.15 K. This range resulted in the highest oil recovery factor, as shown in Figure 13. These findings suggest that using this fluid in enhanced oil recovery methods can significantly improve the oil recovery factor compared to traditional fluids. Generally, the recovery factor increases as the fluid's temperature increases, but the oil recovery factor starts to decline once the volume fraction reaches 5% at temperatures higher than 353.15 K.

The effect of the inlet temperature of the nanofluid on the oil recovery factor depends on the nanofluid volume fraction used. When the VF was 4%, increasing the IT of the NF caused a slight increase in the ORF compared to a VF of 5%, as shown in Figure 13. The higher temperature reduced the oil's viscosity, making it easier to extract from the reservoir. Moreover, when the VF of the NF was 4%, the effect of the IT on the ORF was more pronounced. This is because the higher VF of the NF improved the heat transfer and enhanced convection, reducing the oil's viscosity and improving its recovery. Therefore, it is essential to consider these factors to accurately predict the effect of the IT of the NF on the ORF.

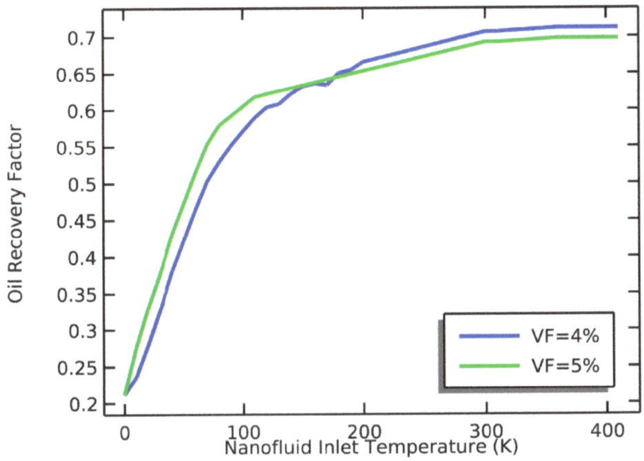

Figure 13. The effect of the inlet temperature of the NF on the ORF at VFs of 4% and 5%.

3.6. Comparison of ORF of Water and NF

The simulation was performed twice, comparing the effects of water-SiO_2 and water on increasing the ORF. The results in Figure 14 indicate that the use of the NF (water-SiO_2) led to a more significant increase in the ORF than water flooding. The mixture of water-SiO_2 produced the highest ORF, which was 30% higher than that of water flooding. In the case of water-SiO_2 flooding, NPs increased OR in the PM compared to pure water flooding for several reasons.

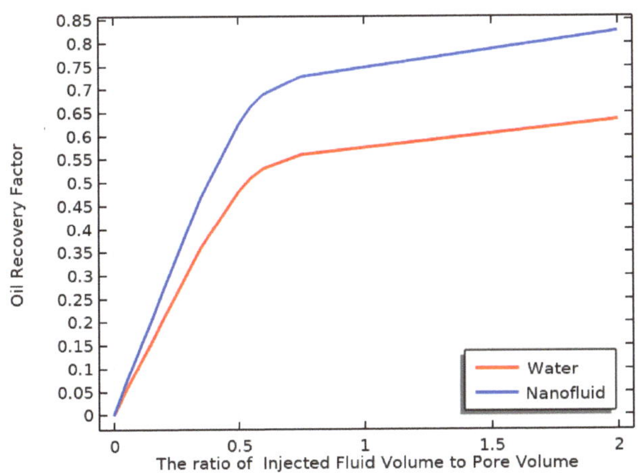

Figure 14. Comparison of ORF of water and nanofluid.

The addition of NPs to water resulted in the increased viscosity of the NF, as demonstrated in Figures 7b and 8b. This led to better sweep efficiency and decreased mobility of the injected fluid, ultimately facilitating the movement of oil through the PM and toward the production well. Additionally, NPs improved the TC of the injected water, as evidenced in Figures 5 and 6. This helped reduce the oil's viscosity and improve its mobility within the reservoir.

When NPs absorb rock and modify it, they alter its surface properties. By using a water-SiO_2 NF, the NPs enhanced the rock surface's ability to be wetted by water, allowing for the easier displacement of oil from the pores by water. The NPs also decreased the interfacial tension between the oil and the water, causing oil droplets to detach more easily from the rock surface and move toward the production well. When using a water-SiO_2 NF with a VF of 4% at 353.15 K, OR in the PM was higher compared to pure water flooding due to these mechanisms. However, the specific effects of NF flooding can vary depending on factors such as the reservoir rock type, injection rate and pressure, and properties of the NF.

4. Conclusions

In summary, the injection of an NF into a 3D porous medium for EOR has been modeled. The FEM has been used to compute the governing equations, Darcy's law, mass conservation, concentration, and energy equations, coded in COMSOL software. The inlet temperature, VF, and pressure were shown to influence the nanofluid's TPs in the PM. Accurately considering these TP changes in modeling the EOR process is crucial for predicting the ORF. This study presented a mathematical model that can describe the changes in the density, viscosity, and TC of nanofluids with variations in the temperature and VF. This information is crucial for developing efficient and cost-effective EOR methods using NFs. The findings of this study provide a foundation for further research in this field and can contribute to the advancement of technology in the oil and gas industry.

The results can be summarized as follows:

- Nanofluid flooding in a 3D porous medium for the EOR process has been modeled.
- The proposed model addresses the modifications in thermophysical properties concurrently with fluctuating temperatures and volume fractions facilitated by innovative empirical correlations.
- An increase in volume fractions led to agglomeration, which reduced the effectiveness of nanofluids in EOR. However, an increase in the inlet temperature resulted in an increase in the thermal conductivity of the nanofluid due to increased Brownian motion. This phenomenon helped to break up agglomerates and improved the oil recovery factor.
- The amplification of NF density due to the heightened temperature and VF fostered oil recovery by diminishing the mobility ratio and augmenting displacement efficiency.
- The increase in the volume fraction led to an increase in the viscosity of the nanofluid. In contrast, increasing the inlet temperature decreased oil viscosity, making it easier for extraction and increasing oil production.
- Notably, the simulation results were analyzed and compared using an NF and pure water. Adding the SiO_2 NPs with a VF of 4% and an NF inlet temperature of 353.15 K resulted in 30% more oil production than water alone for flooding.
- A significant decrease in the ORF was observed when the VF was more than 4% and the inlet temperature exceeded 353.15 K.

Finally, this study is valuable to the field as it models the injection of nanofluids into a 3D porous medium, which can improve oil recovery. The developed mathematical model accurately accounts for changes in the properties of nanofluids, including the density, viscosity, and thermal conductivity, with variations in the temperature and volume fraction. The results are useful for engineering applications, particularly in optimizing the use of nanofluids for enhanced oil recovery. This study emphasizes the impact of aggregation on nanofluid effectiveness and shows how higher inlet temperatures can disperse agglomerates and enhance oil recovery.

Author Contributions: A.A.-Y. wrote the manuscript's main text, and A.A.-Y., D.L.C.C., H.S. and M.S.M. managed the project (methodology, software, validation, formal analysis, and data maintenance). D.L.C.C., H.S. and M.S.M. supervised the project, and H.S. funded the project. M.Z., A.H., Z.M.A.M. and A.S.A. reviewed the manuscript. All authors read and agreed to the published version of the manuscript.

Funding: This research was funded by Universiti Teknologi PETRONAS - Universiti Malaysia Pahang Matching Grant (Cost Center: 015MC0-033).

Institutional Review Board Statement: Not applicable.

Informed Consent Statement: Not applicable.

Data Availability Statement: The data presented in this study are available from the corresponding author upon reasonable request.

Conflicts of Interest: The authors declare no conflict of interest.

Nomenclature

The following abbreviations are used in this manuscript:

Greek Letters

α	Thermal diffusivity
ΔP	Pressure drop
ΔT	Temperature difference
ϵ	Porosity
$\kappa_{r,nf}$	Relative permeability of the NF
$\kappa_{r,o}$	Relative permeability of the oil
λ	Pore-size distribution index
μ	Dynamic viscosity
ρ	Density
θ	Contact angle

Latin Letters

B_c	Square root of the permeability
C	NP concentration
C_p	Specific heat capacity
D	Diffusivity
D_c	Diffusion coefficient
g	Acceleration due to gravity
h	Film thickness
K	Absolute permeability
L	Characteristic length
P	Pressure
P_c	Capillary pressure
Pr	Prandtl number
Re	Reynolds number
S_e	Effective saturation
S_{nf}	NF saturation
S_o	Oil saturation
S_{rnf}	Residual saturation of the NF
S_{ro}	Oil residual saturation
T	Temperature
T_{in}	Inlet temperature of the NF
T_{init}	Initial temperature of the system
t	Time
u	Flow velocity
u_{nf}	Input velocity of the inlet fluid
VF	Volume Fraction

Dimensionless Numbers

Bi	Biot number
Nu	Nusselt number
Pe	Peclet number
Pr	Prandtl number
Re	Reynolds number
Sc	Schmidt number
Sh	Sherwood number

References

1. Shi, F.; Wu, J.; Zhao, B. Preparation and investigation of intelligent polymeric nanocapsule for enhanced oil recovery. *Materials* **2019**, *12*, 1093. [CrossRef] [PubMed]
2. Bear, J. *Dynamics of Fluids in Porous Media*; American Elsevier Publishing Company, Inc.: New York, NY, USA, 1972.
3. Yan, G.; Li, Z.; Galindo Torres, S.A.; Scheuermann, A.; Li, L. Transient two-phase flow in porous media: A literature review and engineering application in geotechnics. *Geotechnics* **2022**, *2*, 32–90. [CrossRef]
4. Sun, X.; Zhang, Y.; Chen, G.; Gai, Z. Application of nanoparticles in enhanced oil recovery: A critical review of recent progress. *Energies* **2017**, *10*, 345. [CrossRef]
5. Zhao, W.; Tang, H.; Lu, F.; Hu, S.; Liu, T.; Li, N.; Song, R. Mathematical model for oil recovery prediction of polymer microsphere conformance control based on the stream tube method. *Materials* **2023**, *16*, 1476. [CrossRef]
6. Al-Yaari, A.; Sakidin, H.; Zainuddin, N.; Hashim, I. Unsteady Nanofluid Flow Over Exponentially Stretching Sheet with Vertical Throughflow. In *Proceedings of the 6th International Conference on Fundamental and Applied Sciences: ICFAS 2020*; Springer: Berlin/Heidelberg, Germany, 2021; pp. 595–609.
7. Foroozesh, J.; Kumar, S. Nanoparticles behaviors in porous media: Application to enhanced oil recovery. *J. Mol. Liq.* **2020**, *316*, 113876. [CrossRef]
8. Irfan, S.A.; Shafie, A.; Yahya, N.; Zainuddin, N. Mathematical modeling and simulation of nanoparticle-assisted enhanced oil recovery—A review. *Energies* **2019**, *12*, 1575. [CrossRef]
9. Kamal, M.S.; Adewunmi, A.A.; Sultan, A.S.; Al-Hamad, M.F.; Mehmood, U. Recent Advances in Nanoparticles Enhanced Oil Recovery: Rheology, Interfacial Tension, Oil Recovery, and Wettability Alteration. *J. Nanomater.* **2017**, *2017*, 2473175. [CrossRef]
10. Monfared, A.D.; Ghazanfari, M.H.; Jamialahmadi, M.; Helalizadeh, A. Potential Application of Silica Nanoparticles for Wettability Alteration of Oil-Wet Calcite: A Mechanistic Study. *Energy Fuels* **2016**, *30*, 3947–3961. [CrossRef]
11. Zafar, M.; Sakidin, H.; Sheremet, M.; Dzulkarnain, I.B.; Hussain, A.; Nazar, R.; Khan, J.A.; Irfan, M.; Said, Z.; Afzal, F.; et al. Recent Development and Future Prospective of Tiwari and Das Mathematical Model in Nanofluid Flow for Different Geometries: A Review. *Processes* **2023**, *11*, 834. [CrossRef]
12. Zafar, M.; Sakidin, H.; Sheremet, M.; Dzulkarnain, I.; Nazar, R.M.; Hussain, A.; Said, Z.; Afzal, F.; Al-Yaari, A.; Khan, M.S.; et al. The Impact of Cavities in Different Thermal Applications of Nanofluids: A Review. *Nanomaterials* **2023**, *13*, 1131. [CrossRef]
13. Yan, G.; Li, Z.; Bore, T.; Torres, S.A.G.; Scheuermann, A.; Li, L. A lattice Boltzmann exploration of two-phase displacement in 2D porous media under various pressure boundary conditions. *J. Rock Mech. Geotech. Eng.* **2022**, *14*, 1782–1798. [CrossRef]
14. Zafar, M.; Sakidin, H.; Dzulkarnain, I.; Farkhanda, A. *Numerical Investigations of Nano-Fluid Flow in Square Porous Cavity: Buongiorno's Mathematical Model*; Springer: Singapore, 2022; Volume 14, pp. 739–748.
15. Su, J.; Chai, G.; Wang, L.; Cao, W.; Yu, J.; Gu, Z.; Chen, C. Direct numerical simulation of pore scale particle-water-oil transport in porous media. *J. Pet. Sci. Eng.* **2019**, *180*, 159–175. [CrossRef]
16. Gharibshahi, R.; Jafari, A.; Haghtalab, A.; Karambeigi, M.S. Application of CFD to evaluate the pore morphology effect on nanofluid flooding for enhanced oil recovery. *RSC Adv.* **2015**, *5*, 28938–28949. [CrossRef]
17. Ma, Y.; Yan, G.; Scheuermann, A. Discrete bubble flow in granular porous media via multiphase computational fluid dynamic simulation. *Front. Earth Sci.* **2022**, *10*, 947625. [CrossRef]
18. Liu, X.; Civan, F. Characterization and prediction of formation damage in two-phase flow systems. In Proceedings of the SPE Production Operations Symposium, OnePetro, Oklahoma City, OK, USA, 21–23 March 1993.
19. Ju, B.; Fan, T.; Ma, M. Enhanced oil recovery by flooding with hydrophilic nanoparticles. *China Particuology* **2006**, *4*, 41–46. [CrossRef]
20. Sepehri, M.; Moradi, B.; Emamzadeh, A.; Mohammadi, A.H. Experimental study and numerical modeling for enhancing oil recovery from carbonate reservoirs by nanoparticle flooding. *Oil Gas Sci. Technol. D'Ifp Energies Nouv.* **2019**, *74*, 5. [CrossRef]
21. Ju, B.; Fan, T. Experimental study and mathematical model of nanoparticle transport in porous media. *Powder Technol.* **2009**, *192*, 195–202. [CrossRef]
22. El-Amin, M.F.; Sun, S.; Salama, A. Enhanced oil recovery by nanoparticles injection: Modeling and simulation. In Proceedings of the SPE Middle East Oil and Gas Show and Conference, OnePetro, Manama, Bahrain, 18–21 March 2019.
23. Feng, Y.; Cao, L.; Shi, E. A numerical investigation of enhanced oil recovery using hydrophilic nanofluids. *J. Sustain. Energy Eng.* **2017**, *5*, 67–97. [CrossRef]

24. Yu, J.; Berlin, J.M.; Lu, W.; Zhang, L.; Kan, A.T.; Zhang, P.; Walsh, E.E.; Work, S.N.; Chen, W.; Tour, J.M. Transport Study of Nanoparticles for Oilfield Application. In Proceedings of the SPE International Conference on Oilfield Scale, OnePetro, Aberdeen, UK, 14–15 May 2014.
25. Rahmani, A.R.; Bryant, S.; Huh, C.; Athey, A.; Ahmadian, M.; Chen, J.; Wilt, M. Crosswell magnetic sensing of superparamagnetic nanoparticles for subsurface applications. *SPE J.* **2015**, *20*, 1067–1082. [CrossRef]
26. Gruesbeck, C.; Collins, R. Entrainment and deposition of fine particles in porous media. *Soc. Pet. Eng. J.* **1982**, *22*, 847–856. [CrossRef]
27. Cullen, E.; O'Carroll, D.M.; Yanful, E.K.; Sleep, B. Simulation of the subsurface mobility of carbon nanoparticles at the field scale. *Adv. Water Resour.* **2010**, *33*, 361–371. [CrossRef]
28. Zhang, T.; Murphy, M.; Yu, H.; Huh, C.; Bryant, S.L. Mechanistic model for nanoparticle retention in porous media. *Transp. Porous Media* **2016**, *115*, 387–406. [CrossRef]
29. An, C.; Alfi, M.; Yan, B.; Cheng, K.; Heidari, Z.; Killough, J.E. Modeling of magnetic nanoparticle transport in shale reservoirs. In Proceedings of the SPE Reservoir Simulation Symposium, OnePetro, Houston, TX, USA, 23–25 February 2015.
30. Abdelfatah, E.R.; Soliman, M.M.; Khattab, H.M. Improving Heavy Oil Recovery by Nanofluid Injection: The factors Affecting and Mathematical Modelling. *J. Pet. Min. Eng.* **2014**, *17*, 88–89.
31. Abdelfatah, E.; Pournik, M.; Shiau, B.J.B.; Harwell, J. Mathematical modeling and simulation of nanoparticles transport in heterogeneous porous media. *J. Nat. Gas Sci. Eng.* **2017**, *40*, 1–16. [CrossRef]
32. Khalilinezhad, S.S.; Cheraghian, G.; Karambeigi, M.S.; Tabatabaee, H.; Roayaei, E. Characterizing the role of clay and silica nanoparticles in enhanced heavy oil recovery during polymer flooding. *Arab. J. Sci. Eng.* **2016**, *41*, 2731–2750. [CrossRef]
33. Salama, A.; Negara, A.; El Amin, M.; Sun, S. Numerical investigation of nanoparticles transport in anisotropic porous media. *J. Contam. Hydrol.* **2015**, *181*, 114–130. [CrossRef]
34. Esfe, M.H.; Esfandeh, S. 3D numerical simulation of the enhanced oil recovery process using nanoscale colloidal solution flooding. *J. Mol. Liq.* **2020**, *301*, 112094. [CrossRef]
35. Esfe, M.H.; Esfandeh, S.; Hosseinizadeh, E. Nanofluid flooding in a randomized heterogeneous porous media and investigating the effect of capillary pressure and diffusion on oil recovery factor. *J. Mol. Liq.* **2020**, *320*, 113646. [CrossRef]
36. Al-Yaari, A.; Ching, D.L.C.; Sakidin, H.; Muthuvalu, M.S.; Zafar, M.; Alyousifi, Y.; Saeed, A.A.H.; Bilad, M.R. Thermophysical Properties of Nanofluid in Two-Phase Fluid Flow through a Porous Rectangular Medium for Enhanced Oil Recovery. *Nanomaterials* **2022**, *12*, 1011. [CrossRef]
37. Esfe, M.H.; Hosseinizadeh, E.; Esfandeh, S. Flooding numerical simulation of heterogeneous oil reservoir using different nanoscale colloidal solutions. *J. Mol. Liq.* **2020**, *302*, 111972. [CrossRef]
38. Al-Yaari, A.; Ching, D.L.C.; Sakidin, H.; Muthuvalu, M.S.; Zafar, M.; Alyousifi, Y.; Saeed, A.A.H.; Haruna, A. Optimum Volume Fraction and Inlet Temperature of an Ideal Nanoparticle for Enhanced Oil Recovery by Nanofluid Flooding in a Porous Medium. *Processes* **2023**, *11*, 401. [CrossRef]
39. Jafari, A.; Hasani, M.; Hosseini, M.; Gharibshahi, R. Application of CFD technique to simulate enhanced oil recovery processes: Current status and future opportunities. *Pet. Sci.* **2020**, *17*, 434–456. [CrossRef]
40. Multiphysics, C. *Introduction to Comsol Multiphysics®*; COMSOL Multiphysics: Burlington, MA, USA, 1998; Volume 9, p. 32.
41. Afsharpoor, A.; Ma, K.; Mateen, K.; Duboin, A.; Jouenne, S.; Cordelier, P. Micro-scale experiment and cfd modeling of viscoelastic polymer; trapped oil displacement and deformation at the dead-end. In Proceedings of the SPE Improved Oil Recovery Symposium, OnePetro, Tulsa, OK, USA, 12–16 April 2014.
42. Das, L.; Habib, K.; Saidur, R.; Aslfattahi, N.; Yahya, S.M.; Rubbi, F. Improved thermophysical properties and energy efficiency of aqueous ionic liquid/mxene nanofluid in a hybrid pv/t solar system. *Nanomaterials* **2020**, *10*, 1372. [CrossRef] [PubMed]
43. Ewing, R.E. *The Mathematics of Reservoir Simulation*; SIAM: Philadelphia, PA, USA, 1983.
44. Russell, T.F.; Wheeler, M.F. Finite element and finite difference methods for continuous flows in porous media. In *The Mathematics of Reservoir Simulation*; SIAM: Philadelphia, PA, USA, 1983; pp. 35–106.
45. Singh, S.; Ghosh, S.K. A unique artificial intelligence approach and mathematical model to accurately evaluate viscosity and density of several nanofluids from experimental data. *Colloids Surf. Physicochem. Eng. Asp.* **2022**, *640*, 128389. [CrossRef]
46. Maxwell, J.C. *A Treatise on Electricity and Magnetism*; Clarendon Press: Oxford, UK, 1873; Volume 1.
47. Murshed, S.; Leong, K.; Yang, C. A combined model for the effective thermal conductivity of nanofluids. *Appl. Therm. Eng.* **2009**, *29*, 2477–2483. [CrossRef]
48. Zuo, L.; Krevor, S.; Falta, R.W.; Benson, S.M. An experimental study of CO_2 exsolution and relative permeability measurements during CO2 saturated water depressurization. *Transp. Porous Media* **2012**, *91*, 459–478. [CrossRef]
49. El-Amin, M.; Kou, J.; Sun, S.; Salama, A. Numerical modeling of nanoparticles transport with two-phase flow in porous media using iterative implicit method. *arXiv* **2013**, arXiv:1310.4769.
50. Diaz-Viera, M.; Lopez-Falcon, D.; Moctezuma-Berthier, A.; Ortiz-Tapia, A. COMSOL implementation of a multiphase fluid flow model in porous media. In Proceedings of the COMSOL Conference, Budapest, Hungary, 24 November 2008; p. 7.
51. Lide, D.R. *CRC Handbook of Chemistry and Physics*; CRC Press: Boca Raton, FL, USA, 2004; Volume 85.
52. Esfe, M.H.; Karimipour, A.; Yan, W.M.; Akbari, M.; Safaei, M.R.; Dahari, M. Experimental study on thermal conductivity of ethylene glycol based nanofluids containing Al_2O_3 nanoparticles. *Int. J. Heat Mass Transf.* **2015**, *88*, 728–734. [CrossRef]

53. Jin, F. Principles of enhanced oil recovery. In *Physics of Petroleum Reservoirs*; Springer: Berlin/Heidelberg, Germany, 2017; pp. 465–506.
54. Minakov, A.V.; Guzei, D.V.; Pryazhnikov, M.I.; Filimonov, S.A.; Voronenkova, Y.O. 3D pore-scale modeling of nanofluids-enhanced oil recovery. *Pet. Explor. Dev.* **2021**, *48*, 956–967. [CrossRef]
55. Jafarbeigi, E.; Kamari, E.; Salimi, F.; Mohammadidoust, A. Experimental study of the effects of a novel nanoparticle on enhanced oil recovery in carbonate porous media. *J. Pet. Sci. Eng.* **2020**, *195*, 107602. [CrossRef]
56. Natalya, S.A.; Kadja, G.T.; Azhari, N.J.; Khalil, M.; Fajar, A.T. Two-dimensional (2D) Nanomaterials for Enhanced Oil Recovery (EOR): A Review. *FlatChem* **2022**, *34*, 100383. [CrossRef]

materials

MDPI

Article

Removal of Iron and Copper Ions and Phenol from Liquid Phase by Membrane Based on Carbonaceous Materials

Joanna Przybyl [1], Aleksandra Bazan-Wozniak [1], Faustyna Poznan [1], Agnieszka Nosal-Wiercińska [2], Judyta Cielecka-Piontek [3] and Robert Pietrzak [1,*]

1 Department of Applied Chemistry, Faculty of Chemistry, Adam Mickiewicz University in Poznań, Uniwersytetu Poznańskiego 8, 61-614 Poznań, Poland; asia.krason@amu.edu.pl (J.P.); aleksandra.bazan@amu.edu.pl (A.B.-W.); faupoz@st.amu.edu.pl (F.P.)

2 Department of Analytical Chemistry, Institute of Chemical Sciences, Faculty of Chemistry, Maria Curie-Skłodowska University in Lublin, Maria Curie-Skłodowska Sq., 3, 20-031 Lublin, Poland; agnieszka.nosal-wiercinska@mail.umcs.pl

3 Department of Pharmacognosy, Faculty of Pharmacy, Poznan University of Medical Sciences, Rokietnicka 3, 60-806 Poznan, Poland; jpiontek@ump.edu.pl

* Correspondence: pietrob@amu.edu.pl; Tel.: +48-61829-1560

Abstract: The present work reports an effective method for the removal of inorganic and organic pollutants using membranes based on different carbonaceous materials. The membranes were prepared based on cellulose acetate (18 wt. %), polyvinylpyrrolidone as a pore-generating agent (2 wt. %) and activated carbon (1 wt. %). Activated carbons were developed from residues after extraction of the mushroom *Inonotus obliguus* using microwave radiation. It has been demonstrated that the addition of activated carbon to the membranes resulted in alterations to their physical properties, including porosity, equilibrium water content and permeability. Furthermore, the chemical properties of the membranes were also affected, with changes observed in the content of the surface oxygen group. The addition of carbon material had a positive effect on the removal of copper ions from their aqueous solutions by the cellulose–carbon composites obtained. Moreover, the membranes proved to be more effective in the removal of copper ions than iron ones and phenol. The membranes were found to show higher effectiveness in copper removal from a solution of the initial concentration of 800 mg/L. The most efficient in copper ions removal was the membrane containing urea-enriched activated carbon.

Keywords: cellulose acetate membrane; physicochemical properties; carbonaceous adsorbents; microwave oven; organic and inorganic pollutants

1. Introduction

In recent years, the study of the effects of pollution on the environment and human health has received increasing attention. The negative effects of heavy metals and phenol in water on both the human body and the environment have been known for years. Consumption of water with increased concentrations of iron and copper ions causes health problems such as disorders of the nervous and cardiovascular systems, abnormal metabolism or cancer [1,2]. Conversely, phenol, an organic aromatic compound, has been demonstrated to exert a detrimental impact on the functioning of the human body, manifesting as skin irritation, cardiac, renal and hepatic impairment, visual impairment and metabolic dysfunction [3]. Furthermore, elevated phenol content and iron and copper ions in the hydrosphere have a negative effect on aquatic organisms, causing, among other things, oxidative stress [4]. The above facts serve to confirm the validity of developing new and modifying existing methods to effectively remove these pollutants from both ground and surface waters [5,6].

Membrane processes represent one of the most popular methods for contaminant removal in chemical engineering [7,8]. At the synthesis stage, membranes can be modified by the addition of a pore-generating agent (e.g., polyvinylpyrrolidone) or materials (e.g., activated carbons) to improve the removal efficiency [9,10]. Modified membranes are obtained using the phase inversion method, the most popular of the flat sheet membrane fabrication methods. This method involves the formation of a mixture comprising polymer, blowing agent, additional components and solvent, which is then spread on a substrate. After the solvent has evaporated, the mixture is immersed in a coagulation bath, such as deionised water [11,12].

At the stage of membrane synthesis, it is possible to modify the membranes with, for example, amines, metal oxides or activated carbons. This is discussed in detail in references [13,14]. Activated carbons are frequently employed as adsorbents for contaminants, both in the liquid and gaseous phases [15]. Due to their large specific surface area and microporous structure, these materials are effective adsorbents of compounds such as dyes, metal ions or organic compounds [16,17].

This paper presents the results of a study on the preparation, characterisation and efficiency in the removal of phenol and iron and copper ions from the liquid phase of membrane composites obtained on the basis of cellulose acetate and carbonaceous materials. The membranes were modified with the blowing agent polyvinylpyrrolidone. In order to test the influence of the type of carbonaceous material on the properties of the resulting cellulose–carbon composite, carbonaceous material was added during preparation. Tests were conducted on both pure membranes and membranes that had been modified with char, activated carbon derived from it and nitrogen-modified activated carbon. Compared to our previous work, the novel aspect of this research is the use of residues from the extraction of the fruiting body of the chaga mushroom (*Inonotus obliguus*). This precursor is a plentiful waste product with no discernible use. Furthermore, this study introduces a novel approach to synthesising membranes based on carbon materials, which enables the production of cost-effective adsorbents. This strategy also enhances environmental sustainability and provides a foundation for further research in the field of sustainability.

2. Materials and Methods

2.1. Materials

Cellulose acetate (acetyl content 39.8%, CA) was procured from Sigma Aldrich (Darmstadt, Germany) and employed as a membrane material. N,N-dimethylformamide (>99.8%, DMF) was sourced from Avantor Performance Materials Poland S.A. (Gliwice, Poland) and utilized as a solvent. Polyvinylpyrrolidone (>99%, PVP, 10,000 g/mol) as a pore former was provided by Sigma Aldrich (Sigma–Aldrich, Darmstadt, Germany). The carbonaceous materials added to the membranes were char (C), activated carbon (AC) and nitrogen-enriched activated carbon (NAC). The precursor carbon material was derived from the residue resulting from the methanol extraction of the fruiting body of the chaga mushroom (*Inonotus obliquus*). The fruiting body was sourced from a birch trunk. Following drying, the fruiting body was ground into a fine powder. Then, methanol removal was conducted. For this purpose, 2.50 g of powder was weighed and extracted into a reflux cooling bath at 80 °C. Carbon C was obtained by carbonisation of the precursor in a nitrogen atmosphere (150 mL/min) at 500 °C for 60 min (technical nitrogen 4.0, Linde Gaz Poland). Activated carbon (AC) was obtained by physical activation of the precursor with carbon dioxide (250 mL/min) at 800 °C for 30 min (technical CO_2 2.8, Linde Gaz Poland). Nitrogen-enriched activated carbon (NAC) was prepared by mixing 40% urea (Chempur, Piekary Śląskie, Poland) with char (C) in a mass ratio of 1:1, followed by activation with CO_2 (250 mL/min) for 15 min at 500 °C. Carbonisation and activation processes were carried out in a microwave oven (Phoenix, CEM Corporation, Matthews, IL, USA).

2.2. Preparation of CA Membranes

Casting solutions of CA 18 wt. %, 2 wt. % of PVP and 1 wt. % of activated carbon were prepared by mixing the ingredients in a flask. The resulting casting solution was then left to rest for approximately 24 h to allow for the complete release of bubbles. Subsequently, the casting solution was cast onto a glass plate using a stainless-steel knife to create a casting film with a thickness of 300 μm. The film was then exposed to the atmosphere for 40 s and subsequently immersed in a coagulation bath of deionised water. The as-prepared cast solution films were immersed and maintained in a deionised water bath at 25 °C for 24 h, after which the membranes formed were washed for a further 24 h using deionised water. The membranes obtained were designated as M, MC, MAC and MNAC.

2.3. Membrane Structure Characterisation

2.3.1. Porosity and Equilibrium Water Content

The membrane porosity was determined by the mass loss of wet membrane after drying. The membrane sample was mopped with water on the surface and weighed under wet status. Then, the membrane sample was dried until a constant mass was obtained. The membrane porosity [18] ε was evaluated from the following equation:

$$\varepsilon = \frac{W_w - W_d}{\rho \times V} \times 100\% \quad (1)$$

where W_w is the mass of a wet membrane sample (g), W_d is the mass of dry state membrane sample (g), ρ pure water density (g/dm^3) and V- is the volume of a membrane in wet state (L).

The equilibrium water content [19] (EWC) was determined by Equation (2):

$$EWC = \frac{W_w - W_d}{W_w} \times 100\% \quad (2)$$

Two parallel series of measurements were conducted, and the resulting data were presented as the arithmetic mean of the values obtained in each series.

2.3.2. Contact Angle

The contact angle between water and the membrane was directly measured using a contact angle measuring instrument (G10, KRUSS, Halle, Germany). In order to evaluate the membrane's hydrophilicity, deionised water was used as a probe liquid in all measurements. In order to minimise the experimental error, the contact angle was measured at five random locations for each sample, and the average was reported.

2.3.3. Surface Oxygen Groups

The surface properties were characterised using potentiometric titration experiments conducted with the aid of the 809 Titrando equipment manufactured by Metrohm (Opacz-Kolonia, Polnad). The instrument was configured to collect equilibrium pH data. The materials were weighed and placed in a container at a temperature of 25 °C. The container was then filled with 50 mL of 0.01 M $NaNO_3$ and allowed to equilibrate overnight with the electrolyte solution. To eliminate the influence of atmospheric CO_2, the suspension was continuously saturated with N_2. The carbon suspension was continuously stirred throughout the course of the measurements. The volumetric standards NaOH (0.1 M) or HCl (0.1 M) were employed as titrants in accordance with the methodology described in reference [20]. For each sample, two parallel determinations were conducted.

2.3.4. Membrane Performance Characterisation

The water permeability of the membranes was quantified in a stainless-steel cell, with an effective membrane area of 19.6 cm^2. Prior to testing, the membranes were subjected to deionised water at 3 bar for approximately 1.5 h. The pure water flux was then measured at 3 bar, 23 ± 1 °C and a crossflow velocity of 0.22 m/s. The pure water flux was calculated using the following equation:

$$J_w = \frac{V}{A \times \Delta t} \tag{3}$$

where J_w (L/(m^2×h)) is the pure water flux, V (L) is the volume of permeated water, A (m^2) is the effective membrane area and Δt (h) is the permeation time.

The experiments were conducted using compressed nitrogen gas and three types of solutions: phenol, copper and iron ions. The initial concentrations of the solutions were varied (15 and 25 mg/L for phenol, 800 and 1000 mg/L for copper ions, 12 and 20 mg/L for iron ions). The selection of concentrations is informed by previous research [19,21]. All measurements were conducted at 3 bar in triplicate. The final concentrations of the solutions were analysed using a double-beam UV–VIS spectrophotometer (Agilent, Santa Clara, CA, USA) at 506 nm for phenol [22], 620 nm for copper ions and 487 nm for iron ions. The rejection of this compound was calculated using Equation (4):

$$R = \left(1 - \frac{C_p}{C_f}\right) \times 100\% \tag{4}$$

where C_p and C_f (mg/mL) were phenol, copper and iron concentrations in the permeate and the feed solutions, respectively.

The membrane resistance was evaluated in accordance with Darcy's law [22] by calculating the resistance in the series of models as follows (5):

$$\mathrm{J} = \frac{\Delta P}{\mu R_t} \tag{5}$$

where J (L/(m^2×h)) is the permeate flux, ΔP is the transmembrane pressure (TMP), μ is the dynamic viscosity of permeate (Pa × s) and R_t is the total filtration resistance (m/s). The resistance in the series of models combines various resistances causing flux decline as follows (6):

$$\mathrm{R}_t = R_m + R_p + R_c \tag{6}$$

The total filtration resistance, designated as R_t, is the sum of various resistances, including that of the membrane itself, R_m, pore-blocking resistance, R_p, and cake resistance, R_c. The intrinsic membrane resistance (R_m) can be estimated from the initial pure water flux (3). The fouling resistance (R_p) is a consequence of pore plugging and the irreversible adsorption of foulants on the membrane pore wall or surface. The cake resistance (R_c) induced by the formation of a cake layer on the membrane surface was calculated from the water flux after pure water washing [23,24].

The behaviour of the detail membrane in the context of fouling was studied in accordance with the following methodology. Initially, the pure water flux of the membrane J_{w1} (L/(m^2×h)) was evaluated at 3 bar. Subsequently, an aqueous solution of phenol (15 and 25 mg/L), copper ions (800 and 1000 mg/L) and iron ions (12 and 20 mg/L) was introduced into the ultrafiltration system. Following a filtration period of 30 min, the membrane was flushed with pure water for a further 10 min, after which the pure water flux of the membrane, designated J_{w2} (L/(m^2×h)), was measured. Two parallel series of measurements were conducted, and the resulting data were presented as the arithmetic mean of the values obtained in each series. The flux recovery ratio (FRR) was calculated using Equation (7) in order to evaluate the antifouling properties of the membrane.

$$\mathrm{FRR} = \frac{J_{w2}}{J_{tw1}} \times 100\% \tag{7}$$

2.4. Carbonaceous Materials Characterisation

The Thermo Scientific FLASH 2000 Elemental Analyzer was utilized for analysing the elemental composition (C, N, H and S) of all carbon materials. The elemental oxygen content was determined by the difference method. Ash content was assessed by subjecting the samples to combustion in a microwave muffle furnace (Phoenix model, CEM Corporation, Matthews, IL, USA) at 815 °C for 60 min.

The prepared carbon materials underwent characterisation for their porous properties utilizing nitrogen adsorption/desorption isotherms measured at 77 K using an AutosorbiQ instrument, provided by Quantachrome Instruments (Boynton Beach, FL, USA). The detailed procedure for carrying out the measurements was previously presented in our article [17].

The SEM images of biocarbons were captured using the scanning electron microscope Quanta 3D FEG (FEI, Field Electron and Ion Co., Hillsboro, OR, USA). pH values and the content of oxygen functional groups of adsorbents were determined following the procedure outlined in our earlier article [25]. Additionally, iodine adsorption analysis was conducted for the obtained carbon materials [25].

3. Results

3.1. Chracterisastion of the Carbon Materials

Table 1 presents the results of elemental analysis for the carbon materials obtained. A review of the data presented in Table 1 reveals that sample C exhibits the lowest elemental carbon content, which also results in the highest oxygen content for the same material. In contrast, the percentages of hydrogen and nitrogen were 2.1 wt. % and 2.9 wt. %, respectively. Further analysis of the data presented in Table 1 revealed that the activation of the precursor with carbon dioxide and the impregnation of sample C with urea resulted in a notable alteration of the carbon materials obtained. The NAC sample was characterised by its highest concentration of elemental carbon (86.9 wt. %). The carbon obtained by direct activation of the starting material with carbon dioxide exhibited a carbon content that was over 10 wt. % lower than that of NAC activated carbon. In the case of AC and NAC samples, an increase in the C^{daf} content was accompanied by an increase in the proportion of hydrogen and oxygen. Modification with urea allowed obtaining an adsorbent containing in its structure N^{daf}—4.7 wt. %. Further analysis of the data revealed that the AC and NAC carbonaceous materials exhibited a low sulphur content, with a maximum of 0.1 wt. %. In contrast, the oxygen content of these samples was found to be between 4.8 wt. % and 5.3 wt. %. It is also pertinent to mention that each of the carbon materials obtained exhibits a markedly high ash content. The mineral substance in question has a weight percentage range of 7.8 wt. % to 12.1 wt. %. The high ash content of the obtained coal materials suggests that ash may be present within the pores of the carbon structure.

Table 1. Elemental analysis (wt. %) of the carbon materials.

Carbon Materials	C^{daf}	H^{daf}	N^{daf}	S^{daf}	$O^{daf,1}$	Ash
C	69.2	2.1	2.9	0.2	25.7	12.1
AC	74.6	2.5	3.0	0.1	19.8	14.3
NAC	86.9	2.7	4.7	0.1	5.6	7.8

^{daf}—dry-ash-free basis; ^{1}—determined by difference, method error ≤ 0.3%.

The results of the textural studies (Table 2) indicate that the thermochemical treatment of the residue following the extraction of the fungus *Inonotus obliquus* did not facilitate the development of a porous structure effectively. This is demonstrated by the specific surface areas of the obtained activated carbons, which are 125 (sample C), 749 (sample AC) and 888 m^2/g (sample NAC), respectively. The most developed surface area was characterised by NAC adsorbent, which is the only one with a specific surface area greater than 400 m^2/g.

Table 2. Textural parameters of the carbon materials.

Carbon Materials	Iodine Number (mg/g)	S_{BET} (m^2/g)	Total Pore Volume (cm^3/g)	Micropore Volume (cm^3/g)	Average Pore Diameter (nm)
C	205	125	0.38	0.04	7.32
AC	356	749	0.52	0.39	4.22
NAC	644	888	0.60	0.44	4.15

S_{BET}—surface area

The data presented in Table 2 indicate a correlation between the specific surface area and the iodine numbers obtained for the carbon adsorbents produced. The tested adsorbents exhibited a porous structure comprising small mesopores, as evidenced by the average pore diameter values, which ranged from 4.15 to 7.32 nm. Furthermore, the nitrogen desorption adsorption isotherms, as illustrated in Figure 1, demonstrated the presence of such pores. In accordance with the IUPAC classification, the isotherms depicted in Figure 1 are indicative of type IV. A defining characteristic of the type IV isotherm is the presence of a clearly delineated hysteresis loop, which is associated with capillary condensation in the region of mesopores. The H4-type hysteresis loops visible in the isotherms around the p/p_0 pressure of approximately 0.4 suggest the condensation of nitrogen in the mesopores. This indicates that the tested coals have developed mesoporosity [26].

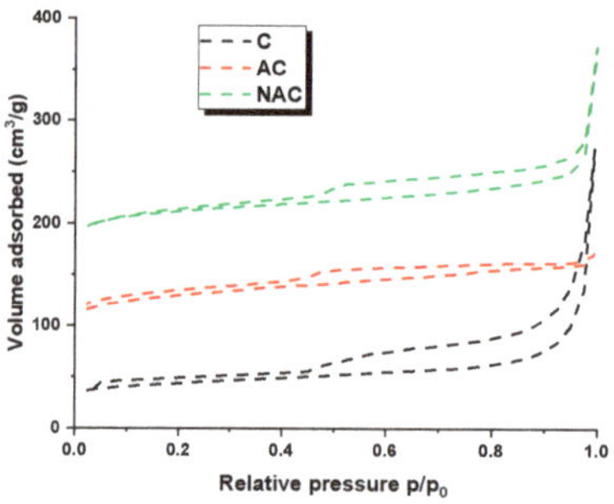

Figure 1. Low-temperature nitrogen adsorption/desorption isotherms for carbon materials obtained.

SEM images of the adsorbent samples are presented in Figure 2. Regarding char samples, the ash content may account for the brighter fragments observed.

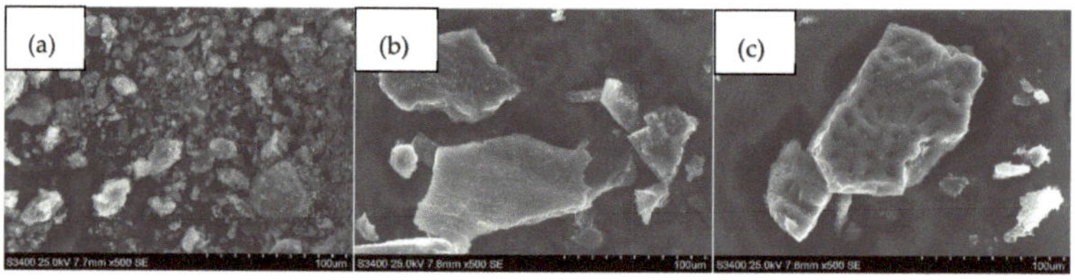

Figure 2. SEM images of the carbon adsorbents obtained: (a) sample C, (b) sample AC and (c) sample NAC.

The data presented in Figure 3 indicate that the obtained adsorbents exhibit comparable acid–base properties. As evidenced by the presented data, the type and quantity of surface oxygen groups are contingent upon the variant of carbon sample production. Regardless of the preparation variant, all coals exhibited acidic and basic groups on their surface. The NAC sample exhibited the highest proportion of both groups. The carbon in question exhibited 2.0 mmol/g of acidic functional groups and 3.5 mmol/g of basic groups on its surface. Furthermore, the data revealed a prevalence of basic groups over acidic ones for all samples. The predominance of basic functional groups is also corroborated by the pH values presented in Figure 3, which range from 7.8 to 8.5.

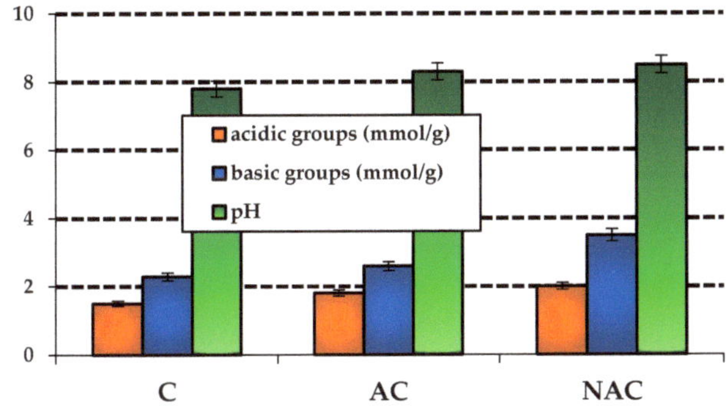

Figure 3. Acid–base properties of the carbon adsorbents.

3.2. Physicochemical Properties and Adsorption Characteristics of the Membranes

Table 3 presents the results for the structural parameters and wetting angle values of the materials under investigation. The data in Table 3 indicate that the char-enriched membrane exhibited the highest values for porosity (68.60%) and equilibrium water content (88.21%). The values for the MNAC membrane and the membrane without added char material were slightly lower. The lowest porosity and equilibrium water content values were determined for the MAC material. The values of the EWC parameter indicate a high susceptibility of the materials tested to drying out, which is a significant factor to be considered when determining the most appropriate storage environment. The analysis of the wetting angles revealed that the MAC material exhibited the most hydrophilic surface character, while the membrane without added carbon material exhibited the least hydrophilic character.

Table 3. Porosity (ε), equilibrium water content (EWC) and contact angle of investigated membranes (standard deviation n = 10).

Membrane	ε (%)	EWC (%)	Contact Angle °
M	50.31 ± 2.61	76.74 ± 4.61	62.80 ± 1.47
MC	68.60 ± 3.79	88.21 ± 5.28	60.08 ± 4.09
MAC	38.01 ± 1.14	80.13 ± 4.17	57.12 ± 2.97
MNAC	62.71 ± 3.19	85.62 ± 4.88	58.24 ± 3.57

Table 4 shows the content of acid and basic oxygen groups on the surface of the extracted materials.

Table 4. Acidic and basic properties of investigated adsorbents (mmol/g).

Membrane	Acidic Groups	Basic Groups	Total Content of Oxygen Groups
M	3.69 ± 0.23	1.66 ± 0.11	5.35 ± 0.34
MC	2.57 ± 0.17	5.05 ± 0.32	7.62 ± 0.37
MAC	5.19 ± 0.32	4.60 ± 0.28	9.79 ± 0.45
MNAC	2.80 ± 0.21	5.19 ± 0.32	7.99 ± 0.39

From the values obtained, it can be seen that the MAC material had the highest number of acid functional groups (5.19 mmol/g). On the surface of the pure membrane, the number of acid groups was 3.69 mmol/g. For the other two membranes, however, the values were much lower at 2.57 mmol/g for MC and 2.80 mmol/g for MNAC. In the case of the alkaline groups, a completely opposite relationship was observed. On the surface of MC and MNAC, the highest content of such groups was found: 5.05 for the activated carbon membrane and 5.19 mmol/g for the nitrogen-enriched activated carbon modified membrane, respectively. A slightly lower amount of 4.60 mmol/g was found for the MAC material. In contrast, the least alkaline functional groups were present on the surface of the membrane without the addition of carbon materials and amounted to 1.66 mmol/g. An analysis of the data presented in Table 4 indicated that the addition of carbon adsorbent to the membrane results in an increase in the number of basic groups compared to the membrane without carbon addition. With regard to acidic groups, no discernible trend was apparent. Consequently, further research should be conducted with the objective of resolving this issue.

Figure 4 illustrates the flow rates determined for the materials under investigation. The highest flow values were observed for the M membrane, with a value of 24.25 $L/m^2 \times h$ prior to filtration of copper ion solutions. A slightly lower value of 23.30 $L/m^2 \times h$ was determined for the MNAC membrane, while the flow values for the MC and MAC membranes were 18.66 and 16.32 $L/m^2 \times h$, respectively. The flow values for the materials used to remove iron ions were similar to each other. The value for the M membrane was 22.48 $L/m^2 \times h$, for the MC material 20.05 $L/m^2 \times h$ and for the MAC and MNAC materials 21.04 and 21.29 $L/m^2 \times h$, respectively. The highest flow rates for MC and MAC were 19.44 and 19.93 $L/m^2 \times h$, respectively, for the materials used in the phenol filtration processes. M and MNAC membranes exhibited slightly lower values for this parameter, with values of 16.73 $L/m^2 \times h$ for M and 18.62 $L/m^2 \times h$ for MNAC.

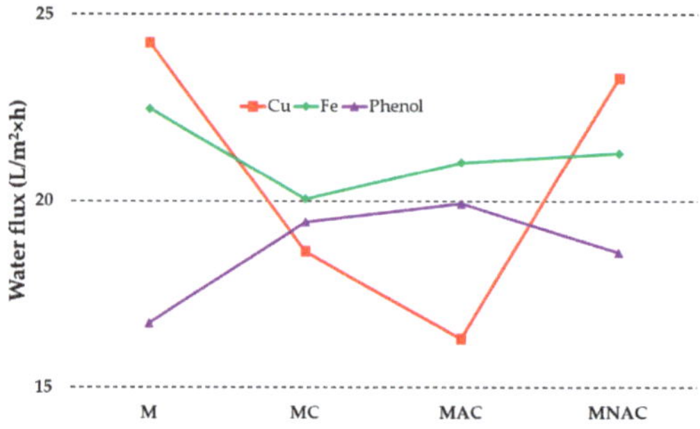

Figure 4. Water flux of the membranes before and after copper, iron and phenol filtrations.

Figure 5 illustrates the efficacy of the materials tested in removing copper ions from aqueous solutions. All of the materials exhibited enhanced efficiency in removing copper ions from solutions with an initial concentration of 800 mg/L. The results indicate that copper ions, regardless of the initial solution concentration, are most effectively removed by the urea-impregnated carbon membrane. It is possible that the observed trend is related to the fact that NAC coal was characterised by the best developed specific surface area and porous structure. The material was found to be highly effective in removing copper ions from aqueous solutions, with 67% of the ions removed from a solution with an initial concentration of 800 mg/L and 60% from a solution with a higher concentration. In contrast, the use of the MAC material resulted in the removal of only 24% of the ions from a solution with an initial concentration of 800 mg/L and 14% from a solution with an initial concentration of 1000 mg/L. In contrast, the use of the MC membrane allowed the removal of 24 and 10% of copper ions, respectively. The membrane without added carbon material was the least effective in removing copper ions from aqueous solutions, removing 21% of the ions from a solution with a lower initial concentration and only 1% from a solution with an initial concentration of 1000 mg/L.

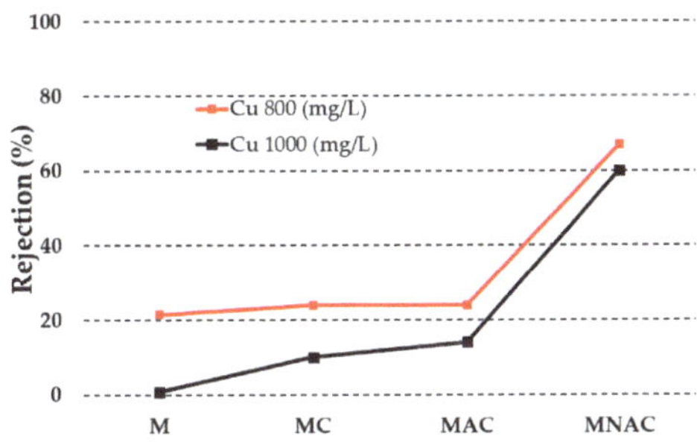

Figure 5. Removal of copper ions by membranes obtained.

The effectiveness of the adsorbents tested for the removal of copper ions can be compared with results already reported in the literature. It should be emphasised that the preparation of the membrane in the form of flat sheets can be considered as a simple process and therefore economical. In addition, the use of carbonaceous adsorbents from waste materials does not significantly increase operating costs. Comparing the results obtained in this article with those reported in [27], where the authors present their findings on the performance of membranes impregnated with polysulfone minerals and clay for Cu(II) ion remediation, it can be concluded that the materials obtained are comparable. Indeed, the authors in [27] used mixing and phase inversion methods to prepare clay-based membranes by mixing bentonite, sepiolite and zeolite in a matrix for the removal of copper ions from aqueous solutions. It was shown that for a final concentration of copper ions equal to 5 mg/L (TMP = 0.5 bar), the following rejections were obtained: 59.2% bentonite/polysulfone, 21.5% sepiolite/polysulfone and 97.0% zeolite/polysulfone. In [28], a membrane with a mixed biochar/polysulfone (PSF) matrix was prepared by incorporating biochar microparticles into the PSF matrix. The biochar was obtained by physical activation of plant waste. It was shown that the rejection was up to 93.7% (Co—12.8 mg/L, TMP = 0.25 bar), which is a much higher value compared to our results.

A quantitative analysis of the data presented in Figure 6 reveals that all the membranes tested exhibited enhanced efficiency in the removal of iron ions from aqueous solutions with an initial concentration of 20 mg/L. The values collected indicate that the membrane without added carbon material exhibited the highest efficiency in the removal of iron ions. The membrane was found to be highly effective in removing iron ions from aqueous solutions. It was observed that 64% of the iron ions were removed from a solution with an initial concentration of 20 mg/L, while 58% were removed from a solution with a concentration of 12 mg/L. In comparison, the results obtained for the other materials were considerably lower. The MC material demonstrated iron ion removal efficiencies of 26% for the lower concentration solution and 39% for the 20 mg/L solution. The use of MAC material resulted in the removal of 30% of the ions from a solution with an initial concentration of 12 mg/L, while 35% of the ions of this metal were removed with MNAC material. In contrast, the use of MAC material removed 34% of the ions from a solution with an initial concentration of 20 mg/L, while 43% of the ions of this metal were removed with MNAC material.

The scientific literature provides evidence that the removal of ions, including iron, by membranes is of interest to many research groups. For example, ref. [29] described the synthesis of graphene oxide and its incorporation into membranes made from recycled PET bottles. The study showed that the rejection rate was 60% (at an initial concentration of C_0 = 0.3 mg/L and TMP pressure = 0.15 bar). It is notable that the incorporation of graphene oxide into PET bottle waste materials as a raw material for the production of membranes for wastewater treatment, as demonstrated in our work, allows the waste to be utilized for the fabrication of effective adsorbents for liquid phase contaminants. In turn, a ceramic membrane containing activated carbon derived from empty oil palm fruit bunches showed an iron ion removal efficiency of over 92% (C_0 = 1.38 mg/L, TMP = 2.0 bar) [30]. Polymer membranes based on polyethersulfone, into which PVP (polyvinylpyrrolidone, a pore-generating agent) was also introduced, exhibited a rejection efficiency of over 90% (C_0 = 5–20 mg/L, TMP = 3.0 bar) [31].

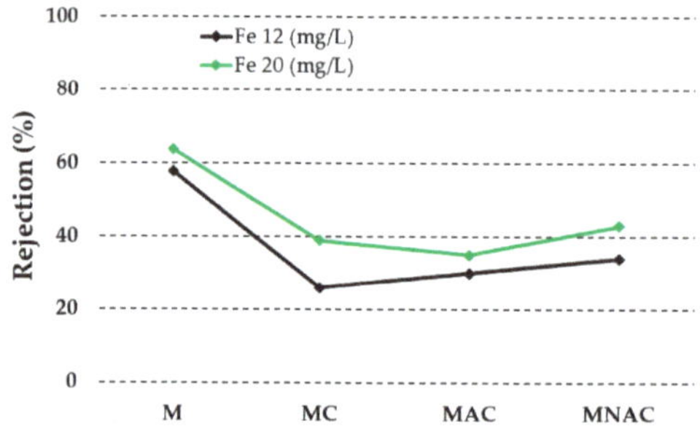

Figure 6. Removal of iron ions by membranes obtained.

As with the removal of copper ions, the membranes tested demonstrated enhanced efficiency in the removal of phenol molecules from solutions with lower initial concentrations (Figure 7). Regardless of the initial concentration of the solution, the membrane without added carbon materials exhibited the highest efficacy (89%). The other materials exhibited significantly lower efficiency in the removal of phenol from aqueous solutions. The use of MC allowed for the removal of 44% of phenol from a solution with an initial concentration of 15 mg/L and 27% from a solution with an initial concentration of 25 mg/L. In contrast, the use of MAC resulted in the removal of 61% and 29% of phenol molecules, respectively. In contrast, MNAC permitted the removal of 51 and 41% of phenol molecules from solutions with initial concentrations of 15 and 25 mg/L, respectively.

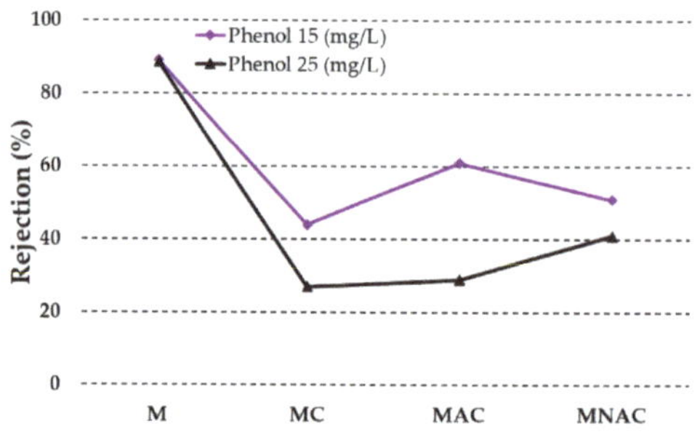

Figure 7. Removal of phenol by membranes obtained.

Given the adverse effects of phenol on human health, numerous research groups are engaged in investigating methods for its removal from wastewater. In [32], researchers employed a composite membrane prepared by coating a mixture of choline chloride and cellulose acetate based on fly ash ceramic substrate for phenol removal. Their findings demonstrated that phenol rejection increased from 56 to 93% with a concomitant increase in the quantity of choline chloride in cellulose acetate. Our previous research [33] indicates that the use of polymeric membranes based on polyethersulfone (PES) modified by the addition of varying quantities of a pore-forming agent (PVP) allows for a significantly higher removal of phenol from the liquid phase (rejection near 100%) compared to the results presented in this paper. Consequently, it can be postulated that the introduction of carbonaceous adsorbents does not have a positive effect on the results obtained when removing this type of contamination.

Figure 8 illustrates the mean values of the renewal rate of the individual membranes following filtration of the solutions tested. The results demonstrate that the highest values for this parameter were observed for the MNAC material, with values of 91% following the filtration of copper ion solutions, 96% following filtration of iron ions and 89% following filtration of phenol. The remaining materials exhibited slightly lower values. The FRR values for M were 82, 85 and 91% after filtration of copper, iron and phenol solutions, respectively. In contrast, the FRR values for MC were 74% after copper filtration, 86% after iron filtration and 92% after phenol filtration. Finally, the FRR values for MAC were 79, 86 and 91% after filtration of copper, iron and phenol solutions, respectively. From the graph obtained, it can be seen that in most cases (for M, MC and MAC materials), the FRR values increase depending on the type of filtered solution. The lowest values were obtained after filtration of copper solutions, followed by increasing values after filtration of ferrous ion solutions, while the highest values were observed after phenol removal processes from aqueous solutions.

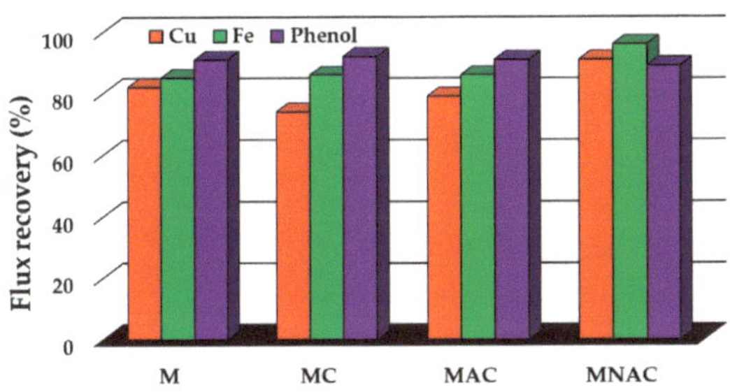

Figure 8. Flux recovery percentage of investigated membranes after filtrations.

Table 5 presents the mean values of the individual resistances determined for the materials obtained following the filtration of copper ion solutions. The data indicate that the highest resistances (both for the membranes, pores and the resulting filter cake) are those of the MAC material, with a total filtration resistance of 25.26×10^{13} L/m^2×h. The MC membrane exhibited slightly lower values, with an R_t of 22.66×10^{13} L/m^2×h. In contrast, the M and MNAC materials exhibited lower resistance values. The total filtration resistances for these membranes are 16.27×10^{13} L/m^2×h for M and 14.32×10^{13} L/m^2×h for MNAC, respectively.

Table 5. Filtration resistance of membranes after filtrations of copper ion solutions (L/m^2×h) (standard deviation n = 10).

Membrane	R_m ($\times 10^{13}$)	R_p ($\times 10^{13}$)	R_c ($\times 10^{13}$)	R_t ($\times 10^{13}$)
M	4.65 ± 0.19	5.82 ± 0.34	5.80 ± 0.31	16.27 ± 1.05
MC	5.79 ± 0.23	7.80 ± 0.57	9.07 ± 0.88	22.66 ± 1.45
MAC	6.64 ± 0.24	8.42 ± 0.68	10.20 ± 0.93	25.26 ± 2.99
MNAC	4.65 ± 0.20	5.15 ± 0.43	4.52 ± 0.42	14.32 ± 0.99

Table 6 presents the average resistance values determined following the iron ion filtration processes. The highest values for individual resistances and total filtration resistance were observed for the MC membrane, with R_t equal to 18.30×10^{13} L/m^2×h. Slightly lower values were found for MAC, with the total filtration resistance equalling 17.50×10^{13} L/m^2×h. The resistance values for M and MNAC are slightly lower than those determined for MC and MAC membranes. The total filtration resistance for M is 15.83×10^{13} L/m^2×h and 15.51×10^{13} L/m^2×h for MNAC.

Table 6. Filtration resistance of membranes after filtrations of iron ion solutions ($L/m^2 \times h$) (standard deviation n = 10).

Membrane	R_m ($\times 10^{13}$)	R_p ($\times 10^{13}$)	R_c ($\times 10^{13}$)	R_t ($\times 10^{13}$)
M	4.87 ± 0.18	5.71 ± 0.53	5.25 ± 0.52	15.83 ± 1.59
MC	5.49 ± 0.24	6.36 ± 0.77	6.45 ± 0.65	18.30 ± 1.78
MAC	5.14 ± 0.21	5.98 ± 0.49	6.38 ± 0.63	17.50 ± 1.68
MNAC	5.09 ± 0.24	5.34 ± 0.47	5.08 ± 0.49	15.51 ± 1.56

The average resistance values determined following filtration of the phenol solutions are presented in Table 7. The values obtained indicate that all materials exhibit comparable resistance values. The Rt value for M is 17.19×10^{13} $L/m^2 \times h$, while for MC it is equal to 17.22×10^{13} $L/m^2 \times h$. In contrast, the total filtration resistance value for MAC is 17.69×10^{13} $L/m^2 \times h$, while that for MNAC is 18.61×10^{13} $L/m^2 \times h$.

Table 7. Filtration resistance of membranes after filtrations of phenol solutions ($L/m^2 \times h$) (standard deviation n = 10).

Membrane	R_m ($\times 10^{13}$)	R_p ($\times 10^{13}$)	R_c ($\times 10^{13}$)	R_t ($\times 10^{13}$)
M	6.59 ± 0.88	7.19 ± 0.94	6.41 ± 0.82	17.19 ± 1.48
MC	5.58 ± 0.79	6.07 ± 0.84	5.57 ± 0.57	17.22 ± 1.47
MAC	5.62 ± 0.81	6.12 ± 0.83	5.95 ± 0.59	17.69 ± 1.54
MNAC	5.80 ± 0.84	6.54 ± 0.87	6.27 ± 0.63	18.61 ± 1.59

4. Conclusions

The carbonaceous adsorbents obtained from the extraction residues of the fungus *Inonotus obliquus* exhibited a specific surface area of 205 to 644 m^2/g and contained both acidic and basic functional groups on their surface, with their aqueous solutions displaying an alkaline character. This study demonstrated that urea impregnation produced an adsorbent with the highest carbon content.

This study demonstrated that the incorporation of char, derived from the carbonisation of the extraction residue of the fungus *Inonotus obliguus*, into the membrane resulted in the highest porosity values and equilibrium water content. Conversely, the membrane enriched with AC material exhibited the highest surface hydrophilicity. Furthermore, the addition of activated carbon, obtained through physical activation of char, led to an increase in the content of acidic oxygen groups on the surface of the membrane under study. In contrast, membranes based on char and nitrogen-enriched activated carbon have a significantly higher number of basic groups on their surface than acidic groups. A membrane with carbon material containing urea in its structure shows the highest efficiency in removing copper ions. In contrast, a membrane without added carbon materials shows the highest efficiency in removing phenol and iron ions from aqueous solutions. Urea-impregnated activated carbon has a positive effect on the renewal rate of membranes. The research findings indicate that the addition of carbonaceous material has an inconclusive effect on the removal efficiency of the pollutants tested. The carbon materials obtained were characterised by a poorly developed specific surface area and porous structure, which suggests that further research should be directed towards the use of carbon adsorbents obtained by chemical activation of waste materials. This method allows a much more effective development of the surface area of carbonaceous material. The next step is to confirm the usefulness of the research carried out. This can be achieved by conducting stability and reusability studies of the obtained materials on model systems and real wastewater. Additionally, a cost analysis of the membranes produced should be carried out to enable a comparison of their production costs with those of commercial products currently available on the market.

Author Contributions: Conceptualization, J.P. and A.B.-W.; methodology, J.P., A.B.-W. and F.P.; validation, J.P., A.B.-W., F.P., J.C.-P. and A.N.-W.; formal analysis, J.P., A.B.-W. and F.P; investigation, J.P. and A.B.-W.; resources, R.P.; data curation, J.P. and A.B.-W.; writing—original draft preparation, J.P. and A.B.-W.; writing—review and editing, A.B.-W., F.P., J.C.-P., A.N.-W. and R.P.; visualization, J.P. and A.B.-W.; supervision, R.P.; funding acquisition, R.P. All authors have read and agreed to the published version of the manuscript.

Funding: This research received no external funding.

Institutional Review Board Statement: Not applicable.

Informed Consent Statement: Not applicable.

Data Availability Statement: The original contributions presented in the study are included in the article, further inquiries can be directed to the corresponding author.

Conflicts of Interest: The authors declare no conflicts of interest.

References

1. Mulungulungu, G.A.; Mao, T.; Han, K. Efficient removal of high-concentration copper ions from wastewater via 2D g-C_3N_4 photocatalytic membrane filtration. *Colloids Surf. A Physicochem. Eng. Asp.* **2021**, *623*, 126714. [CrossRef]
2. Virolainen, S.; Wesselborg, T.; Kaukinen, A.; Sainio, T. Removal of iron, aluminium, manganese and copper from leach solutions of lithium-ion battery waste using ion exchange. *Hydrometallurgy* **2021**, *202*, 105602. [CrossRef]
3. Khoshtinat, F.; Tabatabaie, T.; Ramavandi, B.; Hashemi, S. Phenol removal kinetics from synthetic wastewater by activation of persulfate using a catalyst generated from shipping ports sludge. *Chemosphere* **2021**, *283*, 131265. [CrossRef] [PubMed]
4. Ade, I.A.; Tran, H.N.; Zhang, J.-W.; Wang, Y.-C.; Dat, N.D.; Nguyen, D.T.; Chao, H.-P. Adsorption characteristics of lead, copper, cadmium, methylene blue, phenol, and toluene in water using composite synthesized from titanium dioxides and carbon spheres through hydrothermal method. *J. Water Process Eng.* **2022**, *50*, 103221. [CrossRef]
5. Hamad, H.T. Removal of phenol and inorganic metals from wastewater using activated ceramic. *J. King Saud Univ. Eng. Sci.* **2021**, *33*, 221–226. [CrossRef]
6. Mojiri, A.; Bashir, M.J.K. Wastewater Treatment: Current and Future Techniques. *Water* **2022**, *14*, 448. [CrossRef]
7. Plisko, T.; Karslyan, Y.; Bildyukevich, A. Effect of Polyphenylsulfone and Polysulfone Incompatibility on the Structure and Performance of Blend Membranes for Ultrafiltration. *Materials* **2021**, *14*, 5740. [CrossRef] [PubMed]
8. Cheng, Y.; Ding, H.; Liu, Y.; He, D.; Peng, L.E.; Matsuyama, H.; Hu, M.; Li, H. Fabrication of polyethersulfone/sulfonated polysulfone loose nanofiltration membranes for enhanced selectivity of pharmaceuticals and personal care products and minerals. *Sep. Purif. Technol.* **2024**, *337*, 126466. [CrossRef]
9. Gharbani, P.; Mehrizad, A. Preparation and characterization of graphitic carbon nitrides/polyvinylidene fluoride adsorptive membrane modified with chitosan for Rhodamine B dye removal from water: Adsorption isotherms, kinetics and thermodynamics. *Carbohydr. Polym.* **2022**, *277*, 118860. [CrossRef]
10. Mompó-Curell, R.; Biti, S.; Iborra-Clar, A.; Iborra-Clar, M.I.; Garcia-Castello, E.M.; Fernández-Martín, C. Activated-Carbon-Doped Non-Solvent-Induced Phase-Inversion Membranes: A Comprehensive Study on Synthesis, Characterisation, and Performance Evaluation. *Sustainability* **2024**, *16*, 1150. [CrossRef]
11. Yang, H.; Ye, Q.; Zhou, Y.; Xiang, Y.; Xing, Q.; Dong, X.; Wang, D.; Xu, W. Formation, morphology and control of high-performance biomedical polyurethane porous membranes by water micro-droplet induced phase inversion. *Polymer* **2004**, *55*, 5500–5508. [CrossRef]
12. Zaherzadeh, A.; Karimi-Sabet, J.; Mohammad, S.; Mousavian, A.; Ghorbanian, S. Optimization of flat sheet hydrophobic membranes synthesis via supercritical CO_2 induced phase inversion for direct contact membrane distillation by using response surface methodology (RSM). *J. Supercrit. Fluids* **2015**, *103*, 105–114. [CrossRef]
13. Lazarenko, N.S.; Golovakhin, V.V.; Shestakov, A.A.; Lapekin, N.I.; Bannov, A.G. Recent Advances on Membranes for Water Purification Based on Carbon Nanomaterials. *Membranes* **2022**, *12*, 915. [CrossRef]
14. Ullah, N.; Ali, Z.; Khan, A.S.; Adalat, B.; Nasrullah, A.; Khan, S.B. Preparation and dye adsorption properties of activated carbon/clay/sodium alginate composite hydrogel membranes. *RSC Adv.* **2024**, *14*, 211–221. [CrossRef]
15. Raheel, F.; Rafay, A.; Bibi, B.; Ahmad, S.; Ali, Z.; Saleem, M.; Butt, M.S.; Rehman, A.U.; Irfan, M. Synthesis and Characterization of Activated Carbon and Its Application for Wastewater Treatment. *Mater. Proc.* **2024**, *17*, 4.
16. Obayomi, K.S.; Lau, S.Y.; Danquah, M.K.; Zhang, J.; Chiong, T.; Meunier, L.; Rahman, M.M. Selective adsorption of organic dyes from aqueous environment using fermented maize extract-enhanced graphene oxide-durian shell derived activated carbon composite. *Chemosphere* **2023**, *339*, 139742. [CrossRef]
17. Bazan-Wozniak, A.; Machelak, K.; Nosal-Wiercińska, A.; Pietrzak, R. Microwave Heating for Synthesis of Carbonaceous Adsorbents for Removal of Toxic Organic and Inorganic Contaminants. *Molecules* **2023**, *28*, 6825. [CrossRef] [PubMed]
18. Jasiewicz, K.; Pietrzak, R. The influence of pore generating agent on the efficiency of copper and iron ions removal from liquid phase by polyethersulfone membranes. *Chem. Eng. J.* **2013**, *228*, 449–454. [CrossRef]

19. Krason, J.; Pietrzak, R. Removal of Iron and Copper Ions from the Liquid Phase by Modified Polymeric Membranes. *J. Polym. Environ.* **2018**, *26*, 3237–3242. [CrossRef]
20. Hofman, M.; Pietrzak, R. Adsorbents obtained from waste tires for NO_2 removal under dry conditions at room temperature. *Chem. Eng. J.* **2011**, *170*, 202–208. [CrossRef]
21. Krason, J.; Pietrzak, R. Membranes obtained on the basis of cellulose acetate and their use in removal of metal ions from liquid phase. *Pol. J. Chem. Technol.* **2016**, *18*, 104–110. [CrossRef]
22. Rice, E.W.; Baird, R.B.; Eaton, A.D.; Clesceri, L.S. *APHA 5530 Standard Methods for the Examination of Water and Wastewater*; American Public Health Association, American Water Works Association, Water Environment Federation: Washington, DC, USA, 2012.
23. Di Bella, G.; Trapani, D. A Brief Review on the Resistance-in-Series Model in Membrane Bioreactors (MBRs). *Membranes* **2019**, *9*, 24. [CrossRef] [PubMed]
24. Basile, A.; Gallucci, F. *Membranes for Membrane Reactors: Preparation, Optimization and Selection*; John Wiley & Sons, Ltd.: Chichester, UK, 2011.
25. Bazan, A.; Nowicki, P.; Pietrzak, R. The influence of activation procedure on the physicochemical and sorption properties of activated carbons prepared from pistachio nutshells for removal of NO_2/H_2S gases and dyes. *J. Clean. Prod.* **2017**, *152*, 211–222. [CrossRef]
26. Thommes, M.; Kaneko, K.; Neimark, A.V.; Olivier, J.P.; Rodriguez-Reinoso, F.; Rouquerol, J.; Sing, K.S.W. Physisorption of gases, with special reference to the evaluation of surface area and pore size distribution (IUPAC Technical Report). *Pure Appl. Chem.* **2015**, *87*, 1051–1069. [CrossRef]
27. Hamid, S.A.; Shahadat, M.; Ballinger, B.; Azha, S.F.; Ismail, S.; Ali, S.W.; Ahammad, S.Z. Role of clay-based membrane for removal of copper from aqueous solution. *J. Saudi Chem. Soc.* **2020**, *24*, 785–798. [CrossRef]
28. He, J.; Song, Y.; Chen, J.P. Development of a novel biochar/PSF mixed matrix membrane and study of key parameters in treatment of copper and lead contaminated water. *Chemosphere* **2017**, *186*, 1033–1045. [CrossRef] [PubMed]
29. Agboola, O.; Khalih, A.A.; Oladokun, O.; Ayeni, A.O.; Uzokwe, F.C.; Odunlami, O.; Elehinafe, F.; Yahaya, A.; Fayomi, O.S.I. Statistical analyses of pore radii on the performance of PET-nanocomposite membranes in the removal of iron and anions from Ibeshe River. *S. Afr. J. Chem. Eng.* **2023**, *44*, 89–102. [CrossRef]
30. Sisnayati Said, M.; Nasir, S.; Priadi, D.P.; Faizal, M.; Aprianti, T. Effect of activated carbon made from oil palm empty-fruit bunch and iron oxide powder on the performance of ceramic membrane. *Open Ceram.* **2023**, *13*, 100335. [CrossRef]
31. Jasiewicz, K.; Pietrzak, R. Metals Ions Removal by Polymer Membranes of Different Porosity. *Sci. World J.* **2013**, *2013*, 957202. [CrossRef] [PubMed]
32. Gupta, V.; Raja, C.; Anandkumar, J. Phenol Removal by Novel Choline Chloride Blended Cellulose Acetate-Fly Ash Composite Membrane. *Period. Polytech. Chem. Eng.* **2020**, *64*, 116–123. [CrossRef]
33. Hofman-Bieniek, M.; Jasiewicz, K.; Pietrzak, P. Synthesis of polymer membranes of different porosity and their application for phenol removal from liquid phase. *Korean J. Chem. Eng.* **2014**, *31*, 304–309. [CrossRef]

materials MDPI

Article

Mathematical Model for Oil Recovery Prediction of Polymer Microsphere Conformance Control Based on the Stream Tube Method

Wenyue Zhao [1,2,*], Huo Tang [1], Fan Lu [3], Shuai Hu [4], Tongjing Liu [2], Nannan Li [1] and Renzhi Song [4]

1 College of Mathematics and Computer Science, Chifeng University, 1 Yingbin Road, Hongshan District, Chifeng 024000, China
2 The Unconventional Oil and Gas Institute, China University of Petroleum (Beijing), 18 Fuxue Road, Changping District, Beijing 102249, China
3 Fengdikeng Working Area, No. 5 Oil Production Plant of Petrochina Changqing Oilfield Company, 1189 Jinghuan South Road, Gaoling District, Xi'an 710064, China
4 Irap Branch, Greatwall Drilling Company (GWDC), CNPC, 101 Anli Road, Chaoyang District, Beijing 102249, China
* Correspondence: wenyuezhao@126.com

Abstract: Oil recovery is an essential parameter for reservoir development performance evaluation, but there is no specific research on the theoretical oil recovery prediction model of polymer microspheres (PMs)' conformance control. This research aims to establish an oil recovery prediction model that depends on the definition of oil recovery based on stream tube theory. PMs' enhanced oil recovery mechanism is to plug the pore throat to expand the swept area. The assumption that the stream tube between injection and production wells is trapezoidal is proposed. Based on this premise, the area sweep efficiency equation suitable for the rhombus inverse nine-spot well pattern is established based on the stream tube theory. The vertical sweep efficiency equation is optimized by introducing the equivalent mobility ratio. Additionally, the model's adaptability and validity are studied. The analysis results show that oil recovery decreases with increasing injection rate, injection concentration, and PMs size but increases with the increasing injection period. The theoretical oil recovery is 1.37%, and the actual oil recovery of the field application is 1.22%, with an error of 0.15%. This model has good consistency with the actual physical process of the field application. The oil recovery prediction model can provide oil recovery and optimize PMs' conformance control injection scheme. This study fills the gap in the mathematical model for oil recovery prediction of PMs' conformance control.

Keywords: polymer microspheres' conformance control; oil recovery prediction model; inverted nine-spot rhombus pattern; areal sweep efficiency; stream tube method

1. Introduction

Low-permeability reservoirs are an essential type of reservoir globally, and are mainly distributed throughout Mexico, China, Algeria, Egypt, the United States, and other countries [1,2]. Currently, complex and challenging low-permeability oil fields account for an increasing proportion of those in China, and a considerable number of oil fields are working at a low production rate and production efficiency [3–5]. Therefore, developing low-permeability oil reservoirs is significant for China's energy supply [6,7].

In low-permeability reservoirs, severe heterogeneity and an inconsistent water flooding displacement profile will significantly affect water flooding performance [8]. In the process of water flooding, a quick water breakthrough will occur because the injected water rushes along existing micro-cracks, leading to a shorter stable production period [9]. Unlike the chain-like polymer repeatedly connected by specific structural units, polymer microspheres (PMs) are spherical polymer composite materials with diameters ranging from nanometers to micrometers, a high specific surface area, high reactivity, and other unique physical, chemical, and biological properties [10]. PMs show good conformance control ability as salt tolerance materials [11,12]. In the reservoir migration process, PMs can be plugging, deformation, migration, and second plugging, and have an excellent ability to improve formation heterogeneity and stop or slow down one-way water rush [13]. PMs can not only reduce the water cut rising of water breakthrough well; they also have double functions of conformance control and oil displacement. Increasing water injection volume can achieve a long-term stable oil production period and the ultimate goal of enhanced recovery [14,15]. PMs conformance control technology has been applied to Bohai, Shengli, Changqing, and Jidong oilfields and has achieved good performance of oil incrementation [16–18].

The oil recovery factor is one of the essential parameters to describe the effect of oilfield development performance. The primary means of accessing oil recovery include numerical simulation, laboratory experiments, empirical formulas, and theoretical equations [19–23]. However, for PMs conformance control, oil recovery efficiency cannot be predicted by numerical simulation because numerical simulation technology cannot characterize the enhanced oil recovery mechanism of plugging, deformation, migration, and second plugging. For the laboratory experiment method, due to the apparent differences between the scale of the core and the field, and the relatively simple distribution of core pore-throat, the oil recovery obtained from the laboratory is significantly different from the actual field application effect. Therefore, the mathematical model of oil recovery can only be established based on the definition of oil recovery.

The theoretical equation based on the oil recovery factor's definition is fundamental to oil production forecasting and the development of performance evaluation. The oil recovery factor is determined by oil displacement efficiency, areal sweep efficiency, and vertical sweep efficiency [24]. Oil displacement efficiency at different water cuts can be resolved with an oil–water relative permeability curve and extreme oil displacement efficiency [25–27].

Areal sweep efficiency is the critical parameter to evaluate the displacement performance of area well pattern [28]. Many scholars in China and other countries have conducted corresponding research on areal sweep efficiency. Before 2000, research methods for the area sweep efficiency included laboratory experiments, field data analysis, numerical simulation, probability theory, and dimensional analysis [29–37]. Since 2010, the stream tube method has been used to study the relationship between area sweep efficiency and time [38–45]. The most significant advantage of streamlined simulation is its fast computing speed, high precision, and good visibility [46].

Vertical sweep efficiency is an essential parameter for evaluating the water flooding effect in the thick reservoir. The standard approach to measure vertical sweep efficiency includes a multilayer core flooding experiment and a numerical model [47–51]. DeSouza and Qitai Yu proposed the equation of vertical sweep efficiency calculation considering the mobility ratio, permeability variation coefficient, and the water–oil ratio [24,32]. However, for PMs flooding, the primary enhanced oil recovery mechanism is significantly different from polymer flooding, as PMs cannot increase the mobility ratio but do block the pore throat. Therefore, the vertical calculation method should be optimized to present the influence of throat blocking.

There needs to be more specific research on the theoretical oil recovery prediction model of PMs' conformance control. This research aims to establish an oil recovery prediction model depending on the definition of oil recovery and stream tube theory. PMs' enhanced oil recovery mechanism is to plug the pore throat to expand the swept area. The assumption that the stream tube between the injection and production wells is trapezoidal is proposed. Based on this premise, an area sweep efficiency equation suitable for the rhombus inverse nine-spot well pattern is established based on the stream tube theory. The vertical sweep efficiency equation is optimized by introducing the equivalent mobility ratio. Additionally, the model's adaptability and validity are studied. This model not only provides a faster and more accessible tool to predict and evaluate the production performance of the PMs flooding process, but also could be used to optimize the injection scheme of PMs' conformance control. The study of this paper fills the gap in the mathematical model for the oil recovery prediction of polymer microspheres' conformance control.

2. Theoretical Model Establishment

2.1. Stream Tube Model

The stream tube method assumes that a non-intersecting flow pipe can express the reservoir's physical displacement mechanism in water flooding between injection and production wells [52]. Multiple flow pipes are regularly distributed between injection and production wells, and each flow pipe is independent and parallel without interference [53–55]. Fluid flows in a tube without material exchange. In the conformance control process of PMs, it is assumed that there is a pipe between the injection well and the production well. The PMs block the large throats that allow the injected water to flow into the small channel within the plugging area [56,57]. After multiple bypasses of injected water, the overall flow trend tends to migrate to the production well in a straight line. Therefore, in the microsphere plugging area, the pipe is considered a straight line, and the flow path of injected water is shown in Figure 1. After flowing through the plugging area, the injected water cannot continue to cruise around and still flows to the production wells along the flow trend formed in the water flooding stage. The shape factor is used to consider the shape influence of each flow pipe. The flow pipe between injection and production wells is approximately trapezoidal, and the flow pipe between injection and production wells is shown in Figure 2.

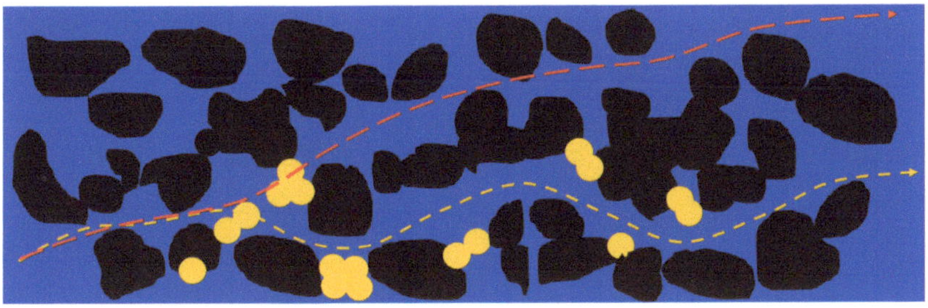

Figure 1. Enhanced oil recovery mechanism schematic diagram of polymer microsphere.

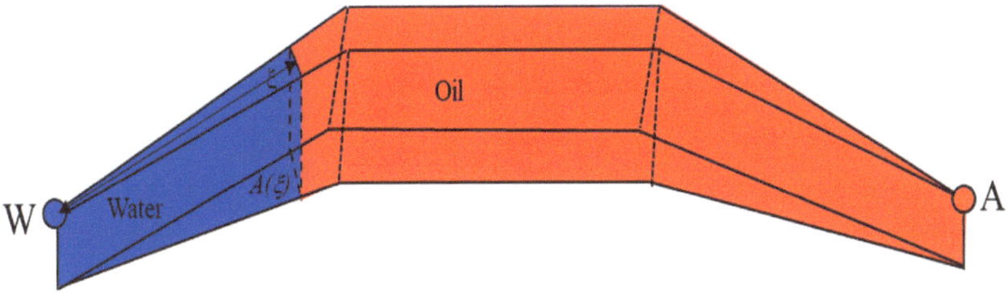

Figure 2. Stream tube model.

2.2. Oil Displacement Efficiency

The ultimate oil recovery factor is the multiplication of sweep efficiency and displacement efficiency, and the mathematical expression is shown in Equation (1) [25]. The sweep coefficient and oil displacement efficiency are two main parameters for predicting oil recovery in water-flooding oilfields.

$$E_R = E_D \cdot E_V \tag{1}$$

where E_R is the oil recovery, E_D is the oil displacement efficiency, and E_V is the volume sweep efficiency.

Volume sweep efficiency refers to the percentage of injected water volume in the swept area compared to the total reservoir volume. Usually, it is expressed by the product of vertical sweep efficiency E_Z and areal sweep efficiency E_A [24]. The expression of the sweep coefficient is:

$$E_V = E_A \cdot E_Z \tag{2}$$

where E_A is areal sweep efficiency, and E_Z is vertical sweep efficiency.

$$E_R = E_D \cdot E_A \cdot E_Z \tag{3}$$

The oil displacement efficiency E_D is the ratio of oil produced to the petroleum geological reserves in the affected area. PMs improve oil recovery by plugging channels to expand the water sweep efficiency. The primary displacement fluid is water, so the oil displacement efficiency is consistent with the water displacement efficiency. According to the reservoir permeability curve combined with the water content formula, the water saturation of the reservoir at different water contents can be calculated, and the oil displacement efficiency can be obtained.

At present, the oil–water relative permeability curve is often expressed in exponential form, under the condition of oil and water two-phase seepage; K_{ro}/K_{rw} and S_w are semi-logarithmic linear relationships [26].

$$\frac{K_{ro}}{K_{rw}} = ae^{-bS_w} \tag{4}$$

According to the partial flow equation, without considering the influence of gravity and capillary force, the moisture content can be expressed as

$$f_w = \frac{1}{1 + \frac{\mu_w}{\mu_o}\frac{K_{ro}}{K_{rw}}} \tag{5}$$

So, the relationship between water content and water saturation is as follows:

$$S_w = \frac{\ln a - \ln\left(\frac{1-f_w}{f_w}\right) - \ln\frac{\mu_o}{\mu_w}}{b} \tag{6}$$

In polymer microsphere flooding, we assume that the injected water enters the original unreached area, and the oil saturation in the unreached area is the initial oil saturation. The oil displacement efficiency formula of the reservoir with different water saturations is expressed as:

$$E_D = \frac{1 - S_w - S_{or}}{1 - S_{wi}} \tag{7}$$

where S_{wi} is the initial water saturation, S_w is the current water saturation, and S_{or} is the residual oil saturation.

Placing Equation (6) into Equation (7), the oil displacement efficiency at different water cuts can be calculated by the following equation:

$$E_D = \frac{1 - \frac{\ln a - \ln\left(\frac{1-f_w}{f_w}\right) - \ln\frac{\mu_o}{\mu_w}}{b} - S_{or}}{1 - S_{wi}} \tag{8}$$

As we know, both vertical sweep efficiency and areal sweep efficiency are essential. They can affect oil recovery significantly when the injected fluid cannot improve the driving efficiency [47].

Among the two factors, areal sweep efficiency is more important than vertical sweep efficiency for the thin thickness of a low permeability reservoir. Therefore, the improvement potential of vertical sweep efficiency is limited, but the possibility of areal sweep efficiency is the opposite. Thus, this paper mainly focuses on the study of areal sweep efficiency.

2.3. Areal Sweep Efficiency

The model assumes that the reservoir is homogeneous. There is a flow pipe between the oil well and the water well; the flow pipe is piston water flooding oil. The compressibility of formation rocks and fluids is not considered. An inverted nine-spot pattern mainly develops low-permeability reservoirs. One well group in the inverted nine-spot pattern can be divided into four injection-production units and 16 seepage units, as shown in Figure 3. Since the rhombic inverse nine-point is a regular symmetric well pattern, the seepage unit's area sweep coefficient can represent the well pattern's sweep coefficient. To obtain the area sweep coefficient of a seepage unit, the seepage unit AWC is divided into four research areas, namely AWG GWE EWH and HWC. These four areas could obtain areal sweep efficiency with the same calculation method. Therefore, we take AWG as an example to deduce the areal sweep efficiency calculation model. The diamond inverted nine-spot well pattern injection–production unit is shown in Figure 4. According to the above assumptions, multiple flow pipes are established between injection wells and production wells. By analyzing the nine-point well pattern's fluid seepage characteristics and symmetry in Figure 3, the seepage unit AWG in the nine-point well pattern is taken as the calculation unit.

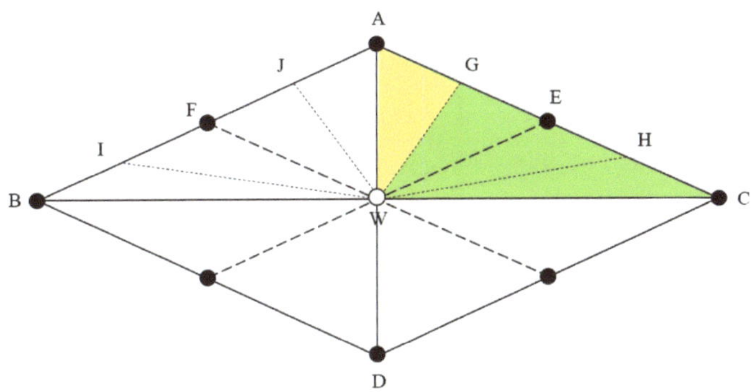

Figure 3. Diamond-shaped reverse nine-spot well pattern.

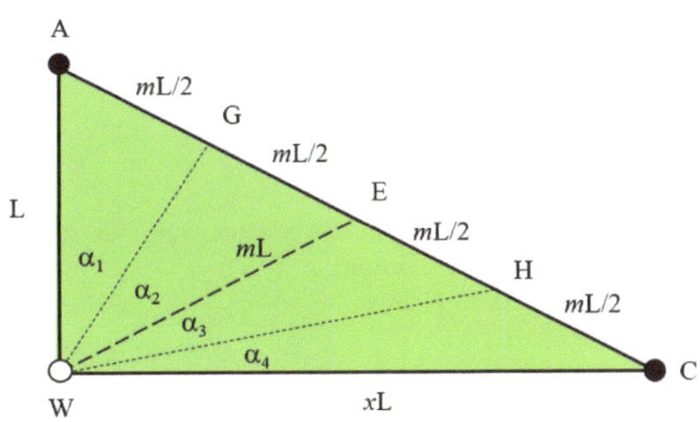

Figure 4. The injection–production unit of the rhombic inverted nine-point well pattern.

In the rhombic inverted nine-spot well pattern calculation unit AWG, the distance between the injection well and the short axis angle well is l, the GWA is $\alpha 1$, and the GAW is $\beta 1$. The flow pipe infinitesimal injection well angle increment EWG $\Delta\alpha$, a production well angle increment FAH $\Delta\beta$, injection well angle variable EWA α, production well angle variable FAW β, a calculation unit, and the flow pipe between injection and production wells are shown in Figure 5.

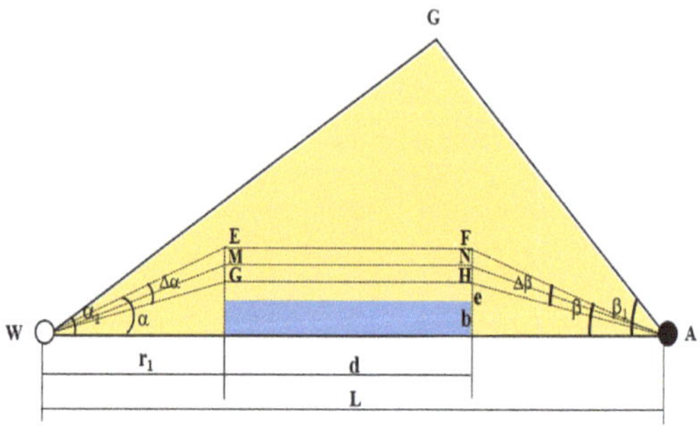

Figure 5. Seepage unit of rhombic inverted nine-spot well pattern, calculation unit of area sweep coefficient, schematic diagram of flow pipe in the calculation unit.

In low-permeability reservoirs, the diamond-shaped reverse nine-point well pattern is used chiefly for reservoir development. According to the symmetry of injection and production units in the well pattern, the ASE can be determined by studying one in four injection and production units. When the microsphere enters and seals off the area with low permeability, it is assumed that the flow line between the injection and production wells is trapezoidal.

Areal sweep efficiency refers to the ratio of the flooded area to the well pattern control area. To calculate the areal sweep efficiency, we must obtain the stream tube's length and cross-sectional area, the oil-water front's position, and the flow equation, considering the starting pressure gradient is necessary.

After being injected into the reservoir, a plugging zone exists in which PMs block the large pore path. We assume that the length of the plugging zone is d, the width of the plugging area is b and the sweep width of subsequent injected water is e. The width of the plugging area can be calculated with Equation (9).

$$b = r_1 \sin \alpha_w \tag{9}$$

where r_1 is the radius of the area in which the permeability increases due to water injection, obtained by well-test interpretation. α_w is the angle of the water drive, obtained by the calculation method of the area spread coefficient in the water drive stage.

After blocking the throat with PMs, the width of the subsequent injection water ripple is:

$$e = \frac{E_{RP}}{E_{RW}} b \tag{10}$$

where E_{RP} is the oil recovery of laboratory polymer microsphere flooding experiment, and E_{RW} is the oil recovery of laboratory water flooding experiment.

The formula of the polymer driving angle α_f is as follows:

$$\alpha_f = \tan(\frac{e}{r_1}) \tag{11}$$

We assume that the injected polymer microspheres will block the whole pore volume. Therefore, the plugging length d can be calculated with the following equation:

$$\frac{Q_{in} C_p C_d}{\rho_p} = 2 r_1 \sin \alpha_w d h \phi (1 - S_w) \tag{12}$$

Currently, the Carman–Kozeny equation is commonly used to calculate the throat's diameter(r_R), shown as Equation (13).

$$r_p = \sqrt{\frac{8K}{\phi}} \tag{13}$$

The ratio of polymer microsphere diameter and throat diameter is demonstrated as follows:

$$\frac{r_P}{r_R} = C_R \tag{14}$$

By bringing Equations (9)–(11), (13) and (14) into Equation (12), and through unit conversion, we obtain the calculation equation of plugging length.

$$d = \frac{Q_{in} C_p r_P^2 C_d}{0.008 K h \rho_p r_1 \sin \alpha_w C_R^2 (1 - S_{wi} - S_{or})} \tag{15}$$

where Q_{in} is the injected volume of PMs, C_p is the concentration of the polymer microsphere, C_d is the expansion ratio of the polymer microsphere in the hydration experiment, R_p is the diameter of PMs, K is reservoir permeability, h is the thickness of formation, ρ_p is the density of polymer microsphere solution, C_R is the ratio of polymer microsphere diameter and throat diameter, S_{wi} is the irreducible brine saturation, r_P is the diameter of polymer microsphere, and S_{or} is the residual oil saturation.

The flow pipe between injection and production wells is trapezoidal, so there are two inflection points. According to the geometric relationship, the length L_1 between the injection well W and the first inflection point E is:

$$L_1 = \frac{r_1}{\cos \alpha} \tag{16}$$

The length L_2 between the injection well W and the second inflection point F is:

$$L_2 = d + \frac{r_1}{\cos \alpha} \tag{17}$$

The length L_3 of any pipe between the injection well and the production well A is:

$$L_3 = d + \frac{r_1}{\cos \alpha} + \frac{l - d - r_1}{\cos \beta} \tag{18}$$

where l is well spacing.

In the flow pipe ξ, the cross-section of the flow tube is shown in Equation (19).

$$A(\xi) = \begin{cases} 2h\xi \tan \frac{\Delta\alpha}{2}, r_w < \xi < \frac{r_1}{\sin \alpha} \\ 2h\frac{r_1}{\sin \alpha} \tan \frac{\Delta\alpha}{2}, \frac{r_1}{\sin \alpha} < \xi < d + \frac{r_1}{\sin \alpha}(1 - \cos \alpha) \\ 2h(d + \frac{r_1}{\sin \alpha}(1 - \cos \alpha) + \frac{l-d}{\cos \beta} - \xi) \tan \frac{\Delta\beta}{2}, d + \frac{r_1}{\sin \alpha}(1 - \cos \alpha) < \xi < d + \frac{r_1}{\sin \alpha}(1 - \cos \alpha) + \frac{l-d}{\cos \beta} - r_w \end{cases} \tag{19}$$

In low-permeability reservoirs, the oil and water phase flow equation is the oil phase and water phase flow when the pressure gradient is started by non-piston water flooding.

The oil phase flow equation is shown as Equation (20).

$$q_t = \frac{\frac{CK}{\mu}(\Delta P - \lambda L_3)}{\int_L \frac{1}{A(\xi)} d\xi} \tag{20}$$

The pseudo tube equation before water channeling is shown as Equation (21).

$$q_t = \frac{\frac{CK}{\mu}(\Delta P - \lambda L_3)}{\frac{1}{2h\tan\frac{\Delta\alpha}{2}}\ln\frac{e}{r_w\sin\alpha} + \frac{d}{2h\frac{e}{\sin\alpha}\tan\frac{\Delta\alpha}{2}} + \frac{1}{2h\tan\frac{\Delta\beta}{2}}\ln\frac{l-d-r_1}{r_w\cos\beta}} \tag{21}$$

In the oil–water phase region, the distance from the oil–water front to the injection well ξ can be calculated with Equation (22):

$$\int_0^{\xi} A(\xi)d\xi = \int_0^{t} \frac{q_t f'_w(S_{wf})}{\phi}dt \tag{22}$$

where $f'(S_{wf})$ is the changing rate of water cut at different water saturations.

When the oil–water front reaches the inflection point E, the time when the oil–water front comes is t_1, the second inflection point F is t_2, and the time when the oil–water front reaches the production well A is t_3.

Assuming that the pipe area at inflection points E and F are equal, there is:

$$\beta = \arctan\frac{r_1\tan\alpha}{l-r_1-d} \tag{23}$$

Combining Equations (18), (21) and (22), the Equation of time t_1 of the oil–water front reaching E at different angles is obtained:

$$t_1 = \frac{\mu\phi\frac{r_1^2}{\cos^2\alpha}}{2CKf'_w(S_{wf})}\frac{\ln\frac{r_1}{r_w\cos\alpha} + \frac{d\cos\alpha}{r_1} + \frac{1}{C_1}\ln\frac{l-d-r_1}{r_w\cos\beta}}{\Delta P - \lambda L_3} \tag{24}$$

The time when the oil–water front reaches F at different angles is t_2:

$$t_2 = \frac{\mu\phi(\frac{r_1^2}{\cos^2\alpha} + \frac{2dr_1}{\cos\alpha})}{2CKf'_w(S_{wf})}\frac{\ln\frac{r_1}{r_w\cos\alpha} + \frac{d\cos\alpha}{r_1} + \frac{1}{C_1}\ln\frac{(l-r_1-d)}{r_w\cos\beta}}{\Delta P - \lambda L_3} \tag{25}$$

The time of oil–water front reaching the production well at different angles t_3:

$$t_3 = \frac{\mu\phi(\frac{r_1^2}{\cos^2\alpha} + \frac{2dr_1}{\cos\alpha} + (\frac{l-r_1-d}{\cos\beta})^2)}{2CKf'_w(S_{wf})}\frac{\ln\frac{r_1}{r_w\cos\alpha} + \frac{d\cos\alpha}{r_1} + \frac{1}{C_1}\ln\frac{(l-r_1-d)}{r_w\cos\beta}}{\Delta P - \lambda L_3} \tag{26}$$

$$\Delta P = P_h - P_f \tag{27}$$

P_h is the bottom hole pressure of the injection well, and P_f is the bottom hole pressure of the production well, $r_w << \xi$.

If $t < t_1$, the position of oil–water front ξ_1 is

$$\xi_1^2 = \frac{\frac{2CKf'_w(S_{wf})}{\mu\phi}(\Delta P - \lambda L_3)}{\ln\frac{e}{r_w\sin\alpha} + \frac{d\cos\alpha}{r_1} + \frac{\sin\alpha}{\sin\beta}\ln\frac{l-d-r_1}{r_w\cos\beta}}t \tag{28}$$

If $t_1 < t < t_2$, the position of oil–water front ξ_2 is

$$\xi_2 = \frac{2CKf'_w(S_{wf})\cos\alpha}{\mu\phi r_1}\frac{\Delta P - \lambda L_3}{\ln\frac{r_1}{r_w\cos\alpha} + \frac{d\cos\alpha}{r_1} + \frac{1}{C_1}\ln\frac{(l-r_1-d)}{r_w\cos\beta}}t + \frac{r_1}{2\cos\alpha} \tag{29}$$

If $t_2 < t < t_3$, the position of oil–water front ξ_3 is

$$\xi_3 = \frac{r_1}{\cos\alpha} + d + \frac{l - r_1 - d}{\cos\beta} - \sqrt{\frac{1}{C_1}(\frac{2CKf'_w(S_{wf})}{\mu\phi}\frac{\Delta P - \lambda L_3}{\ln\frac{r_1}{r_w\cos\alpha} + \frac{d\cos\alpha}{r_1} + \ln\frac{(l-r_1-d)}{r_w\cos\beta}}t - \frac{r_1^2}{\cos^2\alpha} - \frac{2dr_1}{\cos\alpha})} \quad (30)$$

The area of the seepage unit is

$$A_{\Delta AWD} = \frac{ml^2\sqrt{1 - \frac{1}{4m^2}}}{4} \quad (31)$$

When $t < t_1$, the water flooding swept area of a single stream tube is

$$S_i = \frac{1}{2}\xi_1^2\Delta\alpha \quad (32)$$

According to Equation (24), the angle α_0 for water drive in time t can be obtained by

$$E_{A1} = \frac{\int_0^{\alpha_0}\frac{1}{2}\xi_1^2 d\alpha}{A_{\Delta AWD}} \quad (33)$$

When $t_1 < t < t_2$, the angle α_m can be determined by Equation (24). Therefore, the areal sweep efficiency is

$$E_{A1} = \int_0^{\alpha_m}\frac{1}{2}\frac{e^2}{\sin^2\alpha}d\alpha + \int_{\alpha_m}^{\alpha_f}\frac{2e}{\cos\alpha}(d + \frac{e}{\sin\alpha}(1 - \cos\alpha))d\alpha \quad (34)$$

$$E_{A1} = \frac{\sin(\alpha_f - \alpha_m)\xi_1(\alpha_m)\xi_1(\alpha_f) + (\xi_2(\alpha_m) - \frac{r_1}{\cos\alpha_m} + r_1 + d)\cdot r_1\tan\alpha_m}{2A_{\Delta OAD}} \quad (35)$$

When $t_2 < t < t_3$, the angle α_n can be determined by Equation (25). Therefore, the areal sweep efficiency is

$$E_{A1} = \int_0^{\alpha_0}\frac{1}{2}\frac{e^2}{\sin^2\alpha}d\alpha + \int_0^{\alpha_0}\frac{2e}{\cos\alpha}(d + \frac{e}{\sin\alpha}(1 - \cos\alpha))d\alpha + \int_0^{\beta_0}\frac{1}{2}\frac{(l-d)^2}{\cos^2\beta}d\beta \quad (36)$$

$$E_{A1} = \frac{\sin(\alpha_f - \alpha_m)\xi_1(\alpha_m)\xi_1(\alpha_f) + r_1\cdot r_1\tan\alpha_f + (\xi_2(\alpha_m) - \frac{r_1}{\cos\alpha_m} + \xi_2(\alpha_n) - \frac{r_1}{\cos\alpha_n})\cdot r_1(\tan\alpha_m - \tan\alpha_n) + (\xi_2(\alpha_n) - \frac{r_1}{\cos\alpha_n} + \xi_3(0))\cdot r_1\tan\alpha_n}{2A_{\Delta OAD}} \quad (37)$$

When $t > t_3$, the angle α_t can be determined by Equation (26); therefore, the areal sweep efficiency is

$$E_{A1} = \int_0^{\alpha_0}\frac{1}{2}\frac{e^2}{\sin^2\alpha}d\alpha + \int_0^{\alpha_0}\frac{2e}{\cos\alpha}(d + \frac{e}{\sin\alpha}(1 - \cos\alpha))d\alpha + \int_0^{\beta_0}\frac{1}{2}\frac{(l-d)^2}{\cos^2\beta}d\beta \quad (38)$$

$$E_{A1} = \frac{\sin(\alpha_f - \alpha_m)\xi_1(\alpha_m)\xi_1(\alpha_f) + r_1\cdot r_1\tan\alpha_m + (\xi_2(\alpha_m) - \frac{r_1}{\cos\alpha_m} + \xi_2(\alpha_n) - \frac{r_1}{\cos\alpha_n})\cdot r_1(\tan\alpha_m - \tan\alpha_n) + (l - r_1 + d)r_1\tan\alpha_n - (\xi_3(\frac{\alpha_t+\alpha_n}{2}) - d - \frac{r_1}{\cos(\frac{\alpha_t+\alpha_n}{2})})^2(\alpha_n - \alpha_t)}{2A_{\Delta OAD}} \quad (39)$$

According to the established calculation method of the area spread coefficient, the area spread coefficients E_{A2}, E_{A3}, and E_{A4} of ΔGWE, ΔEWH, and ΔHWC are calculated. The injection–production well spacing, injection well angle variable, production well angle variable, and injection–production unit area should be replaced in the calculation.

The area sweep efficiency of the injection–production unit is calculated by the area sweep coefficient of the four seepage units.

$$E_A = \frac{E_{A1} \times A_{\Delta AWD} + E_{A2} \times A_{\Delta GWE} + E_{A3} \times A_{\Delta EWH} + E_{A4} \times A_{\Delta HWC}}{A_{\Delta AWC}} \quad (40)$$

where $A_{\triangle GWE}$, $A_{\triangle EWH}$, $A_{\triangle HWC}$, and $A_{\triangle AWC}$ is triangle area.

2.4. Vertical Sweep Efficiency

The vertical sweep efficiency c is mainly affected by the oil flow ratio (M), the water–oil ratio (F_{WO}), and the permeability coefficient of variation (V_K) [32].

$$Y = 3.33c^{0.77}(1-c)^{-1.23} = \frac{(F_{WO} + 0.4) \times (18.948 - 2.499V_K)}{(M + 1.137 - 0.809V_K) \times 10^{f(V_K)}} \quad (41)$$

where:

$$f(V_K) = -0.6819 + 0.9735V_K + 1.6453V_K^2 \quad (42)$$

The mechanism of EOR by polymer flooding decreases the mobility difference between displaced and displacing fluids to reduce fingering effects. In contrast, PMs can plug high permeability channels and improve sweep efficiency, but they have little effect on displacement efficiency [56,57].

Due to the difference between the primary mechanism of spreading expansion and the polymer, the equivalent oil flow ratio is introduced to characterize the ability of PMs to adjust the profile. For PMs flooding, the resistance coefficient is the ratio of the water flow rate to the microsphere solution, so the resistance coefficient can characterize the flow rate ratio. The plugging ability evaluation experiment provides the flow rate ratio.

In the water flooding stage of laboratory experiment of the PMs' conformance control,

$$Q_w = \frac{K_w A \Delta P_w}{\mu_w l} \quad (43)$$

During the PMs injection' period:

$$Q_p = \frac{K_p A \Delta P_p}{\mu_p l} \quad (44)$$

The equivalent mobility ratio is

$$M_E = \frac{\lambda_w}{\lambda_p} = \frac{\frac{k_w}{\mu_w}}{\frac{k_p}{\mu_p}} = \frac{\Delta P_p}{\Delta P_w} \times \frac{Q_w}{Q_p} \quad (45)$$

The calculation equation of vertical sweep efficiency is as follows:

$$Y = 3.33c^{0.77}(1-c)^{-1.23} = \frac{(F_{WO} + 0.4) \times (18.948 - 2.499V_K)}{(M_E + 1.137 - 0.809V_K) \times 10^{f(V_K)}} \quad (46)$$

3. Adapt Ability Analysis

In this theoretical model, five major parameters adjust the working scheme of polymer microsphere flooding, including injection period, injection rate, injection concentration, injection volume, and polymer microsphere size. These parameters remarkably affect the areal sweep efficiency and vertical sweep efficiency of the polymer microsphere flooding process. A series of sensitivity analyses are introduced to investigate how the parameters may affect the oil recovery of a square inverted nine-spot pattern. The parameters and set values are shown in Table 1, and other support parameters are shown in Table 2.

Table 1. Parameters and design values for sensitivity analysis.

No.	Factors	Parameter Design Value	Default Value
1	Injection period (d)	50, 70, 90, 120, 150	90
2	Injection rate (m^3/d)	25, 30, 35, 40, 45	40
3	Injection concentration (ppm)	2000, 3000, 4000, 5000, 6000	5000
4	Polymer microsphere size (μm)	0.1, 0.3, 0.8, 5, 10, 20	0.1

Table 2. Support parameters for sensitivity analysis.

No.	Parameters Name	Default Value	No.	Parameter Name	Default Value
1	Well spacing (m)	300	12	Expansion ratio	5
2	m	1	13	Irreducible water saturation (f)	0.322
3	C	0.0864	14	Split coefficient of injection volume (f)	0.25
4	Inner zone permeability (10^{-3} μm^2)	30	15	Blocking rate (f)	0.4
5	Outer zone permeability (10^{-3} μm^2)	25	16	Oil displacement efficiency (f)	0.3
6	Crude oil viscosity (MPa)	1.80	17	Vertical sweep efficiency (f)	0.42
7	Porosity (f)	0.2	18	Permeability variation coefficient (f)	0.4
8	Threshold pressure gradient (MPa/m)	0.01	19	Water viscosity (MPa)	0.46
9	Hole diameter (m)	0.1	20	Equivalent mobility ratio	10
10	Polymer microsphere density (g/cm^3)	1.2	21	Recovery of polymer microsphere (f)	0.85
11	Water flooding recovery (f)	0.4	22	Production pressure differential (MPa)	25

3.1. Injection Period

When the effect of the injection period is being investigated, the polymer microsphere injection rate is controlled to 40 m^3/d, and the concentration is headed to 5000 ppm. By applying different PMs injection periods to the oil recovery calculation model, we obtained the oil recovery factor curve corresponding with the injection period, as shown in Figure 6.

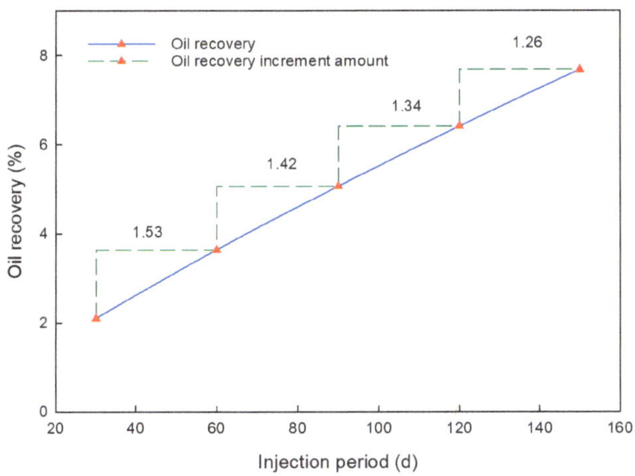

Figure 6. Oil recovery with different injection periods.

According to Figure 6, the recovery factor increases with the injection period of PMs. The oil recovery rises 5.55% from 30 d of the PMs' injection period to 150 d. It can be found that the injection period shows a significant influence on oil recovery. The longer the injection period, the greater the oil recovery. However, as oil recovery contributed, the increased PMs' injection period declined. The oil recovery increment amount decreases from 1.53 to 1.26 as the injection period increases from 30 d to 150 d. This decrease indicates that continuing to lengthen the injection period cannot continue to improve the oil recovery.

3.2. Injection Rate

As for the PMs' injection rate, the oil recovery curve for different injection rates is plotted in Figure 7. The concentration of PMs is controlled to 5000 ppm, and the total injection volume is controlled to 3.6 t.

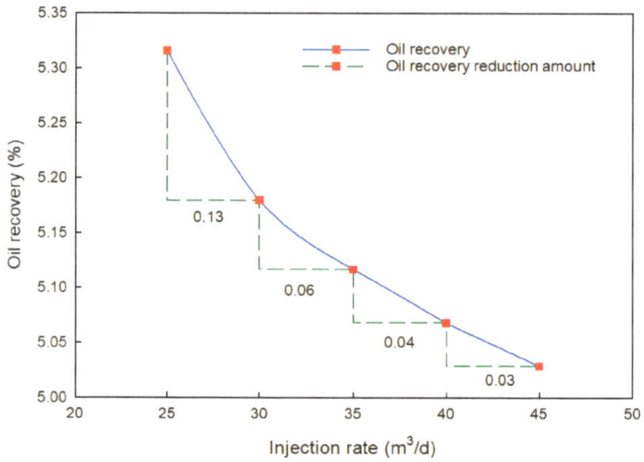

Figure 7. Oil recovery with different injection rates.

Figure 7 shows that oil recovery decreases with the PMs' injection rate. The oil recovery decreases from 5.31% to 5.01% as the injection rate increases from 25 m^3/d to 45 m^3/d. Therefore, a low injection rate is conducive to more significant recovery. Meanwhile, the oil recovery reduction amount shows the same variation trend. The oil recovery reduction amount decreased from 0.13 to 0.03. This indicates that the recovery factor can be increased significantly by lowering the injection rate when in the range of a low injection rate. When the injection rate is high, changing the injection rate to a small range has little effect on the recovery factor.

3.3. Injection Concentration

The concentration of the PMs' dispersed system notably affects recovery performance in the PMs' flooding process. The oil recovery factor–concentration relationship is studied at the same injection volume.

The oil recovery with different injection concentrations is shown in Figure 8. It can be seen that the oil recovery decreases with the increase in injection concentration. The final oil recovery reduces by 3.54%, with the size increasing from 2000 ppm to 6000 ppm. Therefore, low concentration is to the benefit of enhancing oil recovery. However, the best blocking rate of PMs demands the optimal injection concentration. Therefore, the injection concentration could be slightly less than the optimal injection concentration obtained from laboratory experiments to achieve the most enhanced oil recovery performance.

3.4. Polymer Microsphere Size

To investigate the influence of PMs' size on conformance control performance, the injection period, injection rate, and injection concentration remained the same. In this case, the injection period is 90 d, the injection rate is 40 m^3/d, and the injection concentration is 5000 ppm.

Figure 9 shows the effect of PMs' size on oil recovery. According to Figure 9, the oil recovery decreases with the increased PMs' size. The oil recovery reduces by 2.315% as the size increases from 100 nm to 20 μm. The decreasing degree of oil recovery shows significant influence on the PMs' size. The oil recovery reduces by 0.779 as the polymer microspheres' size increases from 100 nm to 300 nm, but it only reduces by 0.254 when the PMs' size increases from 10 μm to 20 μm. When the injected PMs have the same volume, the number of microspheres with smaller particle sizes is much larger than that with larger particle sizes. Therefore, the total volume of the tiny microspheres after hydration and expansion is more extensive. A higher plugging probability and plugging efficiency can be obtained, so smaller PMs show a much-improved oil recovery performance.

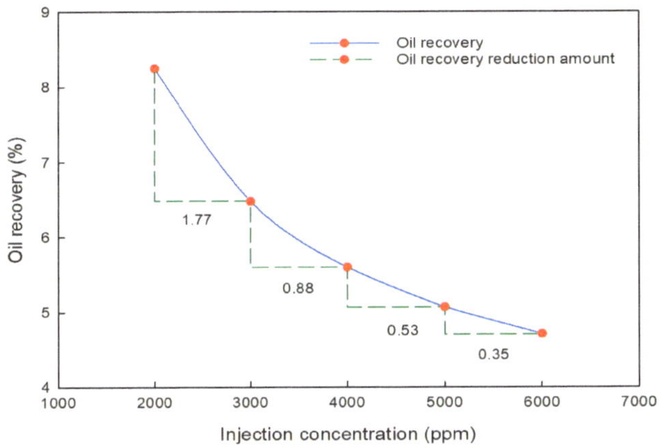

Figure 8. Oil recovery with different injection concentrations of PMs.

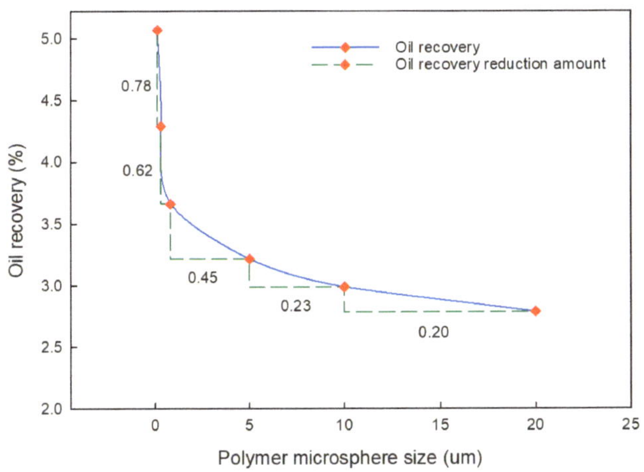

Figure 9. Oil recovery with different sizes of PMs.

Figure 10 shows the blocking distance of different sizes of PMs. From Figure 10, we can find that the blocking length decreases with the increase in PMs' size. The blocking distance decreases from 72.11 m to 44.98 m as the PMs' size increases from 100 nm to 20 μm. Tiny PMs can provide longer blocking distances. Based on the research results, it can be concluded that the smaller the size, the greater the oil recovery within the range of PMs' size selection.

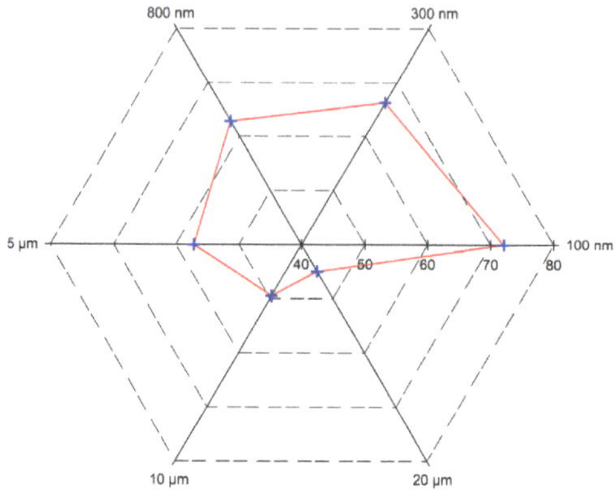

Figure 10. Blocking distance with different sizes of PMs.

4. Comparison with Field Application

Oil production results estimated by the proposed theoretical model are compared with field production data collected from the PMs pilot test. The pilot test was conducted in the Changqing WYQ oil field.

4.1. Reservoir Background

WYQ is a typical low-permeability reservoir located in the Ordos basin. This oil field was put into development in October 1996, using an inverted nine-spot rhombus well pattern. After more than ten years of development, daily oil production decreased rapidly. PMs conformance control technology is chosen to conduct field pilot tests for its ability to improve oil recovery in low-permeability reservoirs.

The porosity and permeability of the selected well group to conduct the pilot test are 12.64% and 28.36 mD. The reservoir temperature is 55 °C, and the formation crude oil viscosity is 1.95 MPa. The initial formation pressure of the WYQ oilfield is 13.2 MPa. Well group I1 has a water injection well and eight oil production wells; all of them are vertical wells, and the well group map is shown in Figure 11. I1 is water injection well, and P1 to P8 are oil production wells.

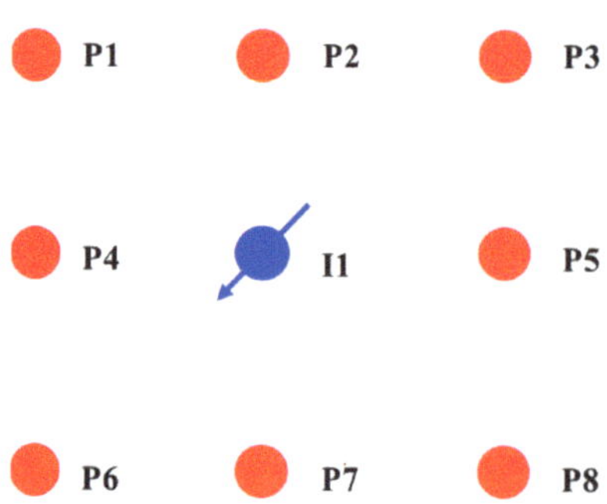

Figure 11. Well location of pilot test well group.

The relative permeability curve is shown in Figure 12; the irreducible water saturation is 0.322, the slope of water cut at the water–oil front is 4.73, and the oil displacement efficiency is 0.42.

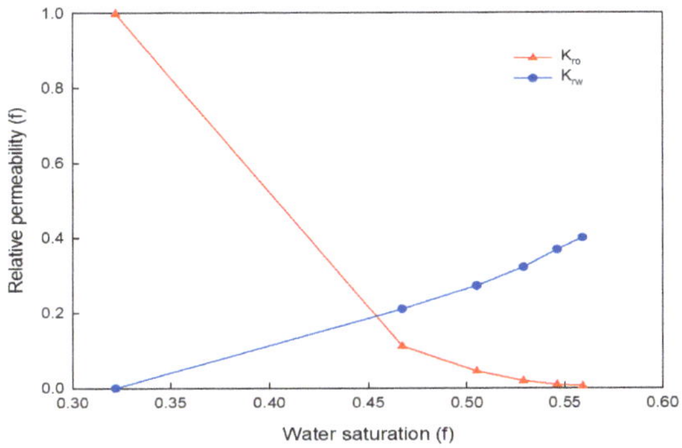

Figure 12. Relative permeability curve of WYQ.

4.2. Field Application Scheme of Polymer Microsphere Flooding

The porosity and permeability of the selected well groups are 12.64% and 28.25 mD, obtained from well testing. The throat diameter calculated with the K-C equation based on the porosity and permeability is 3.36 μm. Therefore, the researchers chose PM800 as the PMs size to conduct the field pilot, and throat diameter ratio is 0.62 at the achieved hydration equilibrium state. With appropriate size, PM800 could smoothly inject into the reservoir, progressively migrate to the deep reservoir, and execute deep plugging. Other physical and injection properties are shown in Table 3.

Table 3. Reservoir and fluid properties.

No.	Parameter Name	Default Value	No.	Parameter Name	Default Value
1	Well spacing (m)	360	12	Expansion ratio	2.6
2	m	1	13	Irreducible water saturation (f)	0.322
3	C	0.0864	14	Split coefficient of injection volume (f)	0.125
4	Inner zone permeability (10^{-3} μm^2)	131	15	Blocking rate (f)	0.8
5	Outer zone permeability (10^{-3} μm^2)	18.32	16	Oil displacement efficiency (f)	0.42
6	Crude oil viscosity (MPa)	1.95	17	Vertical sweep efficiency (f)	0.76
7	Porosity (f)	0.12	18	Permeability variation coefficient (f)	0.37
8	Threshold pressure gradient (MPa/m)	0.01	19	Water viscosity (MPa)	0.46
9	Hole diameter (m)	0.1	20	Equivalent mobility ratio	1.8
10	Polymer microsphere density (g/cm^3)	1.1	21	Recovery of polymer microsphere (f)	0.86
11	Water flooding recovery (f)	0.4	22	Production pressure differential (MPa)	24

Theoretical analysis and experimental evaluation show that PM800 has good adaptability to WYQ and excellent plugging performance. The injection concentration is chosen to be 5000 ppm based on the formula of injection concentration and block rate. Additionally, the injection volume is about 29 tons and the injection period is 160 d. We assume the best injection rate is consistent with the original water injection rate of 42 m^3/d. The injection parameters are shown in Table 4.

Table 4. PMs injection scheme of PM800.

Parameter	Value
Size (nm)	800
Concentration (ppm)	5000
Rate (m^3/d)	42
Injection period (d)	160

4.3. Oil Recovery Comparison

The main parameters of development performance analysis include changing the injection pressure trends of the injection wells, tubing and casing, and the water injectivity presented by the daily water injection rate. In order to study the pressure change trend during and after the PM800 injection, the tubing, casing pressure and the actual injection volume are monitored. The monitored data are shown in Figure 13. Figure 13 shows that during the PM800 injection, there was no significant change in the daily injection rate, and it remains the same as an initial injection rate of 42 m^3/d, indicating that the injection capacity has not decreased.

However, the tubing pressure and casing pressure increase significantly. The increasing pressure indicates that the increasing injection pressure is a crucial indicator parameter for the effectiveness of PMs' conformance control. The average injection pressure rises to 1.6 MPa, which indicates that the PMs formed an effective plugging between the injection and production wells.

Increased sweep efficiency leads to increased oil production. With limited plugging strength, if the injection water breaks through the plugging area, the flow around injection water is not apparent, and the PMs' validity period ends. The validity period starts in July 2015 and ends in February 2017.

In this pilot test of PMs' conformance control, the total oil production volume during the validity period is 8368.5 t. The oil production rate of the selected well groups is shown in Figure 14. According to the former geology study, the reserves of well control of the selected well group is 68.75×10^4 t, and the actual oil recovery calculated with production data is 1.22%. The oil recovery obtained from the proposed model is 1.37%, the oil displacement efficiency is 0.42, the vertical sweep efficiency is 0.76, and the areal sweep efficiency is 0.043. Comparing the two oil recoveries, it can be found that the theoretical oil recovery value is larger than the oil recovery calculated with field data, and the error is 12.3%, i.e., within the acceptable range.

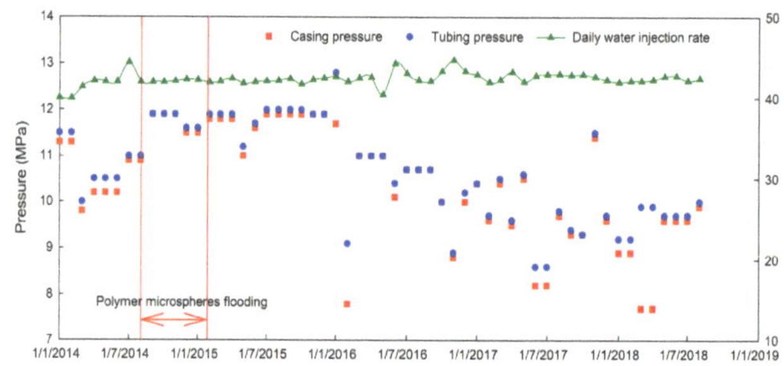

Figure 13. Daily water injection rate, casing pressure and tubing pressure.

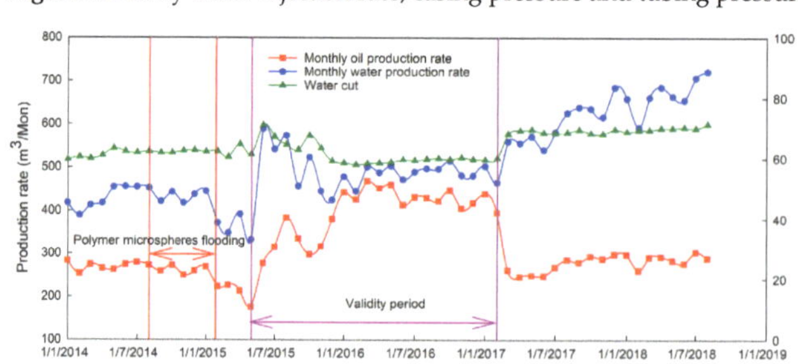

Figure 14. Monthly oil and water production rate and water cut.

5. Conclusions

An effective theoretical model is established to predict the oil recovery of PMs' conformance control. The influences of critical factors, including the injection period, injection rate, injection concentration, and PMs' size, are investigated. Additionally, the model's validity is verified by comparing it with field application data. The following conclusions can be drawn from this study.

(1) The recovery factor calculation model can quickly and accurately evaluate the oil recovery performance of polymer microspheres' conformance control. The equivalent mobility ratio was introduced to optimize the vertical sweep efficiency calculation method. A trapezoidal flow tube is proposed to characterize the streamline shape after PMs' conformance control, and to calculate the areal sweep efficiency.

(2) A series of adaptability analyses on the injection period, injection rate, injection concentration, and PMs' size are conducted in order to reveal how these functional parameters affect results. The results indicate that long-term injection of small size and low concentration benefits higher oil recovery.

(3) The theoretical oil recovery obtained from the proposed model is 1.37%, and the actual oil recovery of field application is 1.22%, with an error of 0.15%. The recovery factor calculated by the theoretical model agrees better with the field application. This model not only provides a faster and more accessible tool to predict and evaluate the production performance of the PMs' flooding process, but also could be used to optimize the injection scheme of PMs' conformance control.

Author Contributions: Conceptualization, W.Z. and T.L.; methodology, W.Z.; validation, H.T.; formal analysis, T.L.; investigation, F.L. and N.L.; data curation, N.L. and R.S.; writing—original draft preparation, W.Z.; writing—review and editing, S.H.; supervision, R.S. All authors have read and agreed to the published version of the manuscript.

Funding: This research received no external funding.

Institutional Review Board Statement: Not applicable.

Informed Consent Statement: Not applicable.

Data Availability Statement: The data presented in this study are available from the corresponding authors upon reasonable request.

Conflicts of Interest: The authors declare no conflict of interest.

References

1. Juan, Z. Study on Gas Channeling Pattern and Optimization Method of CO_2 Injection in C4 + 5 Oil Reservoir in W Oil Region of Northern. Ph.D. Dissertation, Northwest University, Xi'an, China, 2018.
2. Jiayao, L. Study on Secondary Encryption Scheme of CXCY Block in Low Permeability Reservoir. Master's Dissertation, Northeast Petroleum University, Daqing, China, 2020.
3. Yuan, S.; Wang, Q. New Progress and Prospect of Oilfields Development Technologies in China. *Pet. Explor. Dev.* **2018**, *45*, 657–668. [CrossRef]
4. Hu, W.; Wei, Y.; Bao, J. Development of the Theory and Technology for Low Permeability Reservoirs in China. *Pet. Explor. Dev.* **2018**, *45*, 646–656. [CrossRef]
5. Hu, W. The Present and Future of Low Permeability Oil and Gas in China. *Strateg. Study CAE* **2009**, *11*, 29–37.
6. Jia, C. Development Challenges and Future Scientific and Technological Researches in China's Petroleum Industry Upstream. *ACTA Pet. Sin.* **2020**, *41*, 1445–1464.
7. Dang, X. Development Status and Prospect Analysis of Low Permeability Reservoirs in China. *Petrochem. Ind. Technol.* **2018**, *25*, 149.
8. Wang, Z. Analysis of Water Injection Development Process and Technology in Low Permeability Oilfield. *Mod. Chem. Res.* **2019**, *3*, 152–154.
9. Pang, Z.; Li, Y.; Wang, B.; Huang, C.; Duan, W.; Yao, Z. Analysis of Influence Factors of Oil Displacement Efficiency in Ultra-Low Permeability Reservoir. *Unconv. Oil Gas* **2017**, *4*, 76–81.
10. Wang, D.W.; Fang, J.K.; Liu, X.M.; Zhang, J.F.; Shu, D.W.; Zhang, X.Q. Preparation of Colored Polymer Microspheres and Research Progress Thereof in Textile Dyeing and Printing. *J. Text. Res.* **2019**, *40*, 175–182.

11. Chen, X.; Li, Y.; Liu, Z.; Li, X.; Zhang, J.; Zhang, H. Core- and Pore-scale Investigation on the Migration and Plugging of Polymer Microspheres in a Heterogeneous Porous Media. *J. Petrol. Sci. Eng.* **2020**, *195*, 107636. [CrossRef]
12. Yang, H.; Kang, W.; Yin, X.; Tang, X.; Song, S.; Lashari, Z.A.; Bai, B.; Sarsenbekuly, B. Research on Matching Mechanism Between Polymer Microspheres with Different Storage Modulus and Pore Throats in the Reservoir. *Powder Technol.* **2017**, *313*, 191–200 [CrossRef]
13. Wu, T.; Zheng, M.; Zhou, Z.; Yang, H.; Cao, R. New Method for Plugging Performance Evaluation of Polymeric Nanospheres in Low Permeability Reservoir. *Fault-Block Oil Gas F.* **2018**, *25*, 498–501.
14. Sun, Z.; Wu, X.; Kang, X.; Lu, X.; Li, Q.; Jiang, W.; Zhang, J. Comparison of Oil Displacement Mechanisms and Performances Between Continuous and Dispersed Phase Flooding Agents. *Pet. Explor. Dev.* **2019**, *46*, 121–129. [CrossRef]
15. Jia, Y.; Haien, Y.; Cheng, C.; Li, Z.; Hou, G.; Yuan, X.; Zhao, W.; Zhang, Z.; Liu, T. Field Application and Performance Evaluation of Polymer Microsphere Profile Control in Low Permeability Oil Reservoir. SPE-197198-MS. In Proceedings of the Abu Dhabi International Petroleum Exhibition & Conference, Abu Dhabi, United Arab Emirates, 11–14 November 2019.
16. Zhao, Y.; Li, S.; Zhang, W.; Jing, W.; Fang, H. Effect evaluation of polymer microspheres profile control in Chi 46 Chang 8 reservoir *Petrochem. Ind. Appl.* **2018**, *37*, 84–88.
17. Tian, Y.; Yi, Y.; Li, Z.; Bi, T.; Nan, Y. Application and effect analysis of polymer microsphere in Ansai oilfield. *Chem. Eng. Equip.* **2018**, 98–100. [CrossRef]
18. Liao, C. Research on the adaptability of polymer elastic microsphere profile control in Fang169 fault block. *China Pet. Chem. Ind* **2016**, 249.
19. Han, J.; Sun, W.; Bai, Y.; Zhang, Q.; Fan, X. Fluid Seepage Characteristics of Low Permeability Reservoir Based on the Real Sandstone Microscopic Water Flooding Experiment: Taking the Chang 6 Reservoir in Wangpanshan as an Example. *Geol. Sci. Technol. Inf.* **2018**, *37*, 89–95.
20. Yan, W.; Sun, J. Analysis of research present situation of microscopic remaining oil. *Prog. Geophys.* **2016**, *31*, 2198–2211.
21. Mitchell, J.; Chandrasekera, T.C.; Holland, D.J.; Gladden, L.F.; Fordham, E.J. Magnetic resonance imaging in laboratory petrophysical core analysis. *Phys. Rep.* **2013**, *526*, 165–225. [CrossRef]
22. Iglauer, S.; Fernø, M.A.; Shearing, P.; Blunt, M.J. Comparison of residual oil cluster size distribution, morphology and saturation in oil-wet and water-wet sandstone. *J. Colloid Interf. Sci.* **2012**, *375*, 187–192. [CrossRef]
23. Yadali Jamaloei, B.; Asghari, K.; Kharrat, R.; Ahmadloo, F. Pore-scale two-phase filtration in imbibition process through porous media at high- and low-interfacial tension flow conditions. *J. Petrol. Sci. Eng.* **2010**, *72*, 251–269. [CrossRef]
24. Yu, Q.; Zhao, M.; Lin, Z. A Study of the Displacement Efficiency and the Conformance Factor in Waterflooded Sandstone Reservoirs in China (2). *Pet. Explor. Dev.* **1989**, *3*, 46–54.
25. Yu, Q.; Zhao, M.; Lin, Z. A Study of the Displacement Efficiency and the Conformance Factor in Waterflooded Sandstone Reservoirs in China (1). *Pet. Explor. Dev.* **1989**, *2*, 48–52.
26. Du, L.; Chen, Q.; Dai, S. Application of Oil-water Relative Permeability Curve and Dynamic Data to Defining Waterflood Sweep Situation. *Pet. Geol. Recovery Effic.* **2003**, *10*, 45–47.
27. Zhang, L.; Li, T.; Cai, R.; Li, T.; Gai, C.; Fan, Z. Study on Oil Displacement efficiency and Sweep efficiency in Water Flooding Oilfield. *Inn. Mong. Petrochem. Ind.* **2004**, *30*, 135–137.
28. Xie, W.; Ding, Y.; Peng, Z. Research on Sweep Efficiency of Inverted Nine—Spot Pattern in Low Permeability Reservoir *J. Chongqing Univ. Sci. Technol.* **2021**, *23*, 34–38.
29. Zhang, S.; Liu, B.; Zhong, S. Calculation method for areal sweep efficiency of irregular well pattern. *Pet. Geol. Oilfield Dev. Daqing* **2010**, *29*, 49–52.
30. Zhang Lihua, L.Z. Determination of areal sweep efficiency of pattern flooding by use of simulator. *J. China Univ. Pet. Nat. Sci. Ed* **1998**, *12*, 86–98.
31. Fan, J.; Zhang, Z.; Zaladuxin, A.B. A model for the calculation of the sweep efficiency in a heterogeneous oil reservoir. *Acta Pet. Sin.* **1993**, *14*, 92–98.
32. Fassihi, M.R. New correlation for calculation of vertical coverage and areal sweep efficiency. *SPE* **1986**, *1*, 604–606. [CrossRef]
33. Pitts, G.N.; Crawford, P.B. Low areal sweep efficiencies in flooding heterogeneous rock. In Proceedings of the SPE Production Techniques Symposium, Wichita Falls, TX, USA, 14–15 May 1970.
34. Prats, M. The breakthrough sweep efficiency of the staggered line drive. SPE 756G. *Trans. AIME* **1956**, *207*, 361–362. [CrossRef]
35. Slobod, R.L.; Caudle, B.H. X-ray shadowgraph studies of areal sweepout eficiencies. SPE 211G. In Proceedings of the Fall Meeting of the Petroleum Branch of AIME, Houston, TX, USA, 1–3 October 1952.
36. Muskat, M. Flow of homogeneous fluids through porous media. *Soil Sci.* **1937**, *46*, 169. [CrossRef]
37. Shen, F.; Cheng, L.; Huang, S.; Zhao, W.; Liu, X. Calculation of Sweep Efficiency for Water Flooding Development of Conventional Heavy Oil Using the Stream-tube Method. *Oil Drill. Prod. Technol.* **2016**, *38*, 645–649.
38. Lv, D.; Tang, H.; Guo, F. Study of areal sweep efficiency for invert 9-spot pattern of low permeability reservoir. *J. Southwest Pet. Univ. Nat. Sci. Ed.* **2012**, *34*, 147–153.
39. Liu, C.; Fan, Z.; Xu, A. A Calculation Method for the Areal Sweep Efficiency of Heavy oil Reservoirs by Non-piston Like Water Flood in Inverted Nine-spot Pattern. *Oil Drill. Prod. Technol.* **2018**, *40*, 228–233.
40. Sun, Q.; Zhou, H.; Hu, Y.; Shi, H.; Wang, J. Calculation Method of Area Sweep Efficiency of Non-piston Water Flooding for Inverted 9-spot Pattern. *Complex Hydrocarb. Reserv.* **2018**, *11*, 52–56.

41. Zhongyuan, J.; Li, J. Study on Areal Sweep Efficiency of Five-Spot Well Pattern Based on Sreamline Model. *J. Chongqing Univ. Sci. Technol.* **2016**, *18*, 36–40.
42. Shu, J.; Tang , H.; Wang, Y.Y.; Lv, D.L.; Wang, B.; Yang, C. Study on Areal Sweep Efficiency of Square Inverted Nine-spot Well Pattern with Crack. *Fault-Block Oil Gas Field* **2014**, *21*, 360–363.
43. Wang, Q.; Ji, B.; Nie, J. Calculation Methods of Sweep Efficiency Under Different Viscosity Ratios for Polymer Flooding Process. *Oil Gas Geol.* **2014**, *35*, 551–555.
44. Zhou, Y.; Tang, H.; Lv, D.; Tang, H.; Guo, F. Areal Sweep Efficiency of Staggered Well Pattern of Horizontal Wells in Low Permeability Reservoirs. *Lithol. Reserv.* **2012**, *24*, 124–128.
45. Shen, R.; Gao, S.; Hu, Z.; Duan, B. Calclating method and its application of water-flooding swept coefficient for low-permeability oil reservoirs. *Pet. Geol. Oilfield* **2013**, *32*, 60–65.
46. Cao, R.; Ma, M.; Guo, X. Calculation Method of Plane Sweep Coefficient for Water Flooding in Low Permeability Reservoir Based on Flow Tube Model. *Pet. Geol. Recovery Effic.* **2021**, *28*, 100–108.
47. Wu, Z.; Huiqing, L.; Cao, P.; Yang, R. Experimental investigation on improved vertical sweep efficiency by steam-air injection for heavy oil reservoirs. *Fuel* **2021**, *285*, 119–138. [CrossRef]
48. Wang, Z. Study on Sweep Efficiency and Front of Offshore Heterogeneous Thick Oil Layer for Polymer Flooding. Ph.D. Dissertation, China University of Petroleum (Beijing), Beijing, China, 2017.
49. Zhang, J.; Qu, X.; Jin, J.; Ma, K.; Zheng, X. A New Vertical Sweep Efficiency Prediction Model for Water-Flooding Thick Oil Reservoirs. *Spec. Oil Gas Reserv.* **2017**, *24*, 103–106.
50. Lan, Y. Study about the Sweep Coefficient and Oil Displacement Efficiency of Chemical Flooding. Ph.D. Dissertation, Daqing Petroleum Institute, Daqing, China, 2006.
51. Nie, J. Computing Method Research on Sweep Efficiency of Polymer Flooding for Five-Spot Pattern. Master's Dissertation, China University of Geosciences, Beijing, China, 2010.
52. Clermont, J.R. Analysis of incompressible three-dimensional flows using the concept of stream tubes in relation with a transformation of the physical domain. *Rheol. Acta* **1988**, *27*, 357–362. [CrossRef]
53. Bereaux, Y.; Clermont, J.R. The Stream Tube Method and Memory Integral Constitutive Equation Non-newtonian Fluids Using. *Int. Lournal Numer. Methods Fluids* **1995**, *21*, 371–389. [CrossRef]
54. Clermont, J.R.; de La Lande, M.E. Numerical Simulation of Three-Dimensional Duct Flows of Incompressible Fluids by Using the Stream-Tube Method Part I: Newtonian Equation. *Theor. Comput. Fluid Dyn.* **1993**, *4*, 129–149. [CrossRef]
55. Ander, P.; Clermont, J. Experimental and Numerical Study of the swelling of a viscoelastic liquid using the stream tube method and a kinematic singularity approximation. *J. Non-Newton. Fluid* **1990**, *38*, 1–29. [CrossRef]
56. Ramezanpour, M.; Siavashi, M. Application of SiO2–water nanofluid to enhance oil recovery. *J. Therm. Anal. Calorim.* **2019**, *135*, 565–580. [CrossRef]
57. Sun, X.; Alhuraishawy, A.K.; Bai, B.; Wei, M. Combining preformed particle gel and low salinity waterflooding to improve conformance control in fractured reservoirs. *Fuel* **2018**, *221*, 501–512. [CrossRef]

materials

MDPI

Review

Progress in Research and Application of Graphene Aerogel—A Bibliometric Analysis

Bowen Chai [1,2], Wanlin Zhang [3], Yuanyuan Liu [3], Shuang Zhu [1,4,5,*], Zhanjun Gu [1,5,*] and Hao Zhang [3,*]

1 CAS Key Laboratory for Biomedical Effects of Nanomaterials and Nanosafety, Institute of High Energy Physics, Beijing 100049, China
2 School of Science, China University of Geosciences, Beijing 100083, China
3 Aerospace Research Institute of Special Material and Processing Technology, Beijing 100074, China
4 CAS Center for Excellence in Nanoscience, National Center for Nanoscience and Technology of China, Chinese Academy of Sciences, Beijing 100190, China
5 Center of Materials Science and Optoelectronics Engineering, College of Materials Science and Optoelectronic Technology, University of Chinese Academy of Sciences, Beijing 100049, China
* Correspondence: zhus2021@nanoctr.cn (S.Z.); zjgu@ihep.ac.cn (Z.G.); ruizhangh@163.com (H.Z.)

Abstract: In recent years, graphene aerogel (GA) has been widely used as a 3D porous stable network structure material. In order to identify the main research direction of GA, we use the bibliometric method to analyze its hot research fields and applications from the Web of Science database. First, we collected all relevant literature and analyzed its bibliometrics of publication year, country, institution, etc., where we found that China and Chinese Academy of Sciences are the most productive country and institute, respectively. Then, the three hot fields of fabrication, energy storage, and environmental protection are identified and thoroughly discussed. Graphene aerogel composite electrodes have achieved very efficient storage capacity and charge/discharge stability, especially in the field of electrochemical energy storage. Finally, the current challenges and the future development trends are presented in the conclusion. This paper provides a new perspective to explore and promote the related development of GA.

Keywords: graphene aerogel; graphene oxide; composites; nanostructure; fabrication; energy storage; electrode; environmental protection; adsorption

1. Introduction

Graphene was successfully extracted from graphite using mechanical exfoliation by Geim and Novoselov in 2004 [1]. As the first 2D crystalline material composed of single atoms, it is widely used because of its excellent physical and mechanical properties as well as its outstanding electrical conductivity, thermal conductivity, and optical properties [2,3]. At the same time, three-dimensional graphene materials such as graphene aerogel (GA), where graphene sheets are interconnected with each other to retain the unique properties of graphene, has a stable structure that can effectively reduce the problem of greatly reduced physical and chemical properties caused by Π-Π stacking due to the poor dispersion of graphene [4]. Besides, GA is easier to prepare and more efficient in performance compared to carbon nanotube-based 3D architectures [5]. Initially, graphene or graphene oxide was used as precursors, which were formed into graphene hydrogels (GH) by conventional sol-gel molding [6]. Since the internal structure collapses and shrinks in volume under capillary pressure during the normal freeze-drying process thus leading to severely impaired performance, GA is usually formed by supercritical drying to maintain the stability of the internal space [7]. With the development of science and technology, template method and 3D printing are also used to prepare GA with smaller pore size, isotropic, and different types of ordered network structures [8–11]. Meanwhile, a lot of current research has been conducted to achieve the desired performance enhancement through GA compounding with other materials, heteroatom doping, etc. For example, GA is compounded with transition metal nanoparticles to enhance electrical conductivity and maintain stability, or the surface-active sites of GA are increased by doping with N atoms to improve adsorption properties, etc. [12,13]. In terms of functions, GA has excellent properties such as low density, high electrical conductivity, high thermal insulation, large specific surface area, good resilience, and strong adsorption ability, which can effectively be used in important files including resource shortage and environmental pollution. Its high electrical conductivity and stable three-dimensional structure has electrodes in electrochemical energy storage devices for energy applications [14–17], its porous structure and resilience properties enable their application as an adsorbent and solar photocatalytic carrier in environmental remediation [18–20], and its compressibility and sensitivity to gas and liquid for use as sensors [21,22], etc. The emergence of graphene aerogels has greatly facilitated the development and application of graphene-related materials, which can be effectively used in important situations such as resource shortage and environmental pollution. Although GA research has only been developed for more than a decade, it is already feasible to be prepared on an industrial scale and has become one of the popular 3D materials.

Due to the wide range of GA applications, reviews that summarizes its application have been widely reported [23–27]. However, there is still a lack of a work that can provide an overview of the popular research areas from a holistic and objective perspective, which could help researchers to better understand and grasp the key application directions related to GA. Therefore, an objective and comprehensive statistical approach to GA-related research is needed. Based on this, we for the first time use bibliometric analysis in conjunction with the Web of Science repository to summarize and analyze the current published articles and results of GA. We first discuss the current status of research in the field of GA and the extent to which different countries/institutions have contributed to it. Then, according to the objectively obtained keyword mapping, the latest development trend of GA and the research focus including material preparation, energy storage application, and environmental protection application are identified and summarized. This work serves as a focus to show readers the current research hotspots in GA and the feasibility of future development in these three areas. Finally, we discuss the challenges encountered in the current development of GA, as well as the related views on its improvement and the outlook on the application in the future. Our work aims to help readers to grasp a general research status of GA research and further provide insights for guiding future development of this important field.

2. Bibliometric Analysis

2.1. Collection of Data

The data used for this review were retrieved from Web of Science on 4 October 2022. The preliminary results were obtained by advanced search "TS = Graphene Aerogel* not AK = (Foam* or Sponge*)," and then the keywords and abstracts of the retrieved literature were manually filtered to obtain the full WOS core collection of 5705 papers and 234 reviews.

It should be noted that the combination of the word Graphene aerogel first appeared in a paper published by Wang et al. in 2009 [28]. Since then, in many studies on carbon aerogels (CA), many commonalities can be reflected in GA, so they are also included.

2.2. Annual Publications and Countries/Institutes Contributing to GA research

After 13 years of research and development since 2009 (2009–2021), the number of published articles has grown to more than 900 in the last three years, as shown in Figure 1a. Besides, it is still in a steady growth trend, indicating that the field of GA is attracting more and more researchers' attention. Among them, researchers from China contributed the most publications (4295/178,054 citations), followed by the United States (551/36,701 citations). There are also more than 200 publications were Korea (238 publications/8961 citations) and Australia (208 publications/12,672 citations). The criteria for judging contribution need to consider several indicators, such as publication number, citation number, and average citation number at the same time. From this point of view, China provides the highest number of articles while the United States has the highest average number of citations. Furthermore, according to the analysis using VOSviewer, cooperation ties exist between 48 different countries in Figure 1b. In the results of this analysis, China, as the country with the most cooperative relations, cooperates with 43 of them. The next largest country is the United States, which cooperates with 32 of these countries. The rest of the countries also have different collaborations with each other, indicating that extensive collaboration of each country. In addition, when it comes to institute publication, Chinese Academy of Sciences top the list in terms of the number of publications contributed (603) and h-index (96) in Figure 1c. Herein, the h-index reflects the number and level of academic output of researchers. For example, if somebody's h-index is 10, it means that he has 10 articles, and each paper is cited at least 10 times. Lastly, the main disciplines covered in the WOS search are "Chemistry, Materials Science, Science Technology Other Topics, Physics." This phenomenon indicates that the development of research on GA is a joint result of the intersection of multiple disciplines, and most of the research on GA is focused on the properties and applications of the material.

2.3. Research Hotspots

After common bibliometric analysis of GA research performance, we used VOSviewer to analyze the keywords of the articles to determine the hotspots in the GA research field, and selected keywords with more than 20 occurrences. We discarded some keywords like Graphene, Aerogel, Graphene aerogel, etc., which did not show any useful information in Figure 2. The larger the node, the more frequent it appears. The results presented in the figure show that the keywords "Fabrication, Supercapacitor, Adsorption, Performance" appear most frequently. Accordingly, the keywords are subjectively classified into three main research hotspots by Vosviewer. As shown in Figure 2, the green section contains the synthesis, structure, and performance of GA materials. The red section indicates the applications of GA in energy storage, mainly focusing on the applications of GA in electrochemical disciplines such as supercapacitors and electrode materials for battery systems. Finally, the blue section contains GA applications in the environmental protection field, including adsorption and removal of pollutants. In the next chapter, we will focus on each research hotspot of GA.

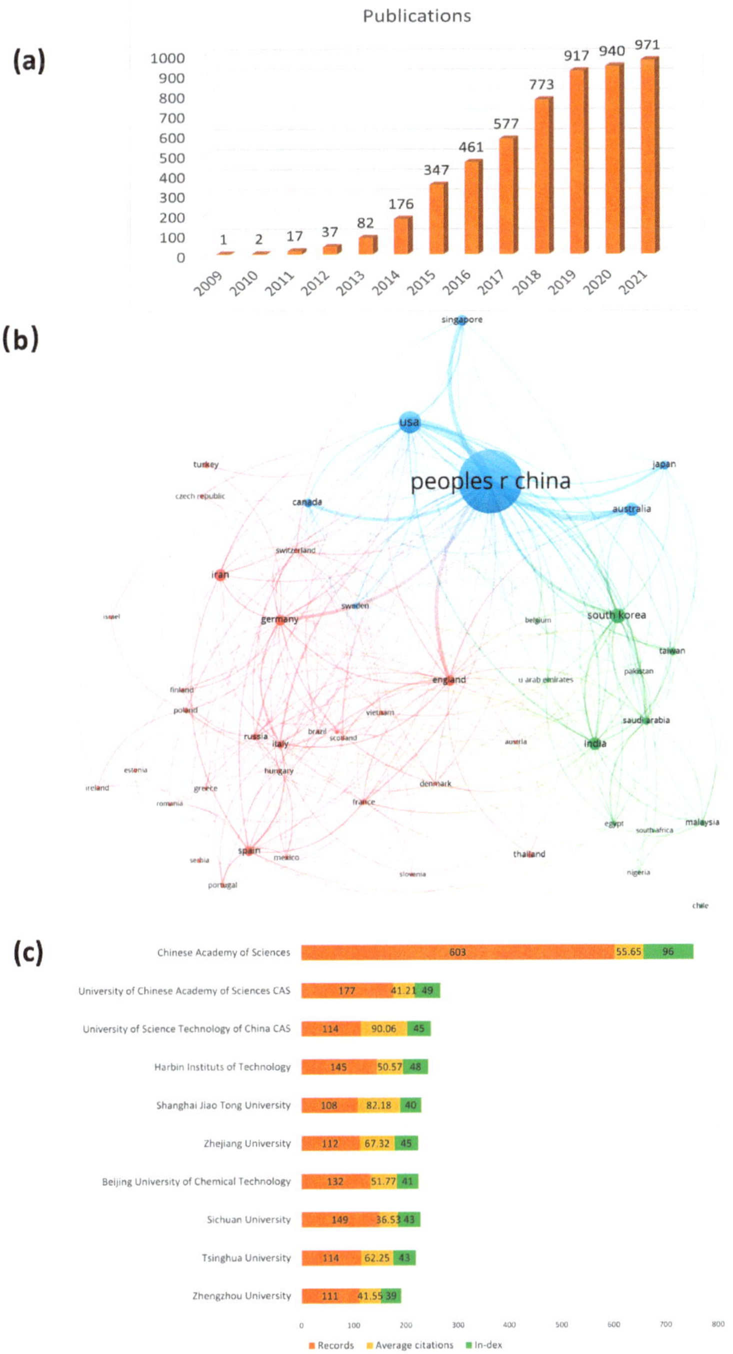

Figure 1. (**a**) Annual number of publications related to GA from 2009 to 2021. (**b**) VOSviewer shows the cooperation between different countries/regions on GA research. (**c**) The most productive institutions based on the total publication.

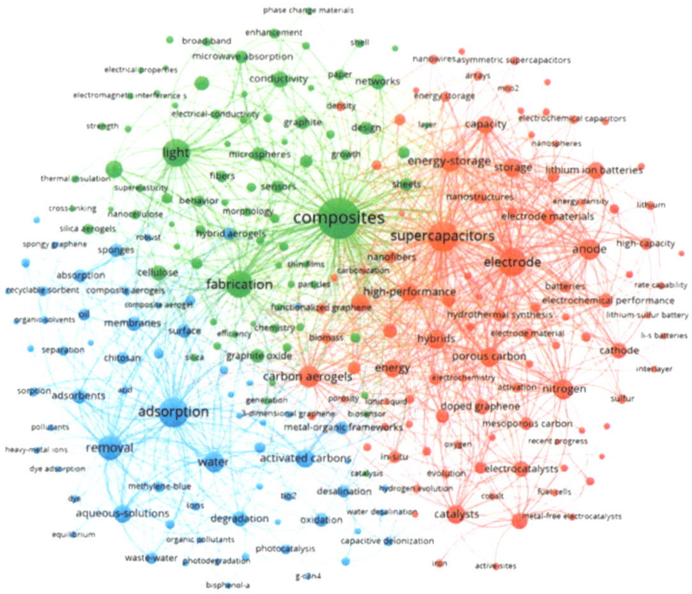

Figure 2. Summary network diagram of GA's bibliometric analysis. Keyword co-occurrence analysis with at least 20 occurrences. The node size represents the occurrence frequency of the keywords. The same color represents relevance.

3. Fabrication of Graphene Aerogels

The fabrication of GAs are mainly classified according to the nodes presented in the green clusters. By employing different methods, diverse properties of GA could be excavated and obtained. At present, the main fabrication methods involved in the formation of graphene aerogels include self-assembly methods, template methods, and 3D printing methods.

3.1. Self-Assembly Method for GA Fabrication

3.1.1. In-Situ Assembly Method

The in-situ assembly method is usually divided into two types: hydrothermal reductive assembly method and chemical reductive assembly method. Graphene oxide (GO) is used as a precursor, which is uniformly and stably dispersed in solution by the electrostatic repulsion of the surface negatively charged functional groups such as epoxy groups and carboxyl groups, which are shown in [29,30] Figure 3a. When the surface functional groups are removed by hydrothermal conditions or chemical reducing agents to eliminate the electrostatic repulsion, the GO layer sheets aggregate with each other. The graphene hydrogels are stacked into three-dimensional structures by the combined effects of van der Waals forces, dipoles, and hydrophilic/hydrophobic interactions, and finally freeze-dried to obtain GA [31].

Hydrothermal assembly utilizes high temperature and pressure conditions in closed conditions where graphene sheets in solution associate with each other to form a stable three-dimensional structure [7,30,32] Figure 3b,c. In this process, the pH value of the solution and the hydrothermal reaction time have a large influence on the structure [30]. First, the oxygen-containing groups on the GO surface are subject to different degrees of ionization by different pH values, which come to influence the final material properties. At low pH (pH < 3) levels, the surface of the synthesized hydrogel adsorbent is positively charged and has a strong adsorption effect on anions. On the contrary, when the pH increase leads to a decrease in the H^+ concentration in the solution, the adsorbent is less protonated, causing a weakening of the adsorption capacity [33,34]. Second, the hydrothermal time affects the degree of GO reduction. Generally, the longer the hydrothermal time, the better the degree of reduction leads to the stronger performance of the formed graphene hydrogel. However, a reduction time that is too long leads to the increase in self-assembled cross-linked sites in the 3D structure and the increase in density due to the decrease in specific surface area and pore volume [35]. Therefore, ensuring a proper hydrothermal time in a low-pH environment can be expected to make the structure and properties of GA reach optimal values.

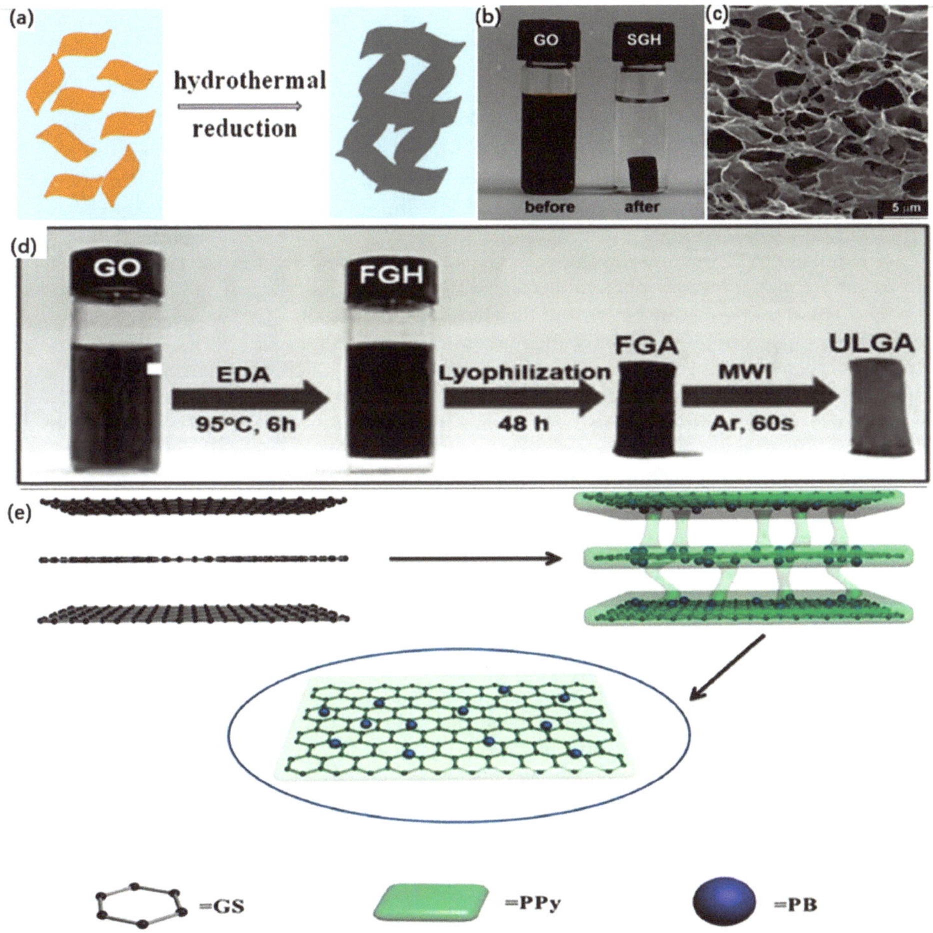

Figure 3. (**a**) The proposed formation mechanism for SGH; (**b**) Photos of GO aqueous dispersion after 12 h of hydrothermal assembly; (**c**) Photos of the internal microstructure of SGH under SEM [30]; (**d**) Illustration of the fabrication process of the ultralight graphene aerogel (ULGA) [7]; (**e**) Schematic diagram of polypyrrole (PPy) used as a crosslinking agent for GA assembly [36].

Compared to the harsh conditions of hydrothermal assembly, the synthesis of graphene hydrogels using chemical reductive assembly requires the addition of reducing agents (commonly used agents are sodium ascorbate, hydroquinone, ethylenediamine, etc.) to chemically reduce GO [31,37,38], which is more convenient to operate and thus widely adopted. Sheng et al. used sodium ascorbate to reduce GO to promote the three-dimensional assembly of graphene sheets, effectively removing the oxygen-containing functional groups between the carbon atom layers and providing a new environmentally friendly pathway for the reductive assembly of GA [31]. In addition, the addition of the reducing agent ethylenediamine (EDA) to the solution of GO resulted in the assembly of monolayers of functionalized graphene hydrogels. The EDA molecules bonded on at the same time reduced the stacking of the lamellae and there was almost no volume change before and after the treatment. the overall structure inside the GA was not disrupted and the integrity was preserved [39]. However, the addition of reducing agents introduces non-carbon impurities and some surface functional groups, so the amount of reducing agents needs to be strictly controlled or de-hybridized [38]. For example, Hu et al. used microwave irradiation to achieve the removal of surface impurities based on the reduction in graphene oxide solutions using EDA, thus reducing the adverse effects on the properties of 3D graphene materials [7], as shown in Figure 3d.

3.1.2. Cross-Linker-Induced Self-Assembly Method

The prominent feature of cross-linker-induced self-assembly lies in the addition of chemical cross-linkers during the synthesis process, and the surfaces where GO and cross-linkers are bound form bond bridges could interlink into a three-dimensional network structure [36,40]. The graphene hydrogel sheets synthesized with the participation of crosslinkers have less interlayer resistance and stronger structure than the in-situ self-assembly method where physical linkage occurs [41]. Simultaneously, the physical and mechanical properties are better, and the electrical conductivity is enhanced by about two orders of magnitude [6]. Usually, the main metal ions (such as Ni^+, Na^+, Co^{3+}, etc.) and organic compounds containing functional groups (mainly include acrylic acid, phenolic resin, and other polyacid esters) can be able to act as cross-linking agents to promote gel formation. Graphene aerogel materials with different properties can be achieved by adding different kinds of cross-linking agents. For example, Cong et al. reported the use of α-$Fe(OH)_3$ nanorods with magnetic Fe and other metal oxides to induce the assembly of graphene sheets into three-dimensional graphene hydrogels used as adsorbents for water purification. The combination of electrostatic attraction, ion exchange between metal hydroxides and heavy ions, surface complexation, and oxygen-containing functional groups showed a strong adsorption capacity [41]. In addition, Wang et al. added metal ions and resorcinol-formaldehyde (RF) mixture to GO solution to generate phenolic resin as a cross-linking agent to form the gel. During the cross-linking process, metal ions interact with GO through bonding forces and promote the protonation of functional groups contained on the surface of the phenolic resin by lowering the pH of the solution to produce hydrogen bonds and thus the mechanical properties of the gel [42]. Different content (0–4 wt%) of RF will form different morphologies of 3D structures. The lower the RF content, the the more the subsequent aerogel becomes thinner, with smaller pores between the sheets and stronger electrical conductivity. Conversely, when RF content is increased, the synthetic aerogel has greater porosity and better adsorption performance [43], and the addition of cross-linker content also has an important effect on the final properties of the material.

3.2. *Template Method for GA Fabrication*

3.2.1. Template-Assisted Chemical Vapor Deposition

Template-assisted chemical vapor deposition (CVD) is another widely used method to catalyze the synthesis of three-dimensional GA [9]. The GA synthesized using the CVD template method possess many advantages, such as fewer stacking interfaces between the flake layers, more complete structure, controlled and uniform pore size, and better electrical conductivity, exhibiting properties closer to those of graphene [44]. The steps of using CVD template method include the following: at the beginning, preparing a foam Ni metal substrate, then growing graphene catalytically on the substrate by CVD, and finally removing the substrate using etchant to obtain 3D graphene. The preparation of 3D graphene by CVD template method was first reported by Chen et al. Carbon generated by methane decomposition was precipitated at 1000 °C on a foam Ni template used as a 3D graphene growth, and the constant folding of graphene was shaped in three dimensions using the different thermal expansion coefficients of Ni and graphene. Then hot HCl (or FeCl) is used to etch the substrate while hot acetone is used to remove the polymethylmethacrylate that prevents the collapse of the graphene network to obtain the final three-dimensional graphene structure [9], as shown in Figure 4a. In addition to methane, Cao et al. and Chen et al. use ethanol as the carbon source, respectively, making the experimental environment safer [45,46], as shown in Figure 4b,c. What's more, the choice of etchant is quite important. Pettes et al. used $Fe(NO_3)_3$ and $(NH_4)_2S_2O_8$ to remove the foam Ni, which shows gentler and less bubbles during etching. As a result, better structural integrity was obtained with better and less defects [47]. Finally, the substrate plays a considerable role as the support structure and catalyst for graphene aerogel forming. The common ones are mainly foam Cu in addition to foam Ni. Kim et al. used foam Cu to prepare graphene aerogels exhibiting good electrical conductivity, but it exhibited poor mechanical properties [48]. In this regard, Ma et al. synthesized graphene aerogel material with the special layered configuration by high temperature CVD method, which can disperse external pressure while having good resilience properties [49].

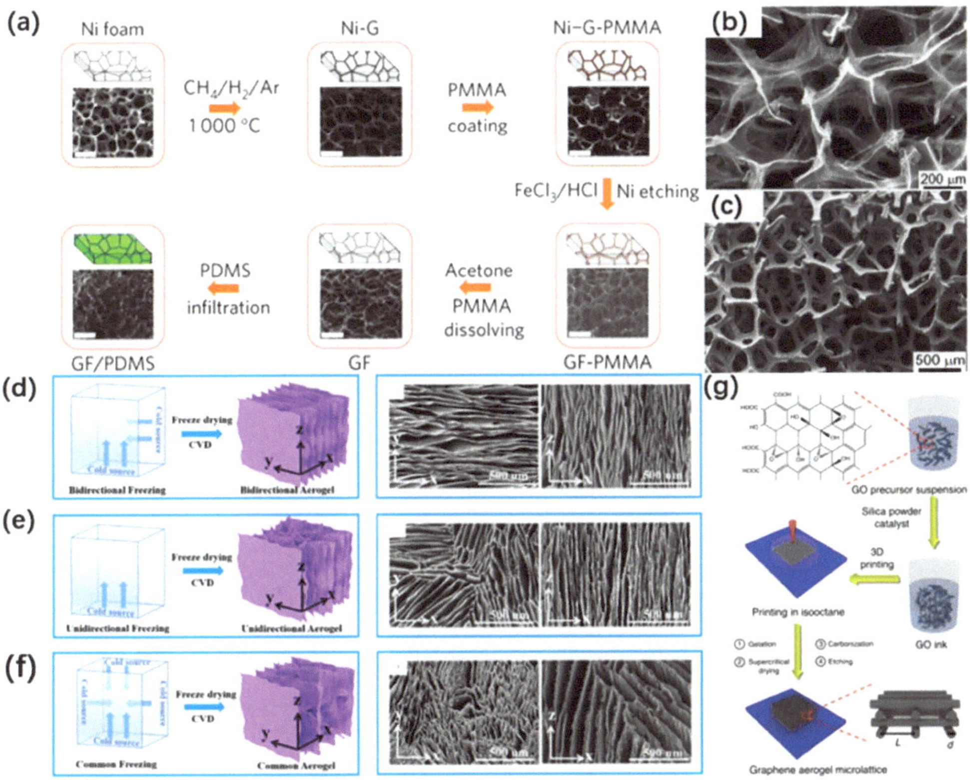

Figure 4. (**a**) CVD growth of 3D graphene using nickel foam as a template [9]; (**b**). SEM image of 3D graphene network grown on Ni foam after CVD; (**c**) SEM image of 3D graphene network after removing Ni foam [46]; (**d**) Schematic diagram and SEM image of GA prepared by two-way freezing casting; (**e**) Schematic diagram and SEM image of GA prepared by unidirectional freezing casting; (**f**) Schematic diagram and SEM image of GA prepared by freezing casting and common freezing casting [50]; (**g**) Schematic diagram of 3D printing GA process [51].

3.2.2. Ice Template Method

Ice templates, also known as freeze casting, is another general method for the preparation of inorganic materials. It mainly refers to the solid phase (ice) generated by the phase separation that occurs during the solidification of the solution as a template [8]. As the temperature rises, the GO in the template is expelled with the sublimation of ice crystals to obtain the desired structural framework. Freeze-drying is also divided into different kinds depending on the method. Generally speaking, the faster the temperature drops, the more ice crystal nuclei are formed, which results in a uniform and small pore size of the sample. On the contrary, if the temperature drops slowly, the ice crystal nuclei near the low-temperature side form and gradually spread to other regions, generating large ice crystals and resulting in a more porous material [52]. In addition, when the reduction level of GO surface is low, the pore wall of the prepared GA is wrinkled and shrunken, and the internal GO lamellae are stacked in a disorderly manner. As the reduction level increases, the pore walls of GA are gradually flattened, and the internal GO lamellae are stacked in a tight and orderly manner [53]. Three main methods are commonly used: unidirectional freezing drying (UDF), non-directional freezing drying (NDF), and air freezing drying (AF) [26]. Firstly, ice crystals with anisotropy grow along the temperature gradient [54], and impurities can be expelled after a certain degree of freezing as the temperature changes [50]. Thus, the aerogel prepared by UFD is purer and the final formed GA has a better performance in terms of electrical and mechanical properties. Secondly, NDF does not grow according to the temperature gradient, so it is better for the overall homogeneity of the material. Finally, UFD and NDF both have smaller pore size and larger porosity for GA prepared under vacuum, so the overall adsorption capacity of UDF and NDF is higher than that of AF [55].

3.3. 3D Printing Methods

3D printing, as a new and unique preparation method, has obvious advantages in GA field [51], as shown in Figure 4g. It not only can the structure be designed according to the desired properties, but the homogeneity of the material can be guaranteed to a greater extent [56]. Compared to traditional GA preparation methods, 3D printing allows the design and processing of aerogels with sizes ranging from tens to hundreds of microns, addressing the key drawbacks of traditional sol-gel and mechanical processing methods. It also uses its controlled tuning to enable nano-scale GA design and precise control. This approach significantly shortens and improves the process from base chemicals or nanomaterials to useful device components, while reducing production costs [57]. This method is mainly based on the flow of GO ink material with viscosity in a certain direction under pressure and the precise deposition of a three-dimensional network structure [15]. Therefore, GO inks play a rather important role in the preparation of GA for 3D printing. Since a low concentration of GO ink does not have the rheological properties required for printing, a high concentration (10–20 mg/mL) of GO ink is beneficial to obtain high viscosity GA. In addition, adding fumed silica powder increases the viscosity of the ink, making it more adaptable for printing [51]. After supercritical drying and carbonization, the residual silica impurities can be removed [32]. Meanwhile, due to the good compatibility of 3D printing, more research on the combination of multiple preparation methods turns out to be more and more. Jiang et al. reported a cross-linking of GO solution into a viscous ink for printing by adding calcium chloride cross-linking agent [58]. Furthermore, Zhang et al. will use nozzle extrusion of GO ink to avoid drying, while using 3D printing and freeze casting methods for preparation. The advantages of both methods were combined to obtain GA with superelasticity and high electrical conductivity [59]. Although only seven years old, the development of 3D printed graphene aerogels has been effective in overcoming the limitations of current aerogel materials and has shown great potential in many applications and new market opportunities.

Synthesis of GA is an important part of GA-related research, and different synthesis methods are used to obtain materials with specified properties to meet the needs of researchers. Therefore, as the first step in GA research, synthesis occupies a fundamental and important role.

4. GA in the Field of Energy Storage

According to the red clusters, we can see that the application of graphene aerogel in the whole energy storage field is mainly focused on functions as an electrode material or a framework for electrode material, as evidenced by energy-storage, electrode, supercapacitors, and batteries. The main research hotspots include battery electrodes and supercapacitor electrodes [60]. For battery electrodes, although there are many other metal battery types such as sodium ion batteries, zinc ion batteries, fuel cells and so on, the most widely studied types are lithium batteries [11]. GA is useful as an electrode material not only because it has excellent electrical conductivity and stable structure, but because it is easy to compound other materials to meet the performance requirements [14].

4.1. Lithium Battery Electrode

Lithium battery is the most widely used battery system in the field of electrochemistry. According to the types of electrodes, lithium batteries are mainly divided into two kinds of lithium-ion batteries (LiBs) and lithium metal batteries [61], as shown in Figure 5a. Due to the good electrical conductivity and energy storage performance of the three-dimensional network porous structure of GA, they are widely used in the negative/positive electrodes of lithium-ion battery, the positive electrode of lithium-sulfur battery, and the lithium-oxygen battery. For example, 3D graphene aerogel can replace graphite anode to solve the defect of low capacity of lithium-ion battery. Meanwhile, the uniform pore distribution and large layer spacing of graphene aerogel can provide more ion channels to improve the efficiency of battery charging and discharging. In addition, the stable structure of graphene aerogel is helpful to solve the problem of service life degradation caused by volume change during the charging and discharging of battery cathode materials. The application of graphene aerogel as electrode materials has greatly improved the performance and safety of lithium batteries.

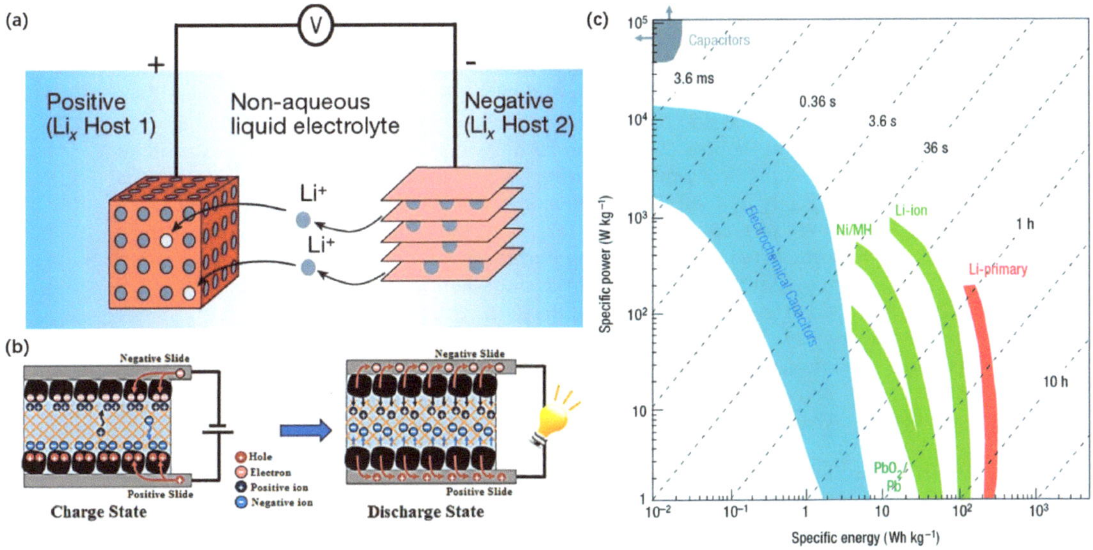

Figure 5. (**a**) Schematic diagram of a rechargeable lithium-ion battery [61]; (**b**) Charging and discharging states of supercapacitors [24]; (**c**) Specific power density against specific energy density, also called a Ragone plot, for various electrical energy storage devices [62].

4.1.1. The LiBs Anode

The anode plays the role of energy storage and release by absorbing or shedding ions during battery charging or discharging [61], and is usually a material with excellent electrical conductivity and ion storage capacity. The reactions involved in charging/discharging are as follows:

Cathode: $Li \leftrightharpoons Li^{+} + e^{-}$
Anode: $6C + XLi^{+} + Xe^{-} \leftrightharpoons LixC_{6}$

The most prominent anode materials for lithium-ion batteries include graphite (A mainly carbon-based material) and lithium titanate [63]. Graphene has attracted much attention as a newly discovered material in the last decades. With a honeycomb network structure, good electrical and thermal conductivity, and high chemical stability, graphene is an ideal electrode material [64], and its specific capacity (740 mAh/g) and electrical conductivity (6000 S/cm) are much higher than those of graphite electrodes with excellent energy-storage capacity. However, the battery life is limited due to the easy stacking of layers during charge/discharge cycles resulting in degraded cycling performance [65]. Therefore, using structurally stable three-dimensional graphene aerogel materials as electrodes is an effective method to solve the stacking problem.

Graphene aerogel-containing anode materials are classified into two categories based on the involvement of GA: pure graphene aerogel anodes and hybrid graphene aerogel composite anodes. Firstly, in the fabrication of pure graphene aerogel anode materials, a good structural design can enhance the electrode performance while ensuring a stable structure. In order to prevent the overlapping of layers as in the case of two-dimensional graphene, graphene oxide can be hydrothermally reduced and annealed under treated filter paper to obtain three-dimensional graphene aerogel electrodes. During the synthesis process, the filter paper acts like a rope to weave the dispersed graphene oxide flakes into a well-connected porous 3DGA. At the same time, the graphene flakes covered with folds are cross-linked with each other to form a porous structure for easy lithium ion embedding [66]. This electrode material provides a stable structure while possessing a good charge/discharge multiplicity. Besides, the choice of pore size also has a large impact on the structure. Generally, structures with microporous or mesoporous pores have a larger specific surface area and more electrochemically active reversible lithium storage sites at the electrode. The interconnection of internal active sites reduces the lithium ion embedding or de-embedding distance and accelerates the charge/discharge cycle. For example, Ren et al. fabricated hierarchical porous graphene aerogels (HPGA) with uniform and tunable mesopores (21 nm and 53 nm) on graphene nanosheets. Attributed to the mesopores capable of providing diffusion pathways for Li ions, HPGA exhibited high specific capacity and excellent cycling stability, i.e., 100 cycles with a capacity of 1100 mAh/g at 0.1 A/g current density [67]. In addition, the electrode capacity can be improved to some extent by introducing surface defects to the graphene aerogel electrode. It is shown that the increase in GA surface defects leads to the increase in active sites to a certain extent, which increases the storage sites for lithium ions and thus increases the capacity. When the defects increase to a certain number, the performance reaches a maximum value of reversible capacity 1430 mAh/g (when the current density is 0.1 A/g) [68]. To increase the GA surface defects, the first step can be taken by introducing heteroatoms. The doping of heteroatoms causes an increase in defects on the surface of graphene materials and the electrochemical activity of the material is improved due to the difference in electronegativity between heteroatoms and carbon atoms. At the same time, the electrochemical activity of the material is improved due to the difference in electronegativity between heteroatoms and carbon atoms. And the lone pair of electrons of heteroatoms as carriers can promote electron migration and ion absorption by changing the spatial structure of π-system on graphene, changing the chemical reactivity of graphene and van der Waals forces between graphene nanosheets [69]. For example, a 3D self-supporting graphene aerogel with N doping and well-controlled nanopore size for lithium-ion battery anode was reported by Yi et al. The N-doped graphene aerogel has a large number of edge sites and defects, and can provide more active sites for ion adsorption, thus enabling fast ion diffusion and charge transfer, revealing excellent energy storage capacity (1147 mAh/g at 0.1 A/g) [70]. At the same time, the dual-doped material shows stronger performance due to the synergistic effect between atoms. This S/N dual-doped graphene aerogels (SNGAs) reversible capacity reaches 1109.8 mAh/g (400 cycles at a current density of 0.8 A/g) when the content of 1.5 wt.% N and 15.11 wt.% S are co-doped in graphene aerogels [71]. Furthermore, after using high-energy photoion (3.8 MeV He) bombardment used to introduce lattice defects in the 3D interconnected network of graphene aerogels (GA), the removal or alteration of the ion beam for the functional groups on the material surface was found. This reveals the complex evolution of defects and the formation of vacancies for the enhancement of Coulomb efficiency and reversibility of graphene [69].

The use of graphene aerogels hybridized with common anode materials including metal/alloys, metal compounds, and silicon-based materials to form composites for use on lithium-ion battery electrodes is another hot research topic in recent years. These materials have good theoretical performance when used as anode alone, but suffer from various shortcomings during use. For example, these materials in ion batteries metal/alloys can charge and discharge by achieving an alloying reaction or a reverse alloying reaction, both of which theoretically have a high specific capacity [72]. However, when many charge/discharge cycles occur, the electrodes suffer from volume expansion leading to reversible capacity drop as well as safety issues. Using the structure of graphene aerogel to stabilize the nanometallic particles is a feasible method at this stage. For instance, Qin et al. designed a 3D graphene-wrapped metal Sn nanoparticles, which can not only effectively buffer the volume expansion, but provide a conductive network to promote electron transfer and improve the charge/discharge multiplicity of the battery [73]. Meanwhile, the cobalt nanoparticles produced by the decomposition of cobalt salt can react with the carbon atoms of graphene to form nanopores to constitute a multilevel porous structure. This structure facilitates ion transfer and electrochemical reactions to enhance cell performance while occupying less space through mechanical compression into thin film electrodes [74], which provides a new enhancement for the application of Li-ion batteries in portable electronics. Similar to metal/alloy electrodes, metal compound electrodes also suffer from lack of conductive network and ion transport channels, crushing or aggregation of materials caused by volume expansion, and contraction leading to batteries damage.

Metal compound electrodes, mainly consisting of two types of metal oxide (MO) electrodes and metal sulfide (MS) electrodes, utilize the Faraday reaction of metal ions to store charge and thus have the potential to achieve ultra-high energy density and large theoretical capacity (500–1000 mAh/g) [75]. It also has advantages such as low preparation cost and environmental friendliness. However, both electrodes lack conductive networks and ion transport channels, and both are prone to volume expansion and contraction-induced material crushing or aggregation during charging and discharging, leading to battery damage [76]. The conductive network of graphene aerogel and the stable 3D structure can give a good solution for the electrodes of metal compounds. GA encapsulates Fe-Co oxide to obtain Fe-Co oxide@Graphene aerogel composite, in which metal oxides are encapsulated into the GA network to prevent stacking between graphene nanosheets, and the good mechanical toughness of graphene aerogel prevents agglomeration between metal oxides. The graphene aerogel with good mechanical toughness also prevents agglomeration between the metal oxides. Importantly, when the material was compared with pure Fe-Co oxide electrodes at the same time, the former was far better than the latter in terms of electrochemical performance during use and stability of the material after multiple cycles [77]. The three-dimensional core-shell MoO_2@few-layered MoS_2 covered by S-doped graphene aerogel (MO/MS/SGA) prepared by Zhang et al. can not only effectively reduce the aggregation of MoO_2, but prevent the MoS_2 stacking. The graphene aerogel structure acts as a flexible matrix and conductive network while effectively relieving the volume buffering of Mo. The performance is good: the capacity is 683 mAh/g when cycled 100 times (current density of 0.1 A/g) [78]. The interplay between the nano-transition metal compounds and graphene aerogel in the composition of the composite is the basis for the excellent performance of the negative electrode.

Silicon is another material system with the highest theoretical specific capacity in the negative electrode of lithium-ion batteries, which is considered as one of the substitutes for carbon materials in commercial batteries [79]. However, silicon-based materials have limitations, which mainly include low electrical conductivity at room temperature, low first efficiency during charging or discharging, unstable SEI film, and poor cycling stability [80]. So far, effective progress has been made in some aspects. For example, a simple freeze-drying strategy was utilized to immobilize silicon nanoparticles on graphene flakes as anode material. The graphene aerogel porous structure acts as a mechanical support and an embedded current collector to ensure that all Si nanoparticles maintain good contact and electrochemical activity during cycling [81]. Peng et al. reported a nascent 3D SiNPs@rGO_1/rGO_2 composite consisting of a 3D conducting graphene backbone and a flexible pre-wrapped graphene coating as well. The special structure of graphene aerogel wrapped with silicon particles is utilized, which can improve the electrical conductivity and has a stable structural buffering volume effect. The bilayer graphene layer Is able to separate the electrolyte and silicon, thus generating a thin and stable SEI film to avoid reversible capacity loss [82].

4.1.2. The Li-S Battery Electrodes

Li-S batteries are expected to replace lithium metal batteries as a new generation of efficient secondary batteries due to their high theoretical specific capacity (1675 mAh/g) and high energy density (2600 Wh/kg) [83]. Besides, this battery system embodies many advantages such as low toxicity, environmental protection, and low cost, and scientists have been enthusiastically researching it. The reactions involved in charging/discharging are as follows:

Cathode: $S_8 + 16Li^+ + 16e^- \leftrightharpoons 8Li_2S$
Anode: $Li \leftrightharpoons Li^+ + e^-$

However, the current development path has been hampered by some obstacles. The main problems include the following: (1) the poor electrical conductivity of cathode sulfur and discharge product Li_2S; (2) the polysulfide (LiPS) produced during charging and discharging is easily dissolved in the electrolyte to form a "shuttle effect" to reduce the battery performance; (3) the large change in electrode volume during cycling, which causes damage to the structure and affects the battery life [84,85]. Graphene aerogel is a good way to solve the above problems as the positive electrode carrier structure. Firstly, graphene aerogel adopts a stable three-dimensional conductive network structure that can effectively promote the electrical conductivity of the cathode as well as prevent the volume expansion of the sulfur element during charging and discharging cycles. Secondly, the porous layered structure not only forms a channel for ion transfer, but also can be used to temporarily separate the LiPS generationed in the electrolyte and increase the sulfur loading to counteract the "dead weight" phenomenon. Thirdly, when sulfur is attached to the aerogel surface, it can provide a larger contact space for the reaction and improve the efficiency for charging and discharging [86]. For example, a GO/CNT Aerogel was proposed as a sulfur body (G/CNT-S) cathode. Thanks to the 3D interconnected porous network structure, the material presents a large surface area (336 m^2/g) and a high electrical conductivity (67 S/m). The porosity is also able to act as a channel for ion diffusion and absorption of LiPS [87]. Meanwhile, the doping of Li-S on graphene aerogel by liquid permeation-evaporation coating as a support cathode can not only avoid the safety problems that occur during the use of lithium metal electrodes, but also the uniformly coated Li-S has a shorter ion/electron transfer pathway on the 3D graphene structure, which can effectively reduce the energy potential barrier [88].

The use of graphene aerogel-encapsulated nano MS/MO as the positive electrode material and the layered porous structure can promote ion diffusion and electron transfer, and also treats LiPS for performance enhancement by the dual effect of adsorption through van der Waals interactions and catalytic conversion of LiPS by MS/MO_2 SnS_2 nanoparticles embedded in graphene aerogel acting as sulfur hosts to form electrodes SnS_2-Nano Dots@GA, which can effectively ensure a small amount of SnS_2 in graphene aerogel with uniform deposition of SnS_2 nanoparticles while still maintaining a high positive sulfur content (75 wt%). Using the adsorption ability of 3D graphene and SnS_2 on LiPS, the material showed chemical stability as well as excellent performance with a capacity of up to 1016 mAh/g after 300 cycles [89]. In addition, nanosheets of other metal sulfides such as MoS_2 with abundant defects wrapped by graphene aerogels also exhibit adsorption properties on LiPS and act as catalysts to accelerate their redox reactions [90]. Similarly, metal oxides, which are also polar nanomaterials, can provide polar-rich active sites for polysulfides as well as enhance the conductivity of electrodes. Meng et al. reported a 3D ordered ternary Fe_3O_4/porous carbon/graphene aerogel (Fe_3O_4/NC/GA). The large specific surface area and porous structure of the graphene aerogel provided the adsorption capacity for LiPS, and the highly conductive Fe_3O_4 embedded in the nanowafer could play a catalytic role to accelerate the conversion of LiPS [91]. Tan et al. used MnO_2-decorated graphene aerogels in the separator for the "shuttle effect" of polysulfides, while the cathode obtained in combination with I/N-doped graphene aerogels was able to effectively immobilize and capture LiPS with long term stability—no sulfite was detected in 200 cycles substances, with intact electrode performance [92].

4.1.3. The Li-O_2 Battery Electrode

As a new electrochemical energy storage device proposed in recent years, Li-O_2 batteries have received widespread attention for their high energy density as well as low cost [84]. Since the negative electrode material is still lithium metal, the main problem for the positive electrode of this battery is how to speed up the oxygen evolution reaction (OER) and oxygen reduction reaction (ORR) that occur during charging and discharging, respectively, the fast reaction rate will make the electrode produce large energy density and high efficiency [93]. The reactions involved in charging/discharging are as follows:

Cathode: $O_2 + 2H_2O + 4e^- \leftrightharpoons 4OH^-$
Anode: $Li \leftrightharpoons Li + e^-$

Many problems such as low cell cycle efficiency and slow reaction kinetics due to the poor conductivity of the reaction product Li_2O_2 need to find a suitable method to solve. The cathode of Li-oxygen battery, as the main research object, is the support for the reaction sites of the three phases (Li^+, O_2 and catalyst). Therefore, the following advantages are required: (1) good electrical conductivity (to facilitate electron transfer); (2) good gas diffusivity (to ensure oxygen concentration at the reaction active site); (3) porous structure, large pore capacity, and high specific surface area (to facilitate the deposition of discharge products); and (4) stable non-decomposition during charging and discharging. Based on these requirements, GA is widely used as the cathode material for lithium-oxygen batteries because of its good electrical conductivity, porosity, light weight, and relatively easy preparation. The special three-dimensional conductive network structure and high specific surface area of GA can accelerate OER and ORR, and its synthesis does not involve other superfluous substances such as adhesive, which can cause further damage to the battery performance [94], so GA has outstanding advantages as the main body of the cathode material. This is mainly reflected in the high specific surface area that enables more active sites, while the high number of pores facilitates electrolyte and oxygen transport. The excellent electrical conductivity of the network structure allows a faster rate of electron transfer as well as facilitating the kinetics of the reaction process [95,96]. At the same time, the porous structure can be used to accommodate Li_2O_2 to reduce deposition and increase the capacity of Li-O_x batteries. For example, Zhao et al. used carbon nanocages prepared from polystyrene sphere@polydopamine (PS@PDA) and then hydrothermally self-assembled coupled into a 3D porous N-doped graphene aerogel (NPGA). The material has a crumpled and loose structure that prevents the stacking of graphene sheets thereby exposing more reactive sites. Compared to the GA without the carbon nanocage structure,

NPGA acts similarly as a catalyst, increasing the ORR onset potential and decreasing the OER onset potential with a faster reaction rate [97]. Not only that, but the team also prepared a stable carbon nanofiber graphene aerogel film (G-CNF film) as the main material for the electrode of Li-O_x battery. As a two-in-one material for positive and negative electrodes, it can not only reduce the formation of negative Li dendrites to increase the battery life, but mitigate the bulk effect to increase the battery capacity and ensure the cycle efficiency [95]. Therefore, in the research related to the application of GA in battery electrode materials, whether as the main framework or composite with other materials, its good conductivity or stable three-dimensional support structure plays a rather critical role.

4.2. Supercapacitor Electrodes

Supercapacitor, also known as Ultracapacitor, is an energy storage device with a large capacity [98]. According to the energy storage mechanism, they are mainly divided into two types: electrochemical double layer capacitors (EDLC), which use the absorption and desorption of ions on the electrode surface to form a double layer of charge, and pseudocapacitors, which generate charge through a Faraday process on the electrode surface [24,99,100], as shown in Figure 5b. As an important dependency of a supercapacitor device, the electrode material is crucial to its performance. The specific surface area and porosity of electrode materials are two important factors to measure the performance of double layer capacitors, but few carbon materials can satisfy both. In particular, redox pseudocapacitors have attracted the attention of researchers due to their ability to provide higher capacity. However, the materials are limited by their own properties, such as poor electrical conductivity, and the long redox reactions during the charging and discharging process tend to cause damage to the material structure, resulting in lower energy density and cycling stability performance [101]. As 3D stable porous carbon materials, graphene aerogels have high electrical conductivity and strong mechanical strength, which provides a fast and convenient channel for the storage and transfer of electrons in EDLC, improving capacitance and efficiency [102]. At the same time, graphene aerogels with high specific surface area and nano-effects in pseudocapacitors can also be effective for traditional materials such as metal oxides or hydroxides with low energy density and cyclic stability [16].

4.2.1. Electrochemical Double Layer Capacitor Electrodes

The main reason for that graphene aerogels can be used as electrode materials in double-layer capacitors is due to their good electrical conductivity and structural properties [103]. When the electrode is immersed in the electrolyte, the structure has a catalytic structure of mesopores and micropores that facilitates the formation of EDLC double layer capacitors [104]. At the same time, the surface macropores are filled while acting as buffer channels for ion diffusion, shortening the ion diffusion time to facilitate ion storage and transfer in the electrolyte [105]. As a representative example, Zhu et al. used direct-ink writing (DIW) to prepare a compressible graphene aerogel network lattice. The GO ink printed graphene aerogel electrodes doped with optimised graphene nanoplatelets have almost constant large specific surface area and lower resistance compared to normal 3D printing to prepare graphene aerogels. The orderly arrangement of the large pore structure promotes the transfer of material and reduces the diffusion resistance in the structure, thus ensuring that the electrode has excellent rate capability and electrochemical stability [106]. At this point, the GA Liquid Crystals (LCs) prepared by Xu et al. have a unique "porous core-dense shell" structure that gives the material good tensile and compressive strength, with a maximum compressive modulus of 3.3 Mpa and can withstand more than 2000 times its own weight without damage. The conductivity is also considerably higher than that of ordinary GA (4.9×10^3 S/m) and remains stable after 100 bending cycles [60]. While in practical application, microporous channels may reduce the rate of ion diffusion in the electrolyte at high loading rates, and the longer the pipe, the slower the diffusion [107]. Therefore, when designing graphene aerogel electrodes, attention should be paid to the length of the pore channels. In addition, Fan et al. reported a unique sheet-like vortex-connected structure of graphene-carbon nanotube aerogel (GCA) as a conducting framework for high theoretical capacity nickel hydroxide. This material improves the performance of the electrode by maximising the exposure of active sites through a convoluted interconnected nanostructure that hinders the stacking of graphene sheets and the generation of mesopores [108].

Graphene aerogel-related composites also offer superior performance when used as electrochemical double-layer capacitor electrode materials. Typically, the electrical conductivity and mechanical stability of metal oxides can be improved when combined with 3D graphene [109]. Tian et al. synthesized a composite aerogel with a three-dimensional layered porous structure by immobilizing MnO_2 on RGO, which was also effective in controlling graphene sheet agglomeration and expanding the electrode capacitance. At a current density of 1 A/g, it exhibited a high specific capacitance of 645 F/g [110]. However, due to its insufficient capacitance retention, Chen et al. further used NiO to replace MnO_2 for the composite with GA. In the course of performing cyclic stability tests, it was found that the specific capacitance of this NiO/GA material remained almost constant and had excellent stability [109]. Composites typically have higher electrical capacity and electrochemical stability than pure GA, which is very important for EDLC, enhancing the efficiency and lifetime of supercapacitors

4.2.2. Pseudocapacitor Electrodes

Compared to EDLC, pseudocapacitor electrode materials transfer electrons and store energy through redox reactions between the electrode surface and the electrolyte, resulting in greater adsorption and thus greater capacitance [111]. Since the reactivity of the electrode surface has a large impact on the pseudocapacitor, it is a good way to improve the performance by improving the active sites on the electrode surface [112]. Three-dimensional graphene has a large specific surface area and can be doped with eroatoms (N, S, B, etc.) to be able to present more active sites for surface reactions and more effectively improve the wettability of electrodes and electrolytes to promote electron transfer and thus increase the capacitance of supercapacitors. For example, 3D nitrogen-doped GA nanomesh (N-GANM) can exhibit a high specific capacitance of 290.03 F/g at a current density of 1 A/g, which is much higher than the 197.5 F/g of ordinary graphene aerogel [113]. Besides, the nitrogen atoms weakened the stacking of graphene sheets and increased the specific surface area of the electrode material, which was able to accelerate the reaction rate with the electrolyte [114]. Compared to single doping, graphene aerogels used as electrodes with multiple doped atoms exhibit better results [115]. Double nitrogen-doped materials possess more nitrogen atoms and are more likely to increase the subversion of the reaction sites to accelerate ion diffusion [116]. Double doping of N and S can improve the electroneutrality of graphene and facilitate the rapid diffusion of ions from the electrolyte into the electrode material to increase the charge/discharge rate and improve the cycling stability [117]. The double doping of N and B can effectively promote the formation of B-N motifs in the carbon matrix, generating local surface polarization and thus increasing the redox sites [118]. In addition to this, the doping of heteroatoms leads to the creation of surface defects and the appearance of more vacancies. The appearance of these vacancies disrupts the original left-right symmetry of the system and allows the electronic structure near the Fermi energy level to change, thus enhancing the conductivity [119]. Generally, metal compounds have high theoretical specific capacitance and wide availability, so recently GA and metal compound nanoparticle composites have also been used as pseudocapacitive electrode materials. Yao et al. used GA as a scaffold and 3D printed graphene/MnO_2 compounded with MnO_2 on its surface using 3D printing technology as an electrode material. It was found that, to a certain extent, the electric capacity increased with increasing MnO_2 loading content and the weight of the electrode material remained almost constant. This shows the ability to substantially increase the loading of the pseudocapacitor material without sacrificing rate capability and weight capacitance, which is essential for the development of practically useful pseudocapacitors [120]. The inherent well-defined porosity and high surface area of metal-organic frameworks (MOFs) and the three-dimensional structure after compounding with GA allow electrolyte ions to diffuse in and out of the network even in thick electrodes with high quality loading, which is the outstanding advantage. Barua et al. generated pseudocapacitance by compounding Co-containing MOF with GA using redox-active ligands composed of hydroquinone units. Strong chemical interfacial engineering between MOF and graphene enhances the ion adsorption energy during electrochemistry and contributes to the charge storage performance [121]. The development of GA in supercapacitors has greatly contributed to the progress in the selection of electrode materials, while realizing a high electric capacity that is beyond the reach of many materials.

The emergence of GA has brought a big change to the energy storage, making battery systems and supercapacitors show high energy conversion efficiency and good development prospects. With its lighter material and larger energy storage capacity, it is expected to replace the traditional carbon material electrode as the new favorite of scientific researchers. However, the practical application of the whole GA material has not been well solved, and a perfect system to promote the practice is needed. There is still a long way to go to realize the application in energy storage field.

According to the keywords presented in the blue clusters, the main research hotspot are Adsorption and Removal, which clearly shows another major application area of GA: environmental protection. Since the 3D porous structure has good adsorption and absorption ability as well as large specific surface area for catalyst loading, GA plays a great role in water pollution and air pollution control methods.

5. Application of GA in the Environmental Protection Field

5.1. Adsorption

The porous structure and large specific surface area of aerogels determine their excellent adsorption of pollutants [18]. Moreover, with exceptional resilience it is easy to separate the adsorbed material for recycling and utilization [122]. Therefore, GA is expected to be an efficient adsorbent to absorb pollutants in water and gases.

5.1.1. Adsorption of Wastewater Pollutants

In recent years, the number of offshore oil spills has been increasing and the leaked crude oil has seriously affected the balance of the surrounding ecosystem [123]. The effect of capillary driving force in the nanochannels of graphene materials leads to low diffusion rate of water molecules between graphene sheets with hydrophobic stability [124]. With the combination of oxygen-containing functional groups on the GA surface structure that can interact with hydrophilic substances and arylalkyl groups that interact with hydrophobic substances, they have a strong adsorption capacity for organic substances such as oil with the help of Π-Π interactions and electrostatic gravitational forces [123,125,126]. Therefore, GA is widely used to absorb organic waste from water to purify water.

In 2016, Liu et al. prepared anisotropic graphene aerogels (AGAs) with lipophilic hydrophobicity and efficient adsorption properties using unidirectional freeze casting. The oxygen-containing functional groups on the surface of GA are quite compatible with many organic polymers and have a good adsorption effect on n-heptane, ethanol, acetone, etc. It can adsorb organic masses more than 120–200 times its own and can be easily recovered by combustion, distillation, and extrusion [125] Figure 6a–c. Meanwhile, graphene aerogel-based composites can effectively regulate the pro/hydrophobic properties as adsorption by adding other substances, and have some other environmental protection concepts. For example, Hu et al. prepared a multifunctional composite aerogel using chitosan (CS) as the backbone substrate and reduced graphene oxide nanosheets as the reinforcing agent. The material not only maintains good adsorption capacity and stability in harsh environments but can be easily degraded [127]. Moreover, Sun et al. synthesized a 3D macrostructure graphene-$CuFeSe_2$ aerogel for crude oil adsorption using solar energy drive. The photoexcitation heat generated by the photothermal material $CuFeSe_2$ to heat the crude oil drives its diffusion into the GA, and oil can be retained in the pores of aerogel after the removal of the light source. The method low cost and in line with the green development concept [128].

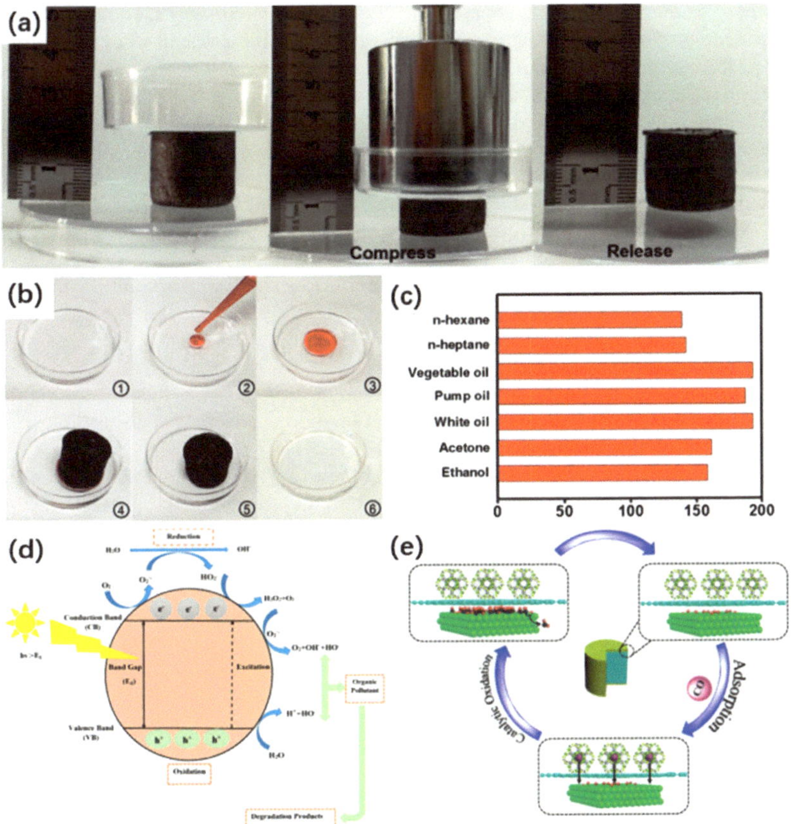

Figure 6. (**a**) Digital images show that AGA-8.3 (height = 18.6 mm, diameter = 21.3 mm, mass = 55 mg) can support 500 g weight and exhibits a good compressive performance; (**b**) Digital images showing the removal of n-heptane that is stained with Sudan Red 5B and floating on water [125]; (**c**) The ability of AGA-4.2 to absorb different organic liquids; (**d**) Photoactivity degradation mechanism of organic pollutants [19]; (**e**) Schematic of the reaction mechanism for CO removal [129].

Pollution caused by heavy metal ions leaking into water bodies when produced by industrial manufacturing can also seriously endanger human health and damage the ecosystem in water [126]. GA with electronegative functional groups on the surface are able to combine with different metal ions, and thus used for metal waste purification. When using GA for adsorption treatment, the adsorption capacity of GA is influenced by various factors (Ph, temperature, surface functional groups, etc.), so needs to be considered. For example, under acidic conditions, the increase in ionization of functional groups on the GA surface leads to an increase in the amount of negative charge, when electrostatic gravity plays an important role in the adsorption of metal ions. When the pH value in the solution gradually increases, the functional groups on the GA surface combine with metal ions to form complexes, and complexation and metal ion exchange become the main mechanism of GA adsorption [130]. Meanwhile, the adsorption capacity of GA increased with the increase in temperature. Since the adsorption reaction requires heat absorption, the temperature decreases during the reaction and the adsorption capacity of GA gradually decreases. In addition, the presence of functional groups helps to expose the surface-active sites of GA, giving it a higher adsorption capacity [131]. A tetraethylenetetramine/polypyrrole/graphene oxide aerogel (TPGA) was prepared by Liang et al. Under acidic conditions, the amino functional group promoted the adsorption of Cr^{2+} up to 408 mg/g [33]. Pan et al. reported that a novel macroporous calcium alginate/graphene oxide composite aerogel (mp-CA/GO) was able to effectively adsorb Pb^{2+}, Cu^{2+}, and Cr^{2+} while achieving remarkable removal rates of 95.4%, 81.2%, and 73.2%, respectively [132]. It can be noted that, based on the strong conductivity of GA, electroabsorption, which removes metal ions from adsorbed water by forming an electrostatic field on the surface of GA electrodes through an applied voltage, is also an efficient way of water purification. Wei et al. reported the use of nitrogen-doped graphene aerogel (NGA) for the removal of lead ions from wastewater by electrosorption. NGA was added to the Pb electrode to improve its mechanical strength and adsorption capacity. It also remained stable at 100 times of adsorption-desorption [133].

5.1.2. Adsorption of Gases

The current applications of GA in gas adsorption include the adsorption of toxic gases (CO, H_2S, NO, etc.) or greenhouse gases (mainly CO_2) and the adsorption-desorption reuse of energy gases (mainly H_2, CH_4).

Leakage of toxic gases is detrimental to both human and environment, so adsorption and subsequent treatment of leaked gases is important for environmental protection [134]. Due to the large degree of environmental hazards, the treatment of toxic inorganic gases such as H_2S and CO by GA includes subsequent auxiliary catalytic conversion to non-toxic substances in addition to adsorption. For example, three-dimensional flexible alkaline graphene aerogel (AGA) was prepared using Na_2CO_3 and GA to adsorb and catalyze the oxidation of H_2S. The macroporous structure of AGA provides large capacity for adsorption of H_2S and the alkaline environment provided by the Na_2CO_3 attached on the large specific surface area promotes the dissociation of H^+ from H_2S. Oxygen molecules are activated as superoxide radicals on the surface of graphene sheets, which promotes the catalytic oxidation of HS^- to sulfur monomers [135]. In addition, Qu et al. took advantage of the large specific surface area and strong adsorption capacity of MOF materials to adsorb CO by synthesizing 3D Ru/GA-HK with HKUST-1 (HK) modified on the GA surface. The macroporous structure of GA provides a pathway for the transport of gas molecules, allowing CO to diffuse between the graphene and Ru layers to form CO_2 using oxygen oxidation. HK loading on the GA surface enhanced the adsorption capacity of CO and increased the catalytic efficiency by $\approx$48.4% [129]. Apart from toxic gases, excessive CO_2 emissions are the main cause of the intensification of the greenhouse effect, and reasonable control of CO_2 content is of great importance [136,137]. Xia et al. compounded MgAl-LDH/RGO aerogel by layer double hydroxide (LDH) with MgAl nanoparticles added to GO, which was used to study the effect on CO_2 capture at high temperature and pressure. Compared to the two-dimensional MgAl-LDH powder, the adsorption efficiency of the three-dimensional porous structure of MgAl-LDH/RGO aerogel was improved by about 160% [138]. Since basic amine groups can react with acidic CO_2 gas by condensation, adsorbents containing amine groups can be used to adsorb CO_2. Therefore, CS-embedded GO aerogels were used for CO_2 adsorption by Hsan et al. The large number of amine groups contained inside the CS attached to the GA surface can have a larger reaction area to react with CO_2, much higher than the adsorption content of pure CS [139]. Leakage of energy gases causes waste of resources, and the treatment of energy gases such as H_2 should ensure the efficiency and stability of adsorption-desorption. Peng et al. reported the shell-core MgH_2@carbon aerogel microspheres for the storage of H_2. Thanks to the bonding between Mg and hydrogen atoms and the microsphere structure that inhibits the agglomeration of the material, the material exhibits a very good adsorption capacity and high cycling stability to meet the requirements of long-term use [140]. After adsorption, the desorption process can be carried out by physical extrusion using the resilience of GA or by creating an acidic environment to increase the number of H ions to replace the heavy metal ions in the active site [141].

5.2. Removal

In addition to adsorption, GA photocatalytic carriers are also a good method for catalytic removal of pollutants. 3D GA has a macroscopic ontological structure and is an ideal carrier for photocatalysts [129]. It can provide a stable backbone structure and large contact area for different photocatalysts. The targets of the scavenging process include a few organic compounds and microorganisms.

Generally, organic compounds are removed by degradation to water and CO_2 with GA-loaded surface catalysts [142]. For instance, a 3D TiO_2/reduced graphene oxide aerogel composite was prepared by Qiu et al. The catalyst TiO_2 grows along the glucose on the surface of GA and interacts to form a stable structure, while the large specific surface area provided by GA accelerates the catalytic degradation of methyl orange (MO) by TiO_2 (up to 90% degradation rate by sunlight irradiation for 300 min). At the end of degradation, GA and TiO_2 can be easily separated by physical manipulation, which is very convenient [143]. At the same time, multiple complexes composed of multiple functional compounds coupled together have higher catalytic properties. $BiVO_4$/RGO aerogel/$CeVO_4$ (BGC) degradation of tetracycline (TC) in visible light was proposed by Liu et al. The high electron mobility inside the material under the joint action of the three compounds enables a more efficient separation of photogenerated electric carriers. And compared to pure $BiVO_4$ and $CeVO_4$, the degradation rate was increased by 66% and 69% using GA as the backbone structural material [144]. In addition, certain heteroatom doping also helps to improve the performance of photocatalysts. Boron-doped graphene aerogels with different concentrations were prepared by Chowdhury et al. for the catalytic degradation of acridine orange (AO). The doping of B opens the band gap of graphene and increases the activity efficiency of the catalyst by $\approx$2 times (compared to the catalyst without GA incorporation) [145]. Additionally, microorganisms (bacteria, fungi, viruses, etc.) in wastewater also have a harmful effect on human health and the environment, and their removal using photocatalysis is also a current research hotspot.

Microbial scavenging mainly uses the products (ions, electrons, free radicals, etc.) obtained from redox reactions induced by active substances to cause damage to cell structures to achieve inactivation, and GA as a carrier template provides a large amount of reaction space and contact area with microorganisms [146]. For example, Zhang et al. used the prepared Ag-AgBr/TiO_2/GA to perform degradation inactivation studies on *E. coli* and *S. aureus*. AgBr nanoparticles are effectively combined with TiO_2 nanocrystals to accelerate the light-induced charge transfer and separation to achieve the decomposition of *E. coli* and *S. aureus*. The backbone structure was provided by GA and the large specific surface area promotes Ag-AgBr light collection and reaction platform [147]. In addition, three-dimensional BiOCl/RGO aerogels prepared by Zhang et al. were used for the catalytic decomposition of oxytetracycline (OTC). The porous structure of three-dimensional graphene with high specific surface area has abundant active sites providing powerful adsorption capacity of BGA composites. Besides, the material has enhanced visible light response and effective separation of electrons and holes, which are important for the antibacterial activity of OTC [148].

In environmental protection, the porous structure and surface functional groups of graphene play an important role. Meanwhile, more and more three-dimensional composites based on graphene aerogels are emerging, demonstrating the performance advantages and wide practicality of graphene aerogels.

6. Conclusions and Outlook

The development of nanotechnology has greatly contributed to the interest in various fields. Surprisingly, the related three-dimensional material GA has made considerable achievements in the past decade, and its three-dimensional stable structure has shown superiority over two-dimensional graphene in many aspects. The review uses bibliometric methods to analyze three hot areas of GA fabrication methods, energy storage applications, and environmental applications.

(1) We firstly discuss the self-assembly method, the template method, and the 3D printing preparation of GA, respectively, and analyze the advantages and limitations of different preparation methods. We found that the assembly method as a commonly used preparation method has the advantages of easy operation and low cost, and is expected to be produced in large quantities on a factory scale. However, the molded GA surface often contains many defects and collapses, which can seriously impair performance. The GA obtained by template method and 3D printing has better results in terms of material structure and uniformity of properties as well as stability in long-term use. However, the relatively high cost of these two methods and the tedious preparation process makes them difficult to meet in industrial scale manufacturing at present.

(2) GA with its excellent conductivity and large specific surface area is used as an electrode for batteries and supercapacitors to store energy. In lithium battery systems, GA is usually used as an electrode material alone or in combination with other electrode materials. Its good electrical conductivity and large specific surface area can effectively improve the transfer efficiency of lithium ions during charging and discharging to achieve rapid charging and discharging. In supercapacitors, the porous structure provided by GA can increase the electrode capacity as well as provide electrochemically active sites for electron attachment after doping with heteroatoms to achieve high energy density. The large specific surface area and high electrical conductivity possessed by GA are its advantages, and it has a huge advantage in the development of energy storage.

(3) In the field of environmental protection, the hydrophobic properties of the oxygen-containing functional groups on the surface of GA combined with the porous structure enable effective adsorption of organic substances, and its good resilience can effectively achieve multiple cycles of adsorption-desorption under the action of mechanical pressure. Meanwhile the large surface area provides loading for photocatalysts to improve the efficiency of catalytic degradation of pollutants, which will play a very significant role in promoting the improvement of environmental problems. The development and progress of GA is closely related to the development of these three major research fields. As a visualization method, the bibliometric analysis presents objective and comprehensive content. The summary of the key sections will be presented to the reader more clearly and quickly.

Currently, more and more studies on GA are being reported in a wider range of applications, for example, as electromagnetic shielding materials to reduce electromagnetic interference and as biosensors for sensitive detection of cells [149–153]. However, there are some challenges in the development of GA. The preparation of controlled structures with ordered arrangements and the loss of performance during use still pose limitations to its development, and the fundamental reason is the lack of most of the research between the transition from 2D graphene to 3D GA. Therefore, a more detailed study of the mechanism from both theoretical and experimental aspects is necessary. Challenges come along with opportunities, and it is foreseeable that GA will have more room for performance in the future.

Author Contributions: Conceptualization, S.Z. and Z.G.; methodology, B.C. and S.Z.; software, B.C.; Validation, S.Z.; formal analysis, B.C.; investigation, B.C. and W.Z.; resources, Z.G.; data curation, B.C.; writing—original draft preparation, B.C. and S.Z.; writing—review and editing, W.Z., Y.L., Z.G. and H.Z.; visualization, B.C.; supervision, S.Z., Z.G. and H.Z.; project administration, Z.G.; funding acquisition, Y.L, Z.G. and H.Z. All authors have read and agreed to the published version of the manuscript.

Funding: Defense Industrial Technology Development Program: JCKY2019204B044; Beijing Nova Program from Beijing Municipal Science & Technology Commission: Z201100006820108; Young Elite Scientists Sponsorship Program by CAST: 2021QNRC001; National Natural Science Foundation of China: 22102166 and 52075510; Strategic Priority Research Program of Chinese Academy of Sciences: XDB36000000; Directional Institutionalized Scientific Research Platform relies on Beijing Synchrotron Radiation Facility of Chinese Academy of Sciences; Beijing Natural Science Foundation: 2222087.

Institutional Review Board Statement: Not applicable.

Informed Consent Statement: Not applicable.

Conflicts of Interest: The authors declare no conflict of interest.

References

1. Novoselov, K.S.; Geim, A.K.; Morozov, S.V.; Jiang, D.; Zhang, Y.; Dubonos, S.V.; Grigorieva, I.V.; Firsov, A.A. Electric field effect in atomically thin carbon films. *Science* **2004**, *306*, 666–669. [CrossRef] [PubMed]
2. Geim, A.K. Graphene: Status and Prospects. *Science* **2009**, *324*, 1530–1534. [CrossRef] [PubMed]
3. Stankovich, S.; Dikin, D.A.; Dommett, G.H.B.; Kohlhaas, K.M.; Zimney, E.J.; Stach, E.A.; Piner, R.D.; Nguyen, S.T.; Ruoff, R.S. Graphene-based composite materials. *Nature* **2006**, *442*, 282–286. [CrossRef]
4. Dong, Q.C.; Xiao, M.; Chu, Z.Y.; Li, G.C.; Zhang, Y. Recent Progress of Toxic Gas Sensors Based on 3D Graphene Frameworks. *Sensors* **2021**, *21*, 3386. [CrossRef] [PubMed]
5. Nardecchia, S.; Carriazo, D.; Ferrer, M.L.; Gutierrez, M.C.; del Monte, F. Three dimensional macroporous architectures and aerogels built of carbon nanotubes and/or graphene: Synthesis and applications. *Chem. Soc. Rev.* **2013**, *42*, 794–830. [CrossRef] [PubMed]
6. Worsley, M.A.; Pauzauskie, P.J.; Olson, T.Y.; Biener, J.; Satcher, J.H.; Baumann, T.F. Synthesis of Graphene Aerogel with High Electrical Conductivity. *J. Am. Chem. Soc.* **2010**, *132*, 14067–14069. [CrossRef]
7. Hu, H.; Zhao, Z.B.; Wan, W.B.; Gogotsi, Y.; Qiu, J.S. Ultralight and Highly Compressible Graphene Aerogels. *Adv. Mater.* **2013**, *25*, 2219–2223. [CrossRef]
8. He, Y.Q.; Liu, Y.; Wu, T.; Ma, J.K.; Wang, X.R.; Gong, Q.J.; Kong, W.N.; Xing, F.B.; Liu, Y.; Gao, J.P. An environmentally friendly method for the fabrication of reduced graphene oxide foam with a super oil absorption capacity. *J. Hazard. Mater.* **2013**, *260*, 796–805. [CrossRef]
9. Chen, Z.P.; Ren, W.C.; Gao, L.B.; Liu, B.L.; Pei, S.F.; Cheng, H.M. Three-dimensional flexible and conductive interconnected graphene networks grown by chemical vapour deposition. *Nat. Mater.* **2011**, *10*, 424–428. [CrossRef]
10. Garcia-Tunon, E.; Barg, S.; Franco, J.; Bell, R.; Eslava, S.; D'Elia, E.; Maher, R.C.; Guitian, F.; Saiz, E. Printing in Three Dimensions with Graphene. *Adv. Mater.* **2015**, *27*, 1688–1693. [CrossRef]
11. Zhang, X.T.; Sui, Z.Y.; Xu, B.; Yue, S.F.; Luo, Y.J.; Zhan, W.C.; Liu, B. Mechanically strong and highly conductive graphene aerogel and its use as electrodes for electrochemical power sources. *J. Mater. Chem.* **2011**, *21*, 6494–6497. [CrossRef]
12. Fu, G.T.; Yan, X.X.; Chen, Y.F.; Xu, L.; Sun, D.M.; Lee, J.M.; Tang, Y.W. Boosting Bifunctional Oxygen Electrocatalysis with 3D Graphene Aerogel-Supported Ni/MnO Particles. *Adv. Mater.* **2018**, *30*, 1704609. [CrossRef] [PubMed]
13. Sui, Z.Y.; Meng, Y.N.; Xiao, P.W.; Zhao, Z.Q.; Wei, Z.X.; Han, B.H. Nitrogen-Doped Graphene Aerogels as Efficient Supercapacitor Electrodes and Gas Adsorbents. *ACS Appl. Mater. Interfaces* **2015**, *7*, 1431–1438. [CrossRef]
14. Hao, P.; Zhao, Z.H.; Tian, J.; Li, H.D.; Sang, Y.H.; Yu, G.W.; Cai, H.Q.; Liu, H.; Wong, C.P.; Umar, A. Hierarchical porous carbon aerogel derived from bagasse for high performance supercapacitor electrode. *Nanoscale* **2014**, *6*, 12120–12129. [CrossRef]
15. Tsang, A.C.H.; Zhang, J.T.; Hui, K.N.; Hui, K.S.; Huang, H.B. Recent Development and Applications of Advanced Materials via Direct Ink Writing. *Adv. Mater. Technol.* **2022**, *7*, 31. [CrossRef]
16. Wu, Z.Y.; Liang, H.W.; Chen, L.F.; Hu, B.C.; Yu, S.H. Bacterial Cellulose: A Robust Platform for Design of Three Dimensional Carbon-Based Functional Nanomaterials. *Acc. Chem. Res.* **2016**, *49*, 96–105. [CrossRef] [PubMed]
17. Yao, B.; Peng, H.R.; Zhang, H.Z.; Kang, J.Z.; Zhu, C.; Delgado, G.; Byrne, D.; Faulkner, S.; Freyman, M.; Lu, X.H.; et al. Printing Porous Carbon Aerogels for Low Temperature Supercapacitors. *Nano Lett.* **2021**, *21*, 3731–3737. [CrossRef]
18. Yang, Q.X.; Lu, R.; Ren, S.S.; Chen, C.T.; Chen, Z.J.; Yang, X.Y. Three dimensional reduced graphene oxide/ZIF-67 aerogel: Effective removal cationic and anionic dyes from water. *Chem. Eng. J.* **2018**, *348*, 202–211. [CrossRef]
19. Hasanpour, M.; Hatami, M. Photocatalytic performance of aerogels for organic dyes removal from wastewaters: Review study. *J. Mol. Liq.* **2020**, *309*, 113094. [CrossRef]
20. Tian, Y.; Du, C.; Yong, S.H.; Zhou, X.; Zhou, C.; Yang, S.Y. Catalysis-involved 3D N-doped graphene aerogel achieves a superior solar water purification rate and efficiency. *Chem. Eng. J.* **2023**, *453*, 139793. [CrossRef]
21. An, B.X.; Ma, Y.; Li, W.B.; Su, M.; Li, F.Y.; Song, Y.L. Three-dimensional multi-recognition flexible wearable sensor via graphene aerogel printing. *Chem. Commun.* **2016**, *52*, 10948–10951. [CrossRef] [PubMed]
22. Gong, J.Y.; Tang, W.Y.; Xia, L.J.; Fu, Z.; Zhou, S.J.; Zhang, J.J.; Zhang, C.H.; Li, L.; Ji, H.; Xu, W.L. Flexible and weavable 3D porous graphene/PPy/lignocellulose-based versatile fibrous wearables for thermal management and strain sensing. *Chem. Eng. J.* **2023**, *452*, 139338. [CrossRef]

23. Sun, Z.X.; Fang, S.Y.; Hu, Y.H. 3D Graphene Materials: From Understanding to Design and Synthesis Control. *Chem. Rev.* **2020** *120*, 10336–10453. [CrossRef]
24. Korkmaz, S.; Kariper, I.A. Graphene and graphene oxide based aerogels: Synthesis, characteristics and supercapacitor applications *J. Energy Storage* **2020**, *27*, 101038. [CrossRef]
25. Zhi, D.D.; Li, T.; Li, J.Z.; Ren, H.S.; Meng, F.B. A review of three-dimensional graphene-based aerogels: Synthesis, structure and application for microwave absorption. *Compos. Pt. B-Eng.* **2021**, *211*, 108642. [CrossRef]
26. Sun, Y.R.; Yang, M.X.; Yu, F.; Chen, J.H.; Ma, J. Synthesis of Graphene Aerogel Adsorbents and Their Applications in Water Treatment. *Prog. Chem.* **2015**, *27*, 1133–1146. [CrossRef]
27. Raagulan, K.; Kim, B.M.; Chai, K.Y. Recent Advancement of Electromagnetic Interference (EMI) Shielding of Two Dimensional (2D) MXene and Graphene Aerogel Composites. *Nanomaterials* **2020**, *10*, 702. [CrossRef]
28. Wang, J.; Ellsworth, M.W. Graphene Aerogels. In Proceedings of the 1st International Symposium on Emerging Materials for Post-CMOS Applications Held at the 215th Meeting of the Electrochemical-Society, San Francisco, CA, USA, 25–29 May 2009 pp. 241–247.
29. Chen, W.F.; Yan, L.F. In situ self-assembly of mild chemical reduction graphene for three-dimensional architectures. *Nanoscale* **2011**, *3*, 3132–3137. [CrossRef]
30. Xu, Y.X.; Sheng, K.X.; Li, C.; Shi, G.Q. Self-Assembled Graphene Hydrogel via a One-Step Hydrothermal Process. *ACS Nano* **2010** *4*, 4324–4330. [CrossRef]
31. Sheng, K.X.; Xu, Y.X.; Li, C.; Shi, G.Q. High-performance self-assembled graphene hydrogels prepared by chemical reduction of graphene oxide. *New Carbon Mater.* **2011**, *26*, 9–15. [CrossRef]
32. Gao, H.C.; Xiao, F.; Ching, C.B.; Duan, H.W. High-Performance Asymmetric Supercapacitor Based on Graphene Hydrogel and Nanostructured MnO2. *ACS Appl. Mater. Interfaces* **2012**, *4*, 2801–2810. [CrossRef] [PubMed]
33. Liang, Q.W.; Luo, H.J.; Geng, J.J.; Chen, J.D. Facile one-pot preparation of nitrogen-doped ultra-light graphene oxide aerogel and its prominent adsorption performance of Cr(VI). *Chem. Eng. J.* **2018**, *338*, 62–71. [CrossRef]
34. Shu, D.; Feng, F.; Han, H.L.; Ma, Z.F. Prominent adsorption performance of amino-functionalized ultra-light graphene aerogel for methyl orange and amaranth. *Chem. Eng. J.* **2017**, *324*, 1–9. [CrossRef]
35. Goldstein, A.P.; Mickelson, W.; Machness, A.; Lee, G.; Worsley, M.A.; Woo, L.; Zettl, A. Simultaneous Sheet Cross-Linking and Deoxygenation in the Graphene Oxide Sol-Gel Transition. *J. Phys. Chem. C* **2014**, *118*, 28855–28860. [CrossRef]
36. Tong, X.; Jia, W.J.; Li, Y.M.; Yao, T.J.; Wu, J.; Yang, M. One-step preparation of reduced graphene oxide/Prussian blue/polypyrrole aerogel and their enhanced photo-Fenton performance. *J. Taiwan Inst. Chem. Eng.* **2019**, *102*, 92–98. [CrossRef]
37. Wu, J.; Tao, K.; Zhang, J.; Guo, Y.Y.; Miao, J.M.; Norford, L.K. Chemically functionalized 3D graphene hydrogel for high performance gas sensing. *J. Mater. Chem. A* **2016**, *4*, 8130–8140. [CrossRef]
38. Wan, W.C.; Zhang, F.; Yu, S.; Zhang, R.Y.; Zhou, Y. Hydrothermal formation of graphene aerogel for oil sorption: The role of reducing agent, reaction time and temperature. *New J. Chem.* **2016**, *40*, 3040–3046. [CrossRef]
39. Li, J.H.; Li, J.Y.; Meng, H.; Xie, S.Y.; Zhang, B.W.; Li, L.F.; Ma, H.J.; Zhang, J.Y.; Yu, M. Ultra-light, compressible and fire-resistant graphene aerogel as a highly efficient and recyclable absorbent for organic liquids. *J. Mater. Chem. A* **2014**, *2*, 2934–2941. [CrossRef]
40. Lim, M.B.; Hu, M.; Manandhar, S.; Sakshaug, A.; Strong, A.; Riley, L.; Pauzauskie, P.J. Ultrafast sol-gel synthesis of graphene aerogel materials. *Carbon* **2015**, *95*, 616–624. [CrossRef]
41. Cong, H.P.; Ren, X.C.; Wang, P.; Yu, S.H. Macroscopic Multifunctional Graphene-Based Hydrogels and Aerogels by a Metal Ion Induced Self-Assembly Process. *ACS Nano* **2012**, *6*, 2693–2703. [CrossRef]
42. Wang, X.; Lu, L.L.; Yu, Z.L.; Xu, X.W.; Zheng, Y.R.; Yu, S.H. Scalable Template Synthesis of Resorcinol-Formaldehyde/Graphene Oxide Composite Aerogels with Tunable Densities and Mechanical Properties. *Angew. Chem.-Int. Edit.* **2015**, *54*, 2397–2401 [CrossRef] [PubMed]
43. Worsley, M.A.; Olson, T.Y.; Lee, J.R.I.; Willey, T.M.; Nielsen, M.H.; Roberts, S.K.; Pauzauskie, P.J.; Biener, J.; Satcher, J.H.; Baumann, T.F. High Surface Area, sp(2)-Cross-Linked Three-Dimensional Graphene Monoliths. *J. Phys. Chem. Lett.* **2011**, *2*, 921–925 [CrossRef] [PubMed]
44. Chen, Z.P.; Xu, C.; Ma, C.Q.; Ren, W.C.; Cheng, H.M. Lightweight and Flexible Graphene Foam Composites for High-Performance Electromagnetic Interference Shielding. *Adv. Mater.* **2013**, *25*, 1296–1300. [CrossRef] [PubMed]
45. Dong, X.C.; Wang, X.W.; Wang, L.H.; Song, H.; Zhang, H.; Huang, W.; Chen, P. 3D Graphene Foam as a Monolithic and Macroporous Carbon Electrode for Electrochemical Sensing. *ACS Appl. Mater. Interfaces* **2012**, *4*, 3129–3133. [CrossRef] [PubMed]
46. Cao, X.H.; Shi, Y.M.; Shi, W.H.; Lu, G.; Huang, X.; Yan, Q.Y.; Zhang, Q.C.; Zhang, H. Preparation of Novel 3D Graphene Networks for Supercapacitor Applications. *Small* **2011**, *7*, 3163–3168. [CrossRef] [PubMed]
47. Pettes, M.T.; Ji, H.X.; Ruoff, R.S.; Shi, L. Thermal Transport in Three-Dimensional Foam Architectures of Few-Layer Graphene and Ultrathin Graphite. *Nano Lett.* **2012**, *12*, 2959–2964. [CrossRef] [PubMed]
48. Kim, B.J.; Yang, G.; Park, M.J.; Kwak, J.S.; Baik, K.H.; Kim, D.; Kim, J. Three-dimensional graphene foam-based transparent conductive electrodes in GaN-based blue light-emitting diodes. *Appl. Phys. Lett.* **2013**, *102*, 161902. [CrossRef]
49. Ma, Y.X.; Yu, M.; Liu, J.H.; Li, X.J.; Li, S.M. Ultralight Interconnected Graphene-Amorphous Carbon Hierarchical Foam with Mechanical Resiliency for High Sensitivity and Durable Strain Sensors. *ACS Appl. Mater. Interfaces* **2017**, *9*, 27127–27134 [CrossRef]

50. Cao, M.; Li, S.L.; Cheng, J.B.; Zhang, A.N.; Wang, Y.Z.; Zhao, H.B. Fully bio-based, low fire-hazard and superelastic aerogel without hazardous cross-linkers for excellent thermal insulation and oil clean-up absorption. *J. Hazard. Mater.* **2021**, *403*, 123977. [CrossRef]
51. Zhu, C.; Han, T.Y.J.; Duoss, E.B.; Golobic, A.M.; Kuntz, J.D.; Spadaccini, C.M.; Worsley, M.A. Highly compressible 3D periodic graphene aerogel microlattices. *Nat. Commun.* **2015**, *6*, 6962. [CrossRef]
52. Deville, S. The lure of ice-templating: Recent trends and opportunities for porous materials. *Scr. Mater.* **2018**, *147*, 119–124. [CrossRef]
53. Xi, X.T.; Luo, X.Q.; Xia, Y.; Yi, L.F.; Wang, Y.; Song, D.Y.; Song, Y.J.; Wu, J.R.; Zhao, L.J. Ice Crystal Growth Mechanism and Structure-activity Relationships of Graphene Oxide/Poly(vinyl alcohol) Aerogels. *Chin. J. Polym. Sci.* **2022**, *40*, 772–780. [CrossRef]
54. Zhao, N.F.; Li, M.; Gong, H.X.; Bai, H. Controlling ice formation on gradient wettability surface for high-performance bioinspired materials. *Sci. Adv.* **2020**, *6*, eabb4712. [CrossRef] [PubMed]
55. Shao, G.F.; Hanaor, D.A.H.; Shen, X.D.; Gurlo, A. Freeze Casting: From Low-Dimensional Building Blocks to Aligned Porous Structures—A Review of Novel Materials, Methods, and Applications. *Adv. Mater.* **2020**, *32*, 1907176. [CrossRef]
56. Zhu, Y.X.; Qin, J.D.; Shi, G.; Sun, C.; Ingram, M.; Qian, S.S.; Lu, J.; Zhang, S.Q.; Zhong, Y.L. A focus review on 3D printing of wearable energy storage devices. *Carbon Energy* **2022**, *4*, 1242–1261. [CrossRef]
57. Feng, J.Z.; Su, B.L.; Xia, H.S.; Zhao, S.Y.; Gao, C.; Wang, L.K.; Ogbeide, O.; Feng, J.; Hasan, T. Printed aerogels: Chemistry, processing, and applications. *Chem. Soc. Rev.* **2021**, *50*, 3842–3888. [CrossRef]
58. Jiang, Y.Q.; Xu, Z.; Huang, T.Q.; Liu, Y.J.; Guo, F.; Xi, J.B.; Gao, W.W.; Gao, C. Direct 3D Printing of Ultralight Graphene Oxide Aerogel Microlattices. *Adv. Funct. Mater.* **2018**, *28*, 1707024. [CrossRef]
59. Zhang, Q.Q.; Zhang, F.; Medarametla, S.P.; Li, H.; Zhou, C.; Lin, D. 3D Printing of Graphene Aerogels. *Small* **2016**, *12*, 1702–1708. [CrossRef]
60. Xu, Z.; Zhang, Y.; Li, P.G.; Gao, C. Strong, Conductive, Lightweight, Neat Graphene Aerogel Fibers with Aligned Pores. *ACS Nano* **2012**, *6*, 7103–7113. [CrossRef]
61. Tarascon, J.M.; Armand, M. Issues and challenges facing rechargeable lithium batteries. *Nature* **2001**, *414*, 359–367. [CrossRef]
62. Simon, P.; Gogotsi, Y. Materials for electrochemical capacitors. *Nat. Mater.* **2008**, *7*, 845–854. [CrossRef] [PubMed]
63. Bruce, P.G.; Scrosati, B.; Tarascon, J.M. Nanomaterials for rechargeable lithium batteries. *Angew. Chem.-Int. Edit.* **2008**, *47*, 2930–2946. [CrossRef]
64. Cheng, X.B.; Zhang, R.; Zhao, C.Z.; Zhang, Q. Toward Safe Lithium Metal Anode in Rechargeable Batteries: A Review. *Chem. Rev.* **2017**, *117*, 10403–10473. [CrossRef]
65. Yoo, E.; Kim, J.; Hosono, E.; Zhou, H.; Kudo, T.; Honma, I. Large reversible Li storage of graphene nanosheet families for use in rechargeable lithium ion batteries. *Nano Lett.* **2008**, *8*, 2277–2282. [CrossRef] [PubMed]
66. Huang, L.A.; He, Z.S.; Guo, J.F.; Pei, S.E.; Shao, H.B.; Wang, J.M. Self-Assembled Three-Dimensional Graphene Aerogel with an Interconnected Porous Structure for Lithium-Ion Batteries. *ChemElectroChem* **2019**, *6*, 2698–2706. [CrossRef]
67. Ren, L.; Hui, K.N.; Hui, K.S.; Liu, Y.D.; Qi, X.; Zhong, J.X.; Du, Y.; Yang, J.P. 3D hierarchical porous graphene aerogel with tunable meso-pores on graphene nanosheets for high-performance energy storage. *Sci Rep* **2015**, *5*, 14229. [CrossRef]
68. Shan, H.; Xiong, D.B.; Li, X.F.; Sun, Y.P.; Yan, B.; Li, D.J.; Lawes, S.; Cui, Y.H.; Sun, X.L. Tailored lithium storage performance of graphene aerogel anodes with controlled surface defects for lithium-ion batteries. *Appl. Surf. Sci.* **2016**, *364*, 651–659. [CrossRef]
69. Ye, J.C.; Charnvanichborikarn, S.; Worsley, M.A.; Kucheyev, S.O.; Wood, B.C.; Wang, Y.M. Enhanced electrochemical performance of ion-beam-treated 3D graphene aerogels for lithium ion batteries. *Carbon* **2015**, *85*, 269–278. [CrossRef]
70. Li, X.F.; Geng, D.S.; Zhang, Y.; Meng, X.B.; Li, R.Y.; Sun, X.L. Superior cycle stability of nitrogen-doped graphene nanosheets as anodes for lithium ion batteries. *Electrochem. Commun.* **2011**, *13*, 822–825. [CrossRef]
71. Shan, H.; Li, X.F.; Cui, Y.H.; Xiong, D.B.; Yan, B.; Li, D.J.; Lushington, A.; Sun, X.L. Sulfur/Nitrogen Dual-doped Porous Graphene Aerogels Enhancing Anode Performance of Lithium Ion Batteries. *Electrochim. Acta* **2016**, *205*, 188–197. [CrossRef]
72. Zhang, C.J.; Chai, F.L.; Fu, L.; Hu, P.; Pang, S.P.; Cui, G.L. Lithium storage in a highly conductive Cu_3Ge boosted Ge/graphene aerogel. *J. Mater. Chem. A* **2015**, *3*, 22552–22556. [CrossRef]
73. Qin, J.; He, C.N.; Zhao, N.Q.; Wang, Z.Y.; Shi, C.S.; Liu, E.Z.; Li, J.J. Graphene Networks Anchored with Sn@Graphene as Lithium Ion Battery Anode. *ACS Nano* **2014**, *8*, 1728–1738. [CrossRef] [PubMed]
74. Cao, H.L.; Zhou, X.F.; Deng, W.; Liu, Z.P. A compressible and hierarchical porous graphene/Co composite aerogel for lithium-ion batteries with high gravimetric/volumetric capacity. *J. Mater. Chem. A* **2016**, *4*, 6021–6028. [CrossRef]
75. Ji, L.W.; Lin, Z.; Alcoutlabi, M.; Zhang, X.W. Recent developments in nanostructured anode materials for rechargeable lithium-ion batteries. *Energy Environ. Sci.* **2011**, *4*, 2682–2699. [CrossRef]
76. Jiang, T.C.; Bu, F.X.; Feng, X.X.; Shakir, I.; Hao, G.L.; Xu, Y.X. Porous Fe_2O_3 Nanoframeworks Encapsulated within Three-Dimensional Graphene as High-Performance Flexible Anode for Lithium-Ion Battery. *ACS Nano* **2017**, *11*, 5140–5147. [CrossRef]
77. Shao, J.X.; Feng, J.H.; Zhou, H.; Yuan, A.H. Graphene aerogel encapsulated Fe-Co oxide nanocubes derived from Prussian blue analogue as integrated anode with enhanced Li-ion storage properties. *Appl. Surf. Sci.* **2019**, *471*, 745–752. [CrossRef]
78. Zhang, Y.Q.; Tao, H.C.; Ma, H.; Du, S.L.; Li, T.; Zhang, Y.K.; Li, J.H.; Yang, X.L. Three-dimensional MoO_2@few-layered MoS_2 covered by S-doped graphene aerogel for enhanced lithium ion storage. *Electrochim. Acta* **2018**, *283*, 619–627. [CrossRef]
79. Meng, J.K.; Cao, Y.; Suo, Y.; Liu, Y.S.; Zhang, J.M.; Zheng, X.C. Facile Fabrication of 3D SiO_2@Graphene Aerogel Composites as Anode Material for Lithium Ion Batteries. *Electrochim. Acta* **2015**, *176*, 1001–1009. [CrossRef]

80. Entwistle, J.E.; Booth, S.G.; Keeble, D.S.; Ayub, F.; Yan, M.; Corr, S.A.; Cumming, D.J.; Patwardhan, S.V. Insights into the Electrochemical Reduction Products and Processes in Silica Anodes for Next-Generation Lithium-Ion Batteries. *Adv. Energy Mater.* **2020**, *10*, 2001826. [CrossRef]
81. Hu, X.Z.; Jin, Y.; Zhu, B.; Tan, Y.L.; Zhang, S.; Zong, L.Q.; Lu, Z.D.; Zhu, J. Free-Standing Graphene-Encapsulated Silicon Nanoparticle Aerogel as an Anode for Lithium Ion Batteries. *ChemNanoMat* **2016**, *2*, 671–674. [CrossRef]
82. Chang, P.; Liu, X.X.; Zhao, Q.J.; Huang, Y.Q.; Huang, Y.H.; Hu, X.L. Constructing Three-Dimensional Honeycombed Graphene/Silicon Skeletons for High-Performance Li-Ion Batteries. *ACS Appl. Mater. Interfaces* **2017**, *9*, 31879–31886. [CrossRef] [PubMed]
83. Manthiram, A.; Fu, Y.Z.; Chung, S.H.; Zu, C.X.; Su, Y.S. Rechargeable Lithium-Sulfur Batteries. *Chem. Rev.* **2014**, *114*, 11751–11787. [CrossRef] [PubMed]
84. Bruce, P.G.; Freunberger, S.A.; Hardwick, L.J.; Tarascon, J.M. Li-O-2 and Li-S batteries with high energy storage. *Nat. Mater.* **2012**, *11*, 19–29. [CrossRef] [PubMed]
85. Fang, R.P.; Zhao, S.Y.; Sun, Z.H.; Wang, W.; Cheng, H.M.; Li, F. More Reliable Lithium-Sulfur Batteries: Status, Solutions and Prospects. *Adv. Mater.* **2017**, *29*, 1606823. [CrossRef]
86. Hu, G.J.; Xu, C.; Sun, Z.H.; Wang, S.G.; Cheng, H.M.; Li, F.; Ren, W.C. 3D Graphene-Foam-Reduced-Graphene-Oxide Hybrid Nested Hierarchical Networks for High-Performance Li-S Batteries. *Adv. Mater.* **2016**, *28*, 1603–1609. [CrossRef] [PubMed]
87. Shi, H.D.; Zhao, X.J.; Wu, Z.S.; Dong, Y.F.; Lu, P.F.; Chen, J.; Ren, W.C.; Cheng, H.M.; Bao, X.H. Free-standing integrated cathode derived from 3D graphene/carbon nanotube aerogels serving as binder-free sulfur host and interlayer for ultrahigh volumetric-energy-density lithium-sulfur batteries. *Nano Energy* **2019**, *60*, 743–751. [CrossRef]
88. Zhou, G.M.; Paek, E.; Hwang, G.S.; Manthiram, A. High-Performance Lithium-Sulfur Batteries with a Self-Supported, 3D Li2S-Doped Graphene Aerogel Cathodes. *Adv. Energy Mater.* **2016**, *6*, 1501355. [CrossRef]
89. Luo, L.; Chung, S.H.; Manthiram, A. A three-dimensional self-assembled SnS_2-nano-dots@graphene hybrid aerogel as an efficient polysulfide reservoir for high-performance lithium-sulfur batteries. *J. Mater. Chem. A* **2018**, *6*, 7659–7667. [CrossRef]
90. Liu, M.M.; Zhang, C.C.; Su, J.M.; Chen, X.; Ma, T.Y.; Huang, T.; Yu, A.S. Propelling Polysulfide Conversion by Defect-Rich MoS_2 Nanosheets for High-Performance Lithium-Sulfur Batteries. *ACS Appl. Mater. Interfaces* **2019**, *11*, 20788–20795. [CrossRef]
91. Ding, M.; Huang, S.Z.; Wang, Y.; Hu, J.P.; Pam, M.E.; Fan, S.; Shi, Y.M.; Ge, Q.; Yang, H.Y. Promoting polysulfide conversion by catalytic ternary Fe_3O_4/carbon/graphene composites with ordered microchannels for ultrahigh-rate lithium-sulfur batteries. *J. Mater. Chem. A* **2019**, *7*, 25078–25087. [CrossRef]
92. Tan, S.Y.; Wu, Y.F.; Kan, S.T.; Bu, M.M.; Liu, Y.; Yang, L.S.; Yang, Y.H.; Liu, H.T. A combination of MnO_2-decorated graphene aerogel modified separator and I/N codoped graphene aerogel sulfur host to synergistically promote Li-S battery performance. *Electrochim. Acta* **2020**, *348*, 136173. [CrossRef]
93. Wang, H.F.; Wang, X.X.; Li, M.L.; Zheng, L.J.; Guan, D.H.; Huang, X.L.; Xu, J.J.; Yu, J.H. Porous Materials Applied in Nonaqueous Li-O(2)Batteries: Status and Perspectives. *Adv. Mater.* **2020**, *32*, 2002559. [CrossRef] [PubMed]
94. Li, L.Y.; Chen, C.G.; Su, J.M.; Kuang, P.; Zhang, C.C.; Yao, Y.; Huang, T.; Yu, A.S. Three-dimensional MoSx ($1 < x < 2$) nanosheets decorated graphene aerogel for lithium-oxygen batteries. *J. Mater. Chem. A* **2016**, *4*, 10986–10991. [CrossRef]
95. Zhao, C.T.; Yu, C.; Li, S.F.; Guo, W.; Zhao, Y.; Dong, Q.; Lin, X.T.; Song, Z.X.; Tan, X.Y.; Wang, C.H.; et al. Ultrahigh-Capacity and Long-Life Lithium-Metal Batteries Enabled by Engineering Carbon Nanofiber-Stabilized Graphene Aerogel Film Host. *Small* **2018**, *14*, 1803310. [CrossRef]
96. Kim, C.H.J.; Varanasi, C.V.; Liu, J. Synergy of polypyrrole and carbon x-aerogel in lithium-oxygen batteries. *Nanoscale* **2018**, *10*, 3753–3758. [CrossRef]
97. Zhao, C.T.; Yu, C.; Liu, S.H.; Yang, J.; Fan, X.M.; Huang, H.W.; Qiu, J.S. 3D Porous N-Doped Graphene Frameworks Made of Interconnected Nanocages for Ultrahigh-Rate and Long-Life Li-O-2 Batteries. *Adv. Funct. Mater.* **2015**, *25*, 6913–6920. [CrossRef]
98. Wang, G.P.; Zhang, L.; Zhang, J.J. A review of electrode materials for electrochemical supercapacitors. *Chem. Soc. Rev.* **2012**, *41*, 797–828. [CrossRef]
99. Zhang, L.L.; Zhao, X.S. Carbon-based materials as supercapacitor electrodes. *Chem. Soc. Rev.* **2009**, *38*, 2520–2531. [CrossRef]
100. Kotz, R.; Carlen, M. Principles and applications of electrochemical capacitors. *Electrochim. Acta* **2000**, *45*, 2483–2498. [CrossRef]
101. Okhay, O.; Tkach, A. Graphene/Reduced Graphene Oxide-Carbon Nanotubes Composite Electrodes: From Capacitive to Battery-Type Behaviour. *Nanomaterials* **2021**, *11*, 1240. [CrossRef]
102. Chen, W.S.; Yu, H.P.; Lee, S.Y.; Wei, T.; Li, J.; Fan, Z.J. Nanocellulose: A promising nanomaterial for advanced electrochemical energy storage. *Chem. Soc. Rev.* **2018**, *47*, 2837–2872. [CrossRef]
103. Wu, Z.S.; Sun, Y.; Tan, Y.Z.; Yang, S.B.; Feng, X.L.; Mullen, K. Three-Dimensional Graphene-Based Macro- and Mesoporous Frameworks for High-Performance Electrochemical Capacitive Energy Storage. *J. Am. Chem. Soc.* **2012**, *134*, 19532–19535. [CrossRef] [PubMed]
104. Xie, L.J.; Su, F.Y.; Xie, L.F.; Li, X.M.; Liu, Z.; Kong, Q.Q.; Guo, X.H.; Zhang, Y.; Wan, L.; Li, K.X.; et al. Self-Assembled 3D Graphene-Based Aerogel with Co_3O_4 Nanoparticles as High-Performance Asymmetric Supercapacitor Electrode. *ChemSusChem* **2015**, *8*, 2917–2926. [CrossRef] [PubMed]
105. Liu, J.; Zeng, Y.; Zhang, J.; Zhang, H.J.; Liu, J.H. Preparation, Structures and Properties of Three-Dimensional Graphene-Based Materials. *Prog. Chem.* **2019**, *31*, 667–680. [CrossRef]

106. Zhu, C.; Liu, T.Y.; Qian, F.; Han, T.Y.J.; Duoss, E.B.; Kuntz, J.D.; Spadaccini, C.M.; Worsley, M.A.; Li, Y. Supercapacitors Based on Three-Dimensional Hierarchical Graphene Aerogels with Periodic Macropores. *Nano Lett.* **2016**, *16*, 3448–3456. [CrossRef] [PubMed]
107. Fan, W.; Shi, Y.Q.; Gao, W.; Sun, Z.; Liu, T.X. Graphene-Carbon Nanotube Aerogel with a Scroll-Interconnected-Sheet Structure as an Advanced Framework for a High-Performance Asymmetric Supercapacitor Electrode. *ACS Appl. Nano Mater.* **2018**, *1*, 4435–4441. [CrossRef]
108. Wang, C.; Huang, Y.S.; Pan, H.; Jiang, J.Z.; Yang, X.W.; Xu, Z.X.; Tian, H.; Han, S.; Wu, D.Q. Nitrogen-Doped Porous Carbon/Graphene Aerogel with Much Enhanced Capacitive Behaviors. *Electrochim. Acta* **2016**, *215*, 100–107. [CrossRef]
109. Chen, W.; Gui, D.Y.; Liu, J.H. Nickel oxide/graphene aerogel nanocomposite as a supercapacitor electrode material with extremely wide working potential window. *Electrochim. Acta* **2016**, *222*, 1424–1429. [CrossRef]
110. Tian, W.W.; Cheng, D.K.; Wang, S.; Xiong, C.X.; Yang, Q.L. Phytic acid modified manganese dioxide/graphene composite aerogel as high-performance electrode materials for supercapacitors. *Appl. Surf. Sci.* **2019**, *495*, 143589. [CrossRef]
111. Zheng, C.L.; Zhang, J.L.; Zhang, Q.; You, B.; Chen, G.L. Three dimensional Ni foam-supported graphene oxide for binder-free pseudocapacitor. *Electrochim. Acta* **2015**, *152*, 216–221. [CrossRef]
112. Chang, J.H.; Hung, Y.H.; Luo, X.F.; Huang, C.H.; Jung, S.M.; Chang, J.K.; Kong, J.; Su, C.Y. The hierarchical porosity of a three-dimensional graphene electrode for binder-free and high performance supercapacitors. *RSC Adv.* **2016**, *6*, 8384–8394. [CrossRef]
113. Su, X.L.; Fu, L.; Cheng, M.Y.; Yang, J.H.; Guan, X.X.; Zheng, X.C. 3D nitrogen-doped graphene aerogel nanomesh: Facile synthesis and electrochemical properties as the electrode materials for supercapacitors. *Appl. Surf. Sci.* **2017**, *426*, 924–932. [CrossRef]
114. Xu, P.; Gan, Q.M.; Ma, L.; Li, Z.Y.; Zhang, H.; Xiao, H.; Liang, X.; Zhang, T.F.; Tian, X.H.; Liu, C.H. A high surface area N-doped holey graphene aerogel with low charge transfer resistance as high performance electrode of non-flammable thermostable supercapacitors. *Carbon* **2019**, *149*, 452–461. [CrossRef]
115. Wu, Z.S.; Winter, A.; Chen, L.; Sun, Y.; Turchanin, A.; Feng, X.L.; Mullen, K. Three-Dimensional Nitrogen and Boron Co-doped Graphene for High-Performance All-Solid-State Supercapacitors. *Adv. Mater.* **2012**, *24*, 5130–5135. [CrossRef] [PubMed]
116. Meng, Q.; Wan, H.C.; Zhu, W.K.; Duan, T.; Yao, W.T. Naturally Dried, Double Nitrogen-Doped 3D Graphene Aerogels Modified by Plant Extracts for Multifunctional Applications. *ACS Sustain. Chem. Eng.* **2018**, *6*, 1172–1181. [CrossRef]
117. Chen, Y.J.; Liu, Z.E.; Sun, L.; Lu, Z.W.; Zhuo, K.L. Nitrogen and sulfur co-doped porous graphene aerogel as an efficient electrode material for high performance supercapacitor in ionic liquid electrolyte. *J. Power Sources* **2018**, *390*, 215–223. [CrossRef]
118. Zou, X.X.; Wu, D.P.; Mu, Y.F.; Xing, L.Y.; Zhang, W.C.; Gao, Z.Y.; Xu, F.; Jiang, K. Boron and nitrogen Co-doped holey graphene aerogels with rich B-N motifs for flexible supercapacitors. *Carbon* **2020**, *159*, 94–101. [CrossRef]
119. Hou, J.H.; Cao, C.B.; Idrees, F.; Ma, X.L. Hierarchical Porous Nitrogen-Doped Carbon Nanosheets Derived from Silk for Ultrahigh-Capacity Battery Anodes and Supercapacitors. *ACS Nano* **2015**, *9*, 2556–2564. [CrossRef]
120. Yao, B.; Chandrasekaran, S.; Zhang, J.; Xiao, W.; Qian, F.; Zhu, C.; Duoss, E.B.; Spadaccini, C.M.; Worsley, M.A.; Li, Y. Efficient 3D Printed Pseudocapacitive Electrodes with Ultrahigh MnO2 Loading. *Joule* **2019**, *3*, 459–470. [CrossRef]
121. Barua, A.; Mehra, P.; Paul, A. Understanding Integrated Graphene-MOF Nanostructures as Binder- and Additive-Free High-Performance Supercapacitors at Commercial Scale Mass Loading. *ACS Appl. Energ. Mater.* **2021**, *4*, 14249–14259. [CrossRef]
122. Jiang, J.X.; Zhang, Q.H.; Zhan, X.L.; Chen, F.Q. A multifunctional gelatin-based aerogel with superior pollutants adsorption, oil/water separation and photocatalytic properties. *Chem. Eng. J.* **2019**, *358*, 1539–1551. [CrossRef]
123. Guo, Q.L.; Chen, S.; Liu, Z.X.; Yan, J.C.; Liu, H.E. Preparation and performance evaluation of graphene/hydroxypropyl methylcellulose composite aerogel for high viscosity oil adsorption. *J. Environ. Chem. Eng.* **2022**, *10*, 108312. [CrossRef]
124. Ge, J.; Zhao, H.Y.; Zhu, H.W.; Huang, J.; Shi, L.A.; Yu, S.H. Advanced Sorbents for Oil-Spill Cleanup: Recent Advances and Future Perspectives. *Adv. Mater.* **2016**, *28*, 10459–10490. [CrossRef] [PubMed]
125. Liu, T.; Huang, M.L.; Li, X.F.; Wang, C.J.; Gui, C.X.; Yu, Z.Z. Highly compressible anisotropic graphene aerogels fabricated by directional freezing for efficient absorption of organic liquids. *Carbon* **2016**, *100*, 456–464. [CrossRef]
126. Sui, Z.Y.; Meng, Q.H.; Zhang, X.T.; Ma, R.; Cao, B. Green synthesis of carbon nanotube-graphene hybrid aerogels and their use as versatile agents for water purification. *J. Mater. Chem.* **2012**, *22*, 8767–8771. [CrossRef]
127. Hu, J.; Zhu, J.D.; Ge, S.Z.; Jiang, C.W.; Guo, T.Y.; Peng, T.P.; Huang, T.; Xie, L. Biocompatible, hydrophobic and resilience graphene/chitosan composite aerogel for efficient oil-water separation. *Surf. Coat. Technol.* **2020**, *385*, 125361. [CrossRef]
128. Sun, A.Q.; Hou, X.A.; Hu, X.G. Super-performance photothermal conversion of 3D macrostructure graphene-CuFeSe2 aerogel contributes to durable and fast clean-up of highly viscous crude oil in seawater. *Nano Energy* **2020**, *70*, 104511. [CrossRef]
129. Qu, J.F.; Chen, D.Y.; Li, N.J.; Xu, Q.F.; Li, H.; He, J.H.; Lu, J.M. Engineering 3D Ru/Graphene Aerogel Using Metal-Organic Frameworks: Capture and Highly Efficient Catalytic CO Oxidation at Room Temperature. *Small* **2018**, *14*, 1800343. [CrossRef]
130. Liu, J.T.; Ge, X.; Ye, X.X.; Wang, G.Z.; Zhang, H.M.; Zhou, H.J.; Zhang, Y.X.; Zhao, H.J. 3D graphene/delta-MnO_2 aerogels for highly efficient and reversible removal of heavy metal ions. *J. Mater. Chem. A* **2016**, *4*, 1970–1979. [CrossRef]
131. Lu, M.X.; Deng, Y.J.; Luo, Y.; Lv, J.P.; Li, T.B.; Xu, J.; Chen, S.W.; Wang, J.Y. Graphene Aerogel-Metal-Organic Framework-Based Electrochemical Method for Simultaneous Detection of Multiple Heavy-Metal Ions. *Anal. Chem.* **2019**, *91*, 888–895. [CrossRef]
132. Pan, L.H.; Wang, Z.Q.; Yang, Q.; Huang, R.Y. Efficient Removal of Lead, Copper and Cadmium Ions from Water by a Porous Calcium Alginate/Graphene Oxide Composite Aerogel. *Nanomaterials* **2018**, *8*, 957. [CrossRef] [PubMed]

133. Wei, Y.; Xu, L.; Tao, Y.X.; Yao, C.; Xue, H.G.; Kong, Y. Electrosorption of Lead Ions by Nitrogen-Doped Graphene Aerogels via One-Pot Hydrothermal Route. *Ind. Eng. Chem. Res.* **2016**, *55*, 1912–1920. [CrossRef]
134. Zhang, Y.N.; Niu, Q.Y.; Gu, X.T.; Yang, N.; Zhao, G.H. Recent progress on carbon nanomaterials for the electrochemical detection and removal of environmental pollutants. *Nanoscale* **2019**, *11*, 11992–12014. [CrossRef] [PubMed]
135. Pan, Y.K.; Chen, M.Q.; Hu, M.F.; Tian, M.; Zhang, Y.Y.; Long, D.H. Probing the room-temperature oxidative desulfurization activity of three-dimensional alkaline graphene aerogel. *Appl. Catal. B-Environ.* **2020**, *262*, 118266. [CrossRef]
136. Vermeulen, S.J.; Campbell, B.M.; Ingram, J.S.I. Climate Change and Food Systems. *Annu. Rev. Environ. Resour.* **2012**, *37*, 195–222. [CrossRef]
137. Kamran, U.; Park, S.J. Chemically modified carbonaceous adsorbents for enhanced CO_2 capture: A review. *J. Clean Prod.* **2021**, *290*, 125776. [CrossRef]
138. Xia, D.; Li, H.; Mannering, J.; Huang, P.; Zheng, X.R.; Kulak, A.; Baker, D.; Iruretagoyena, D.; Menzel, R. Electrically Heatable Graphene Aerogels as Nanoparticle Supports in Adsorptive Desulfurization and High-Pressure CO_2 Capture. *Adv. Funct. Mater.* **2020**, *30*, 2002788. [CrossRef]
139. Hsan, N.; Dutta, P.K.; Kumar, S.; Bera, R.; Das, N. Chitosan grafted graphene oxide aerogel: Synthesis, characterization and carbon dioxide capture study. *Int. J. Biol. Macromol.* **2019**, *125*, 300–306. [CrossRef] [PubMed]
140. Peng, D.D.; Ding, Z.M.; Zhang, L.; Fu, Y.K.; Wang, J.S.; Li, Y.; Han, S.M. Remarkable hydrogen storage properties and mechanisms of the shell-core MgH2@carbon aerogel microspheres. *Int. J. Hydrogen Energy* **2018**, *43*, 3731–3740. [CrossRef]
141. Chen, X.; Zhou, S.K.; Zhang, L.M.; You, T.T.; Xu, F. Adsorption of Heavy Metals by Graphene Oxide/Cellulose Hydrogel Prepared from NaOH/Urea Aqueous Solution. *Materials* **2016**, *9*, 582. [CrossRef]
142. Qiu, B.C.; Xing, M.Y.; Zhang, J.L. Mesoporous TiO_2 Nanocrystals Grown in Situ on Graphene Aerogels for High Photocatalysis and Lithium-Ion Batteries. *J. Am. Chem. Soc.* **2014**, *136*, 5852–5855. [CrossRef] [PubMed]
143. Sun, X.Q.; Ji, S.D.; Wang, M.Q.; Dou, J.J.; Yang, Z.; Qiu, H.W.; Kou, S.; Ji, Y.Q.; Wang, H. Fabrication of porous TiO_2-RGO hybrid aerogel for high-efficiency, visible-light photodegradation of dyes. *J. Alloys Compd.* **2020**, *819*, 153033. [CrossRef]
144. Dong, S.Y.; Cui, Y.R.; Wang, Y.F.; Li, Y.K.; Hu, L.M.; Sun, J.Y.; Sun, J.H. Designing three-dimensional acicular sheaf shaped BiVO4/reduced graphene oxide composites for efficient sunlight-driven photocatalytic degradation of dye wastewater. *Chem. Eng. J.* **2014**, *249*, 102–110. [CrossRef]
145. Chowdhury, S.; Jiang, Y.Q.; Muthukaruppan, S.; Balasubramanian, R. Effect of boron doping level on the photocatalytic activity of graphene aerogels. *Carbon* **2018**, *128*, 237–248. [CrossRef]
146. Chen, Y.X.; Wang, P.L.; Liang, Y.; Zhao, M.J.; Jiang, Y.Y.; Wang, G.T.; Zou, P.; Zeng, J.; Zhang, Y.S.; Wang, Y. Fabrication of a three-dimensional porous Z-scheme silver/silver bromide/graphitic carbon nitride@nitrogen-doped graphene aerogel with enhanced visible-light photocatalytic and antibacterial activities. *J. Colloid Interface Sci.* **2019**, *536*, 389–398. [CrossRef] [PubMed]
147. Zhang, M.J.; Chen, Y.X.; Chen, B.J.; Zhang, Y.S.; Lin, L.; Han, X.W.; Zou, P.; Wang, G.T.; Zeng, J.; Zhao, M.J. Fabrication of a three-dimensional visible-light-driven Ag-AgBr/TiO_2/graphene aerogel composite for enhanced photocatalytic destruction of organic dyes and bacteria. *New J. Chem.* **2019**, *43*, 5088–5098. [CrossRef]
148. Zhang, J.; Wang, Z.W.; Fan, M.G.; Tong, P.P.; Sun, J.Y.; Dong, S.Y.; Sun, J.H. Ultra-light and compressible 3D BiOCl/RGO aerogel with enriched synergistic effect of adsorption and photocatalytic degradation of oxytetracycline. *J. Mater. Res. Technol.-JMRT* **2019**, *8*, 4577–4587. [CrossRef]
149. Feng, L.; Wei, P.; Song, Q.; Zhang, J.X.; Fu, Q.A.; Jia, X.H.; Yang, J.; Shao, D.; Li, Y.; Wang, S.Z.; et al. Superelastic, Highly Conductive, Superhydrophobic, and Powerful Electromagnetic Shielding Hybrid Aerogels Built from Orthogonal Graphene and Boron Nitride Nanoribbons. *ACS Nano* **2022**, *16*, 17049–17061. [CrossRef]
150. Zou, X.; Chen, Y.Y.; Zheng, Z.Y.; Sun, M.Y.; Song, X.F.; Lin, P.R.; Tao, J.; Zhao, P. The sensitive monitoring of living cell-secreted dopamine based on the electrochemical biosensor modified with nitrogen-doped graphene aerogel/Co_3O_4 nanoparticles *Microchem. J.* **2022**, *183*, 107957. [CrossRef]
151. Cheng, Y.J.; Sun, X.X.; Yang, S.; Wang, D.; Liang, L.; Wang, S.S.; Ning, Y.H.; Yin, W.L.; Li, Y.B. Multifunctional elastic rGO hybrid aerogels for microwave absorption, infrared stealth and heat insulation. *Chem. Eng. J.* **2023**, *452*, 139376. [CrossRef]
152. Liang, L.Y.; Li, Q.M.; Yan, X.; Feng, Y.Z.; Wang, Y.M.; Zhang, H.B.; Zhou, X.P.; Liu, C.T.; Shen, C.Y.; Xie, X.L. Multifunctional Magnetic $Ti_3C_2T_x$ MXene/Graphene Aerogel with Superior Electromagnetic Wave Absorption Performance. *ACS Nano* **2021**, *15*, 6622–6632. [CrossRef] [PubMed]
153. Dong, J.B.; Wen, L.; Yang, H.S.; Zhao, J.Y.; He, C.J.; Hu, Z.K.; Peng, L.; Hou, C.J.; Huo, D.Q. Catalytic Hairpin Assembly-Driven Ratiometric Dual-SignalElectrochemical Biosensor for Ultrasensitive Detection of MicroRNABased on the Ratios of Fe-MOFs and MB-GA-UiO-66-NH2. *Anal. Chem.* **2022**, *94*, 5846–5855. [CrossRef] [PubMed]

materials

MDPI

Article

Performance Study of a Leaf-Vein-like Structured Vapor Chamber

Zhihao Zhou, Xu Wang and Yongmin Zhou *

College of Materials Science and Engineering, Nanjing Tech University, South Puzhu Road No. 30, Nanjing 211816, China; 202061203279@njtech.edu.cn (Z.Z.); 202061203241@njtech.edu.cn (X.W.)
* Correspondence: yongminzhou@njtech.edu.cn; Tel.: +86-139-5183-2092

Abstract: As optoelectronic products continue to advance rapidly, the need for effective heat dissipation has become increasingly crucial due to the emphasis on miniaturization and high integration. The vapor chamber is widely used for cooling electronic systems as a passive liquid–gas two-phase high-efficiency heat exchange device. In this paper, we designed and manufactured a new kind of vapor chamber using cotton yarn as the wick material, combined with a fractal pattern layout of leaf veins. A comprehensive investigation was conducted to analyze the performance of the vapor chamber under natural convection circumstances. SEM showed that many tiny pores and capillaries were formed between the cotton yarn fibers, which are very suitable as the wick material of the vapor chamber. Additionally, experimental findings demonstrated the favorable flow and heat transfer characteristics of the cotton yarn wick within the vapor chamber, which makes the vapor chamber have significant heat dissipation capability, compared to the other two vapor chambers; this vapor chamber has a thermal resistance of only 0.43 °C/W at a thermal load of 8.7 W. In addition, the vapor chamber showed good antigravity capability, and its performance did not show significant changes between horizontal and vertical positions; the maximum difference in thermal resistance at four tilt angles is only 0.06 °C/W. This paper also studied the influence of vacuum degree and filling amount on the performance of the vapor chamber. These findings indicate that the proposed vapor chamber provides a promising thermal management solution for some mobile electronic devices and provides a new idea for selecting wick materials for vapor chambers.

Keywords: vapor chamber; leaf vein; cotton thread; natural convection; heat dissipation

1. Introduction

As information technology, electronics, and related industries continue to progress rapidly, a wide range of electronic products have emerged, such as smartphones, tablet PCs, etc., which are developing in the direction of miniaturization, integration, and lightweight. The increase in heat generation not only has a significant impact on the life of electronic devices and the stability of use, but too high a surface temperature will also affect the user experience. A vapor chamber, commonly referred to as a flat heat pipe [1–3], is a specialized heat sink extensively utilized in electronic products. The vapor chamber has attracted significant attention from researchers due to its unique characteristics compared to heat pipes. These include a large condensation area, uniform surface temperature distribution, and wide range of applications, among others [4–6]. The operational mechanism of a vapor chamber closely resembles that of a heat pipe. At the evaporation end of the vapor chamber, the working fluid within the wick absorbs heat, causing it to heat up and evaporate. The vapor condenses when it comes into contact with the colder wall due to the rapid movement of vapor pressure steam around. Consequently, the working fluid releases its latent heat of vaporization, and through capillary action, the condensed fluid returns to the heat source.

Vapor chambers have been widely explored in recent years due to their superior heat homogenization performance. In order to improve the performance of vapor chambers, many researchers are exploring various new wick materials and structures. Tang et al. [7] manufactured a vapor chamber that utilizes sintered copper powder as the wick material and employs a porous cylindrical structure as a support pillar. Their study indicates that this vapor chamber exhibits a minimum thermal resistance of 0.08 K/W and a maximum heat transfer limit of 300 W/cm^2. Ji et al. [8] utilized porous copper foam as a wick material in the vapor chamber and conducted experiments to determine that the minimum thermal resistance of this vapor chamber can reach 0.09 K/W. Furthermore, it was found to operate effectively under a heat flux density of 216 W/cm^2. Yu et al. [9] utilized an electrodeposition method to fabricate porous copper material, which was subsequently applied in a vapor chamber. Experimental results demonstrated that the vapor chamber with this electrodeposited copper core exhibited excellent heat dissipation capabilities. Peng et al. [10] introduced a unique leaf vein microchannel slot wick system in a vapor chamber. The design of this vapor chamber, created through chemical etching, facilitated enhanced temperature uniformity and minimized thermal resistance. These benefits are attributed to the enhanced fluid flow and increased condensation area resulting from this structural configuration. Wong et al. [11,12] developed a new type of vapor chamber by engraving microgrooves on the condenser-side shell plate and using a sintered core as the liquid wick on the evaporator-side shell plate. Assessment of this novel vapor chamber demonstrated a lower thermal resistance compared to conventional counterparts. Wang et al. [13] investigated and studied the heat dissipation performance of groove-type and foam copper-type vapor chambers. The results showed that the thermal resistance of the foam copper-type vapor chamber is lower than that of the groove-type vapor chamber because the foam copper can provide greater capillary force. Yang et al. [14] integrated a wetting-patterned substrate underneath a nanoscale-structured mesh core, and experimentally evaluated the thermal resistance of the vapor chamber. The measurement findings revealed that the utilization of a wetting-patterned substrate beneath the nanoscale-structured mesh core led to a notable reduction in horizontal thermal resistance. Furthermore, this structure also enables a more uniform surface temperature on the condenser side of the vapor chamber. Liu et al. [15] utilized a metal wire mesh as the wick in the vapor chamber and applied a super hydrophilic surface structure to the mesh. This structure significantly improved the thermal performance of the vapor chamber. Yu et al. [16] fabricated a vapor chamber with a spiral woven mesh using a brazing process, and the maximum heat transfer capacity of the vapor chamber was experimentally determined to be 6 W.

Many researchers are not satisfied with the heat dissipation performance of a single-structured wick vapor chamber and have further employed various methods to enhance its heat dissipation performance. Deng et al. [17] fabricated a composite wick structure vapor chamber and engraved numerous small grooves on the evaporator-side shell plate, and analysis showed that it had good temperature uniformity and thermal performance. Shaeri et al. [18] fabricated an innovative vapor chamber to assess the viability of integrating hydrophobic and hydrophilic properties in the evaporator for enhanced thermal performance. The findings demonstrated that the proposed vapor chamber exhibited a lower thermal resistance compared to conventional designs. Ju et al. [19] developed a novel vapor chamber with two types of core structures designed internally. The single-layer sintered copper powder core effectively reduces the thermal resistance of the vapor chamber. They also compared the performance of three different core structures, including vertical columnar artery, convergent lateral artery, and double-hole system. Patankar et al. [20] devised a methodology for designing vapor chambers specifically tailored to meet the distinct transport constraints and thermal management conditions to be met in mobile electronics. The vapor chamber was intended to improve condenser-side temperature uniformity in an ultrathin shape mainly by controlling the condenser-side core to improve lateral heat transfer. Zhao et al. [21] developed a composite wick vapor chamber by sintering copper mesh and powder. Through experimental analysis, it was determined that the vapor chamber achieved a maximum heat dissipation power of 23.29 W, while maintaining a minimum thermal resistance of 0.33 °C/W. Sudhakar et al. [22] designed a dual-layer wick structure in the vapor chamber. Experimental results showed that, compared to the single-layer wick structure vapor chamber, it could withstand higher heat flux densities. George et al. [23]

designed a vapor chamber without a wick structure in which the internal work fluid flows using the surface tension of the wettability pattern inside the shell plate. The thermal resistance of this vapor chamber was found to be 0.18 K/W at 10 W by experimental test results.

Through billions of years of evolution, nature has provided us with opportunities to learn from its various behaviors and physiological structures across different domains. Inspired by leaf veins, Liu et al. [24] designed and engraved a leaf-vein-like groove network on the condenser side of the vapor chamber, while using a sintered copper core as the wick on the evaporator side. Experimental results demonstrated that this vapor chamber exhibited excellent performance. Peng et al. [25,26] fabricated a new vapor chamber wick structure based on a fractal network of plant veins, comprising a grooved design. The findings demonstrated that the system was characterized by sizeable capillary pressure and high penetration; therefore, the vapor chamber has good performance. However, the design and fabrication of this wick were complex, and the fabrication of fine grooves required costly chemical etching.

Inspired by the water transport capacity of plant leaves, a new biomimetic vapor chamber was designed and manufactured in this study. The vapor chamber used an aluminum alloy as its shell and cotton yarn as the wick material. The fractal structure of leaf veins was imitated to design the distribution of the wick material, and the heat dissipation performance of the vapor chamber under natural convection conditions was studied through experimental testing.

2. Design and Fabrication

2.1. Production of Vapor Chamber

The structure of the internal suction wick of the vapor chamber and the fabrication method of the vapor chamber are described in this section. The upper and lower shells of the vapor chamber formulated in this study were made of 6061 aluminum alloy with $160 \times 80 \times 1$ mm as the dimensions. Anhydrous ethanol was used as the working fluid. To create the cavity structure of the vapor chamber, an aluminum wire with a diameter of 2 mm was bent into a 160×80 mm rectangle as the intermediate frame of the vapor chamber, and then the upper and lower shell plates were bonded together with the middle frame using metal adhesive to form the cavity structure. The surrounding area of the bonding was further sealed with organic silicon sealant. The method used to make the vapor chamber in this study was simple, and the vapor chamber can be disassembled and reused for other studies. This significantly reduces the cost of fabrication and saves resources. The roles of the primary materials are shown in Table 1. The main production process includes cleaning the upper and lower shell plates, arranging and fixing the wick, assembling the upper and lower shell plates, adding liquid-filled tubes, leak detection, liquid-filling, vacuuming, and sealing the liquid-filled tubes.

Table 1. Main production materials.

Materials	Role	Size/Diameter
6061 Aluminum alloy	Upper and lower shell plates	$160 \times 80 \times 1$ mm
Aluminum wire	Intermediate frame	2 mm
Ethanol	Working fluid	/
Stainless steel liquid-filled tube	Liquid filling, vacuuming	2 mm
Metal glue	Fixed, sealed	/
Silicone sealant	Sealed	/

2.2. Structural Design of the Wick

Findings on the transpiration of plants and previous studies on plant leaf veins [27,28] indicate that the fractal structure of leaf veins has a high axial conductivity. This characteristic of the fractal structure allows water transport to the leaf's distal end with low water potential loss. This naturally optimized structure was mimicked to design the wick inside the vapor chamber. The wick structure developed in this study was made of cotton woven from cotton. Cotton is a natural fiber with good moisture absorption. Moreover, the woven cotton thread has good flexibility [29] and can be designed into various desired shapes, which can be utilized as the wick of the vapor chamber. Figure 1 shows the scanning electron microscope image of cotton thread; the surface of the cotton thread is formed by layers of interwoven fibers, which create many tiny pores and capillaries between the fibers. These features are crucial for efficiently transporting the working fluid in the vapor chamber. When the vapor chamber is activated, the strong capillary forces drive the working fluid to flow along the tiny pores formed between the cotton thread fibers toward the heat source, facilitating the internal phase-change heat transfer process.

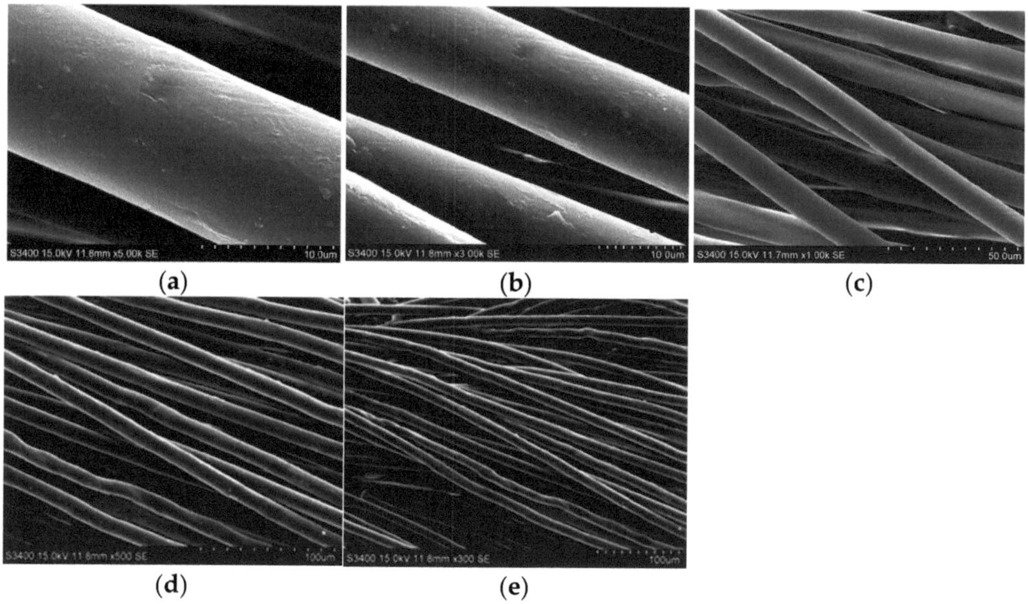

Figure 1. SEM image of the wick: (**a**) 5000 times; (**b**) 3000 times; (**c**) 1000 times; (**d**) 500 times; (**e**) 300 times.

A schematic diagram showing the design process for obtaining the wick is shown in Figure 2a. The fractal points were 2 cm apart, and the fractal angle was 45°. The structure of the wick made by this design is shown in Figure 2b. According to the above method, 24 strands of cotton thread are required to arrange the wick inside the vapor chamber. As the vapor chamber has different lengths and widths, an 8.5 cm cotton yarn was used in the length direction, and a 4.5 cm cotton yarn was used in the width direction, resulting in a 10 × 10 mm intersection area in the center of the shell plate. The heat source is located below the intersection area. Each branch has a strand of cotton thread. The diameter of the cotton thread was 2 mm (made of 19 fine cotton yarn).

According to the above design, the fixation method used a marker pen to draw the fractal network on the shell plate. Then, using adhesive, fine yarn with a diameter of 0.1 mm was fixed at each branching point of the vapor chamber shell plate. The distance between each fixed point was 2 cm. Finally, the cotton yarn was tied to the shell plate according to the fractal structure using fine thread. This process was conducted to ensure the integrity of the internal capillary tube of the wick. Fixing the cotton yarn using dispensing would block the capillary hole and obstruct the working fluid return. Because no suitable method has been found to restore the cotton thread on the shell plate, only this manual method can be used to fix it.

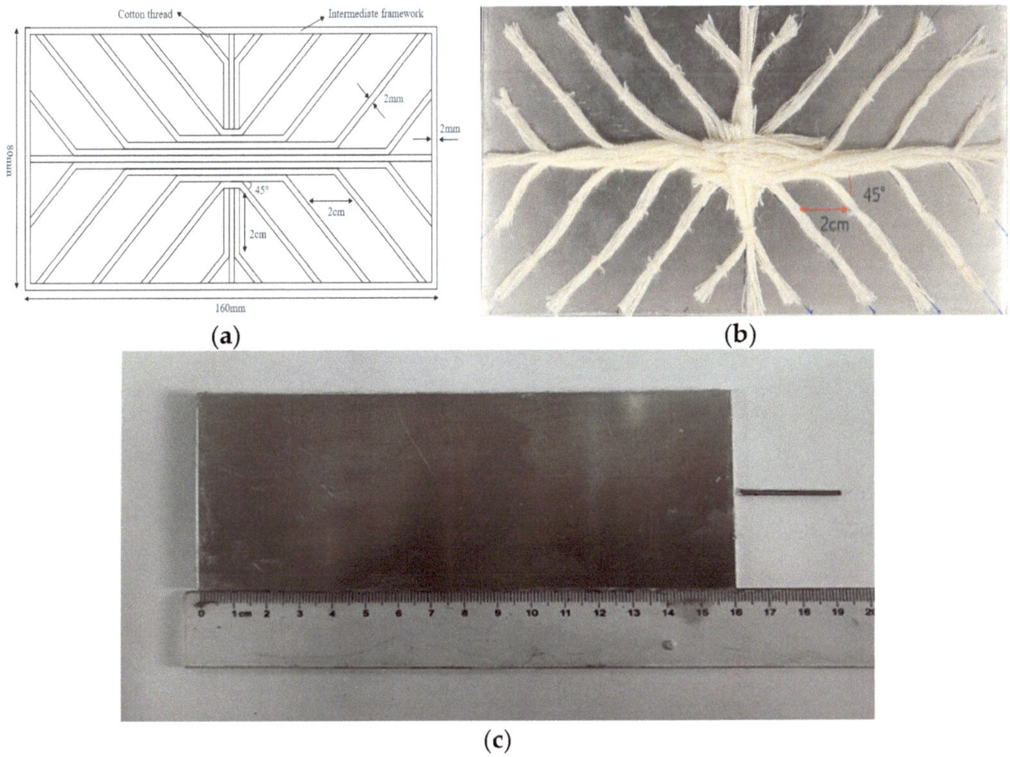

Figure 2. (**a**) Schematic diagram of wick; (**b**) physical diagram of wick; (**c**) finished vapor chamber.

After securing the wick, to prevent deformation under vacuum conditions, support columns were incorporated into the design of the vapor chamber, providing essential structural support. The support column is made of nylon and is a cylinder with a diameter of 5 mm and a length of 2 mm. The upper and lower shell plates, the middle frame, and the liquid-filled tube are assembled using the method described in Section 2.1. The completed finished product is shown in Figure 2c. The working fluid is injected through the filler tube, and the cavity is evacuated using a vacuum pump. Finally, the end of the filler tube is flattened to form a temporary cold weld seal.

3. Experimental Setup and Methods

3.1. Experimental Setup

Figure 3a shows the schematic of the experimental setup used to evaluate the heat transfer performance of the vapor chamber samples designed in this study under natural convection conditions. The heating module was made of a ceramic heating block of $10 \times 10 \times 2$ mm. This module had the advantages of fast heating speed and uniform heat distribution, and its heating power could be adjusted by adjusting the voltage of the connected DC power supply. To minimize heat loss, the heating block was positioned at the central bottom of the vapor chamber and surrounded by insulation cotton. Additionally, a thin layer of thermal grease was applied to the contact surface between the heating block and the vapor chamber, effectively reducing contact thermal resistance. The thermocouples were type K (−20–200 °C) and were calibrated to an uncertainty of ±0.1 °C; this range covered the anticipated temperature fluctuations during the operation of the vapor chamber. The chosen precision level was sufficient to meet experimental needs. Thermocouple T_0 was placed at the edge of the heat source in close proximity to the vapor chamber shell to approximate the heat source's surface temperature and the vapor chamber's evaporator temperature, as shown in Figure 3a. Figure 3b shows the distribution of thermocouples on the condenser surface. The room temperature was approximately around 20 °C during the experiments. However, it is important to note that we did not deliberately control or regulate the room temperature; therefore, the room temperature may fluctuate slightly during the experiment.

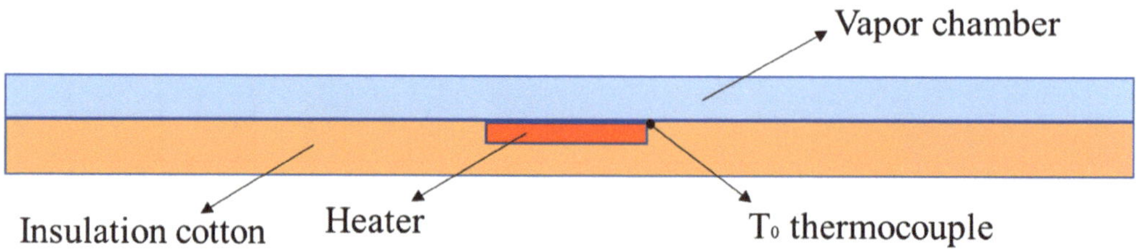

(a)

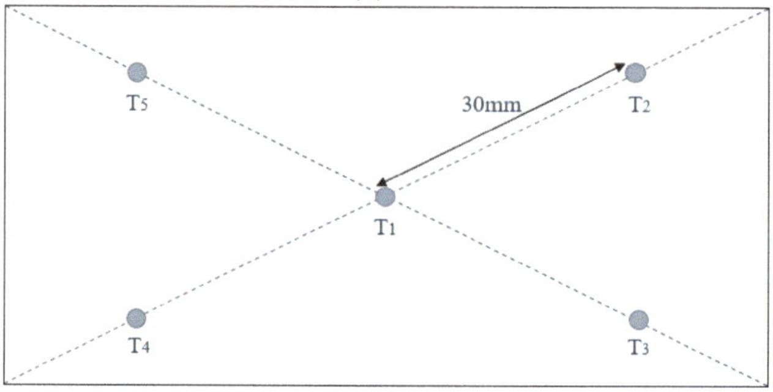

(b)

Figure 3. (**a**) Testing bench of vapor chamber; (**b**) condensing end surface thermocouple distribution map.

3.2. Data Reduction and Uncertainty Analysis

The input power of the heat source is expressed as follows:

$$Q = UI \tag{1}$$

The input power of the heat source was determined by multiplying the readings displayed on the DC power meter, denoted as U and I, which were connected to the heating block.

The maximum temperature difference ΔT_{max} at the surface of the condensing end of the vapor chamber is expressed as $\Delta T_{max} = \max\{T_1, T_2, T_3, T_4, T_5\} - \min\{T_1, T_2, T_3, T_4, T_5\}$, where T_1, T_2, T_3, T_4, T_5 represent the temperatures measured by the thermocouples on the surface of the condensing end.

The thermal resistance R of the vapor chamber is expressed as follows:

$$R = \frac{T_0 - T_{average}}{Q} \tag{2}$$

where T_0 denotes the temperature measured by the thermocouple and $T_{average}$ represents the average temperature of the surface at the condensing end of the vapor chamber; $T_{average}$ can be expressed as $T_{average} = \frac{T_1+T_2+T_3+T_4+T_5}{5}$.

The primary sources of uncertainty in the test system can be attributed to the adjustable DC power supply and the K-type thermocouple. The adjustable DC power supply has an accuracy of 0.5% for current measurements and 0.1% for voltage measurements. The calibration measurement accuracy of the type K thermocouple was 0.1 °C, and the error of the data logger was 0.5%. The measurement uncertainties were analyzed using the standard error analysis method. The maximum relative uncertainties for the input power and thermal resistance were 0.1% and 6.5%, respectively.

4. Results and Discussion

4.1. Performance Comparison of Different Core Structures

This section compares the heat transfer performance among three different vapor chambers. VC1 has no wick inside, VC2 is the vapor chamber designed in this study, and VC3 has a wick made of cotton threads with the same mass as VC2, randomly distributed in the vapor chamber's cavity. The dimensions of the three vapor chambers are 160 × 80 × 4 mm, with a filling amount of 1 g and an absolute pressure of 0.028 MPa in the empty hole.

The three vapor chambers were tested at heating powers of 2.6 W, 4.1 W, 5.8 W, 6.9 W, and 8.7 W, respectively. Figure 4 shows the connection between the surface temperature of the heat source and time for the three types of vapor chambers. From the figure, it can be found that the temperature of the heat source corresponding to all three vapor chambers increases with the increase of heat load. The time taken by VC1 to reach stability is the longest, and the stabilization time is 1380 s, 1440 s, 1500 s, 1560 s, and 1740 s, respectively, for the three heating powers. Furthermore, the surface temperature of the heat source is the highest after stabilization. The stabilization time of VC2 is the shortest, and it is 1260 s, 1320 s, 1380 s, 1380 s, and 1440 s, respectively, for the five heating powers. VC2 has the lowest surface temperature of the heat source after stabilization and VC3 is in between the two. This may be because VC1 has no wick, and the working fluid in the cavity of the vapor chamber can only return to the heat source through gravity and natural flow after being heated and evaporated, resulting in the drying out of the heat source area and seriously damaging the performance of the vapor chamber. Although VC3 has a wick, its random distribution obstructs the flow of working fluid and steam, affecting its performance. The wick of VC2 is made of a leaf vein fractal structure, which is conducive to the flow of working fluid inside and enables the condensate, after being heated and evaporated, to quickly return to the heat source through the wick at the condensation end of the vapor chamber.

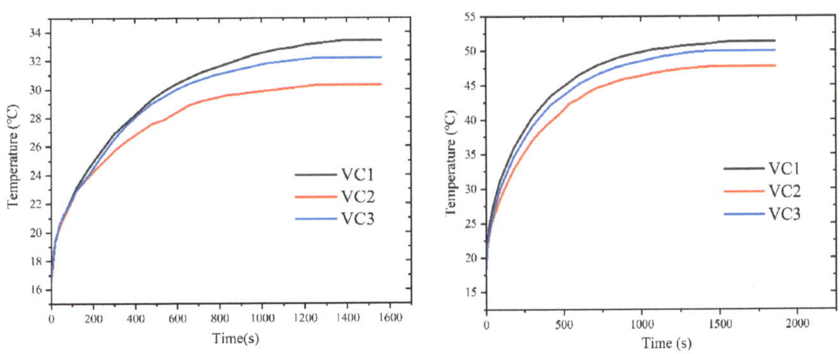

(a) (b)

Figure 4. *Cont.*

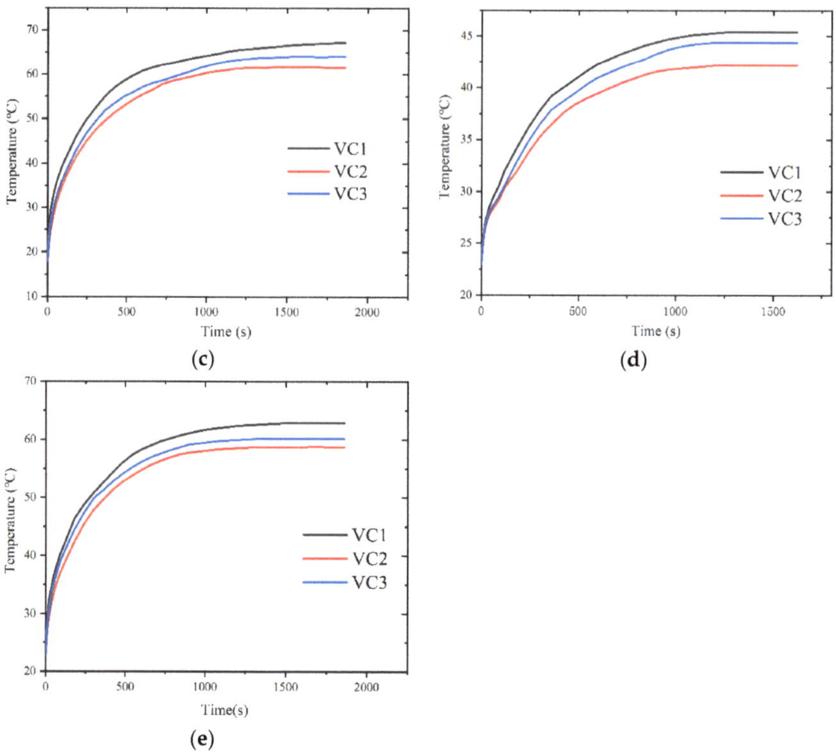

Figure 4. Heat source temperature of different structure of the vapor chamber under different heat generating power: (**a**) 2.6 W; (**b**) 5.8 W; (**c**) 8.7 W; (**d**) 4.1 W; (**e**) 6.9 W.

Figure 5 shows the thermal resistance and condenser surface temperature difference of the three vapor chambers after reaching a steady state. From Figure 5a, it can be seen that the maximum surface temperature difference at the condensing end of the three vapor chambers increases with the growth of the thermal load. The surface temperature difference of VC2 is more significant than the other two under all five heating powers. This may be because the wick of VC2 converges at the center position of the vapor chamber, connecting the upper and lower shell plates. At the same time, the heat source is heated at the bottom, forming another heat transfer path: the heat is transferred to the evaporative end shell plate through the wick and the working fluid inside the wick, which results in a greater surface temperature difference than the other two structures of the vapor chamber. As seen in Figure 5b, the thermal resistance of VC1 is the largest, followed by VC3, and VC2 has the most minor thermal resistance. This is because in VC1, the condensate returns too slowly after the working fluid evaporates, causing drying out of the heat source area and failure of the vapor chamber, resulting in an increase in thermal resistance. This further demonstrates that the vapor chamber with leaf vein fractal structure designed in this study has excellent fluid flow and heat convection performance.

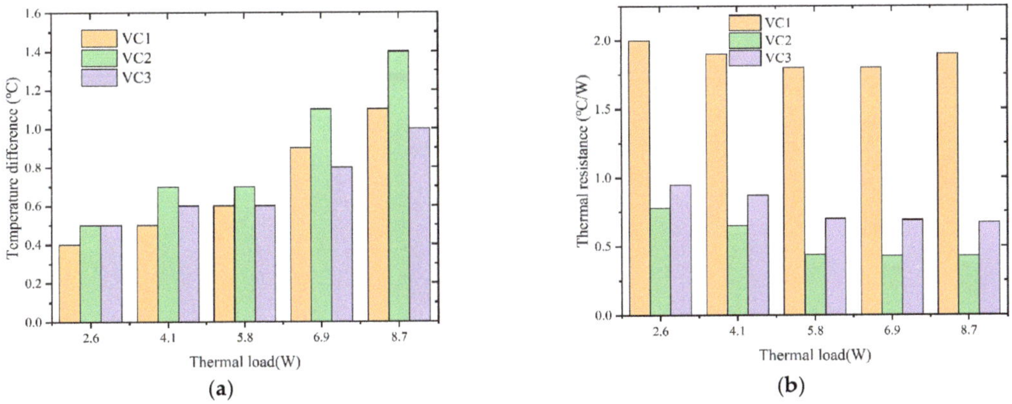

Figure 5. Surface temperature difference and thermal resistance of different structures of vapor chamber under different heat generating power: (**a**) surface temperature difference of vapor chamber; (**b**) thermal resistance of vapor chamber.

4.2. Empty Cavity Pressure's Effect on Vapor Chamber Performance

Figure 6 shows the variation of the temperature of the heat source surface with time at different pressures inside the vapor chamber, which was tested under a heat load of 5.8 W with a filling amount of 1 g. The results indicate that the vapor chamber with lower pressure has a shorter startup time, as indicated by the smaller slope at the beginning and a lower temperature after reaching a steady state. Specifically, the heat source temperature at pressures of 0.1 MPa, 0.044 MPa, 0.028 MPa, and 0.019 MPa were 51 °C, 50.1 °C, 48.1 °C, and 46.4 °C, respectively. This is because as the pressure inside the vapor chamber decreases, the boiling point of the working fluid also decreases, making it easier for the liquid to evaporate. Therefore, at lower pressures, the internal working fluid can participate in the phase-change heat transfer process in a shorter time, allowing the heat generated by the heat source to be promptly transferred from the evaporator side of the vapor chamber to the condenser side and then to the surrounding air through natural convection.

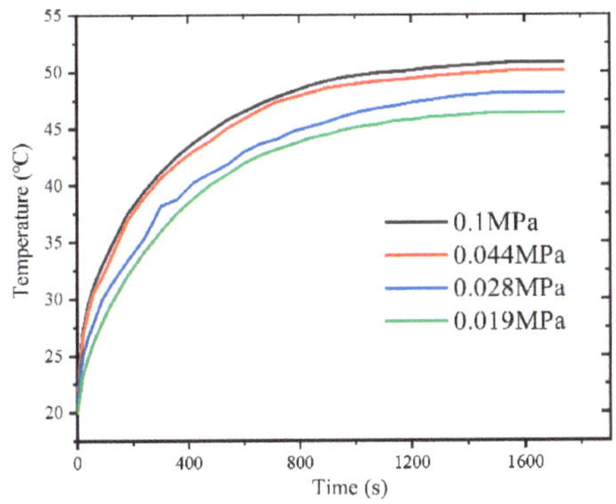

Figure 6. Variation of surface temperature of heat source with time under different pressure of vapor chamber.

Figure 7 shows a vapor chamber's thermal resistance and maximum surface temperature difference under a thermal load of 5.8 W. As seen from the graph, with decreasing pressure, the surface temperature difference of the vapor chamber increases, and the thermal resistance gradually decreases. For example, at atmospheric pressure, the surface temperature difference and thermal resistance of the vapor chamber are 0.5 °C and

1.1 °C/W, respectively. At a pressure of 0.019 MPa, the maximum surface temperature difference of the vapor chamber is 1.2 °C, and the thermal resistance is 0.39 °C/W. This may be due to the decrease in latent heat of the working fluid and the corresponding decline in boiling point as the pressure decreases. Under the same thermal load, the boiling of the working liquid becomes more intense, which enhances the phase-change heat transfer process in the vapor chamber and reduces thermal resistance.

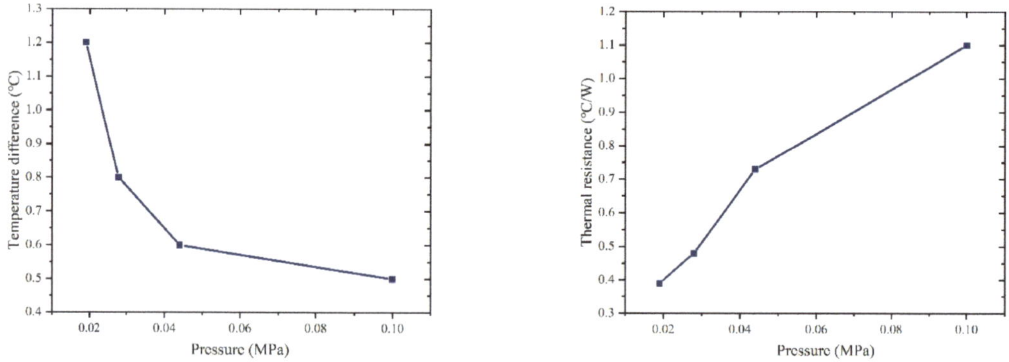

(**a**) (**b**)

Figure 7. The temperature difference of the surface of the vapor chamber at different pressures: (**a**) maximum temperature difference of the surface of the vapor chamber; (**b**) thermal resistance of the vapor chamber.

4.3. Impact of Filling Level on Vapor Chamber Performance

Figure 8 shows the variation of heat source temperature with time for different liquid filling amounts in the vapor chamber at a pressure of 0.028 MPa inside the vapor chamber. The results show that the temperature of the heat source gradually increases with increasing heat load. However, for a constant heat load, the temperature of the heat source decreases with increasing filling levels. For example, at a heat load of 2.6 W, the temperatures of the heat source for filling levels of 0.5 g, 1 g, and 1.5 g are 36.5 °C, 32.5 °C, and 31.3 °C, respectively. Similarly, at a heat load of 5.8 W, the temperatures of the heat source for filling levels of 0.5 g, 1 g, and 1.5 g are 53.1 °C, 48.4 °C, and 46.4 °C, respectively. At a heat load of 8.7 W, the temperatures of the heat source for filling levels of 0.5 g, 1 g, and 1.5 g are 65.9 °C, 60.7 °C, and 59.2 °C, respectively.

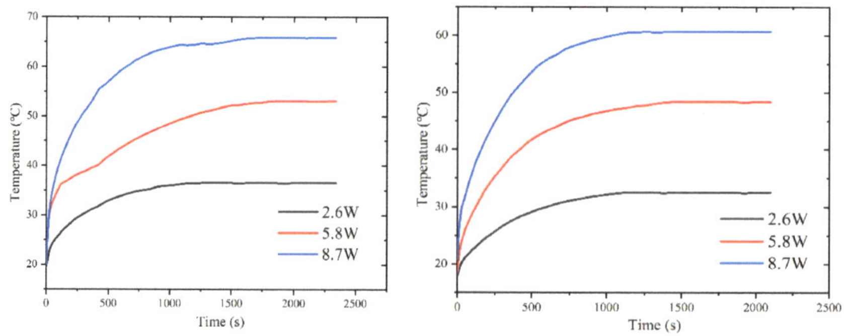

(**a**) (**b**)

Figure 8. *Cont.*

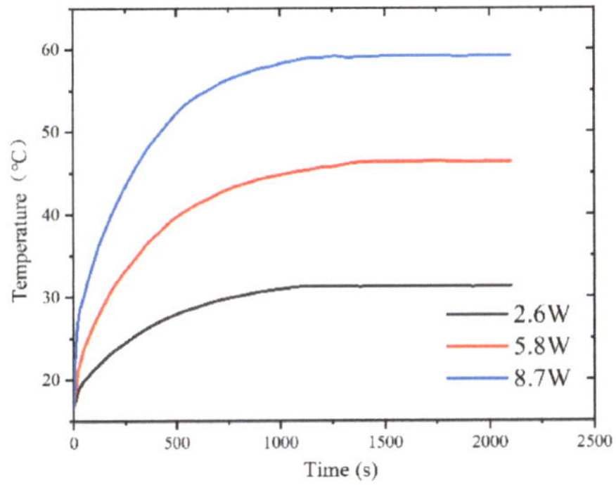

(c)

Figure 8. Variation of heat source temperature with time for different liquid filling amounts: (**a**) 0.5 g; (**b**) 1 g; (**c**) 1.5 g.

However, we observed a significant temperature change in the heat source when the working fluid filling increased from 0.5 g to 1 g. In contrast, when the filling amount increased from 1 g to 1.5 g, the temperature change of the heat source was not significant, especially at higher heat loads. This may be because a low filling amount resulted in the working fluid being unable to return to the evaporation end during the operation of the vapor chamber, leading to partial drying at the heat source and a decrease in heat transfer efficiency. With an increase in filling amount, enough working fluid was participating in the phase-change heat transfer process, thereby improving heat transfer efficiency.

In Figure 9a, the thermal resistance of the vapor chamber shows a different trend with increasing thermal load, depending on the filling amount. In the tested range, the thermal resistance increases with increasing thermal load for vapor chamber with 0.5 g filling amount, while it decreases with thermal load for 1 g and 1.5 g filling amount. The reason for this phenomenon may be that when the filling amount is 0.5 g, the increase in heat load causes partial drying inside the vapor chamber, which results in its inability to function correctly and a rapid rise in thermal resistance. At a heat load of 2.6 W, the thermal resistance of the vapor chamber with a filling amount of 0.5 g was significantly higher than that of the other two samples, indicating that the vapor chamber with a filling amount of 0.5 g had already experienced partial drying at this heat load, leading to poor performance. At high heat loads, the thermal resistance of the vapor chamber with filling amounts of 1 g and 1.5 g decreases with increasing heating power because the working fluid inside the vapor chamber can ensure sufficient liquid evaporation and reflux without drying out. In addition, with increasing heating loads, the boiling phenomenon inside the working fluid intensifies, enhancing its heat transfer performance. Meanwhile, we found that at 2.6 W, the vapor chamber with a 1.5 g filling amount has a higher thermal resistance compared to the vapor chamber with a 1 g filling amount. However, at heat loads of 5.8 W and 8.7 W, the vapor chamber with a 1.5 g filling amount exhibits a smaller thermal resistance than the vapor chamber with a 1 g filling amount. This may be because at low heat loads, the working fluid only partially boils, and the higher filling amount of the vapor chamber results in more working liquid in the wick, increasing the flow resistance of the working fluid and making circulation difficult, thereby reducing the heat transfer efficiency and increasing the thermal resistance. However, when working in a high-power environment, the higher heat flux promotes more working fluid to participate in the phase-change cycle, thereby reducing thermal resistance.

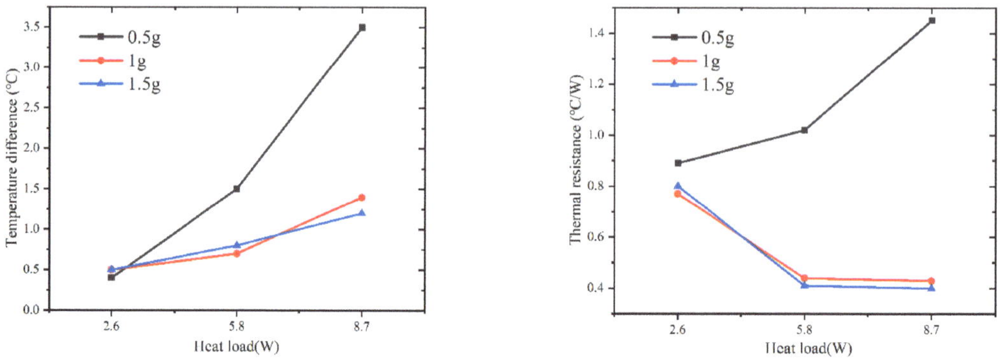

(a) (b)

Figure 9. (**a**) Maximum temperature difference of the surface of the condensing end of the vapor chamber under different liquid filling amounts; (**b**) thermal resistance of the vapor chamber under different liquid filling amounts.

From Figure 9b, it can be seen that the maximum temperature difference on the condensation surface of the vapor chamber increases with increasing heat load. The surface temperature difference of the vapor chamber with a filling amount of 0.5 g increases significantly compared to that of the vapor chambers with filling amounts of 1 g and 1.5 g with increasing heating power. This may be because the filling amount is too small, and the internal working fluid is insufficient to consume the heat generated by the high heat flux, resulting in vapor chamber failure.

4.4. Antigravity Performance

To investigate the antigravity performance of the vapor chamber designed in this paper, the heat transfer performance of the vapor chamber was studied at four different angles. As shown in Figure 10, experiments were conducted on the vapor chamber at four positions: horizontal, inclined at 30°, inclined at 60°, and vertical.

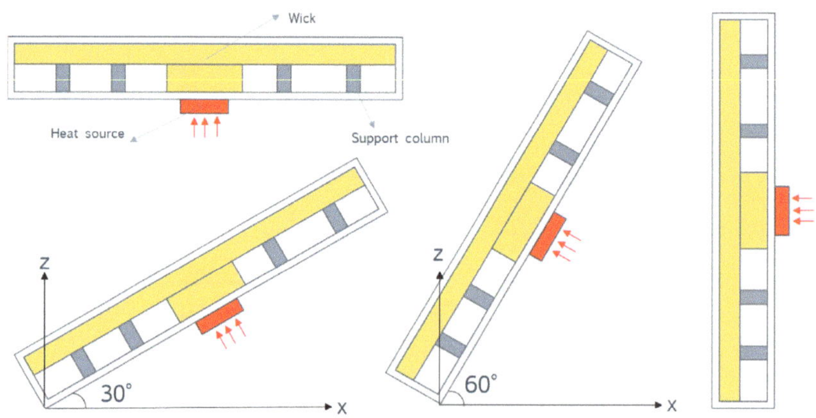

Figure 10. Schematic diagram of the tilt angle of the vapor chamber.

Figure 11a shows the variation of heat source temperature with time at different tilt angles for a vapor chamber with a working fluid filling of 1 g, internal cavity pressure of 0.028 MPa, and a heat generation rate of 5.8 W. It can be observed that the heat source temperature does not change significantly with the change in tilt angle. Figure 11b,c indicate that the vapor chamber's surface temperature difference and thermal resistance are not significantly affected by the tilt angle, which shows that the performance has not changed considerably. This is because the vapor chamber's wick has a full capillary force that allows the working fluid to flow back to the evaporation end, even in the vertical position. Additionally, the tilt angle does not affect the internal vapor flow, which is dominated by molecular forces and can penetrate the entire vapor chamber even in the vertical position. The vapor chamber's strong heat transfer capability relies on the internal vapor flow, proving that the designed vapor chamber in this study has good antigravity performance.

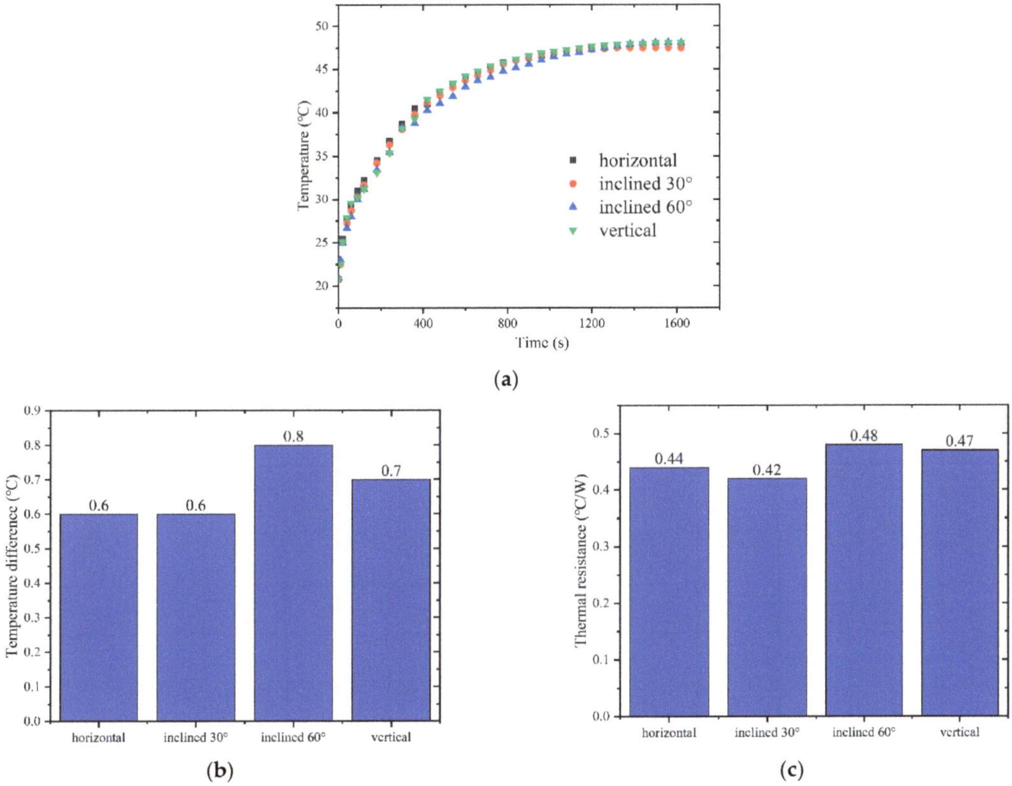

Figure 11. (**a**) Surface temperature of heat source at different tilt angles; (**b**) maximum temperature difference on the surface of vapor chamber at different tilt angles; (**c**) thermal resistance of vapor chamber at different tilt angles.

5. Conclusions

In this study, a new vapor chamber was fabricated with an aluminum alloy shell and a cotton wick arranged in a fractal vein-like structure within the vapor chamber. This novel design provides an effective thermal management strategy for some portable electronic devices and offers a new approach to selecting wick materials in vapor chambers. The experimental results demonstrate that the vein-like structure of the wick has good flow and heat transfer performance and exhibits excellent antigravity performance in both horizontal and vertical positions. The study also explores the effects of cavity pressure and working fluid filling on the performance of the vapor chamber. The main conclusions from this study are as follows:

(1) The designed vapor chamber with woven cotton as the wick structure exhibited excellent fluid flow and thermal convection performance. Compared to the other two vapor chambers, this vapor chamber exhibits better performance, with lower thermal resistance and corresponding heat source temperature.
(2) At the filling amount of 0.5 g, the performance of the vapor chamber is the worst. The performance of the vapor chamber does not differ much between the filling amounts of 1 g and 1.5 g, and within the tested range, the thermal resistance decreases with the increase of heat load.
(3) The thermal resistance of the vapor chamber decreases gradually as the pressure in the cavity decreases, and the temperature of the heat source also decreases accordingly.
(4) The designed vapor chamber demonstrated good antigravity performance, and there was no significant change in its heat dissipation performance at four different tilt angles.

In future work, we will explore more suitable manufacturing methods to produce this type of vapor chamber and conduct in-depth research on the working mechanism of this wick. We will also conduct a more in-depth study of this type of vapor chamber.

Author Contributions: Conceptualization, Z.Z. and X.W.; Methodology, Z.Z. and Y.Z.; Validation, Z.Z., X.W. and Y.Z.; Formal analysis, Z.Z.; Investigation, Z.Z. and Y.Z.; Data curation, Z.Z.; Writing—original draft, Z.Z.; Writing—review & editing, Z.Z. and Y.Z.; Supervision, X.W. and Y.Z. All authors have read and agreed to the published version of the manuscript.

Funding: This research received no external funding.

Conflicts of Interest: The authors declare no conflict of interest.

References

1. Chan, C.W.; Siqueiros, E.; Ling-Chin, J.; Royapoor, M.; Roskilly, A.P. Heat utilisation technologies: A critical review of heat pipes. *Renew. Sustain. Energy Rev.* **2015**, *50*, 615–627. [CrossRef]
2. Shabgard, H.; Allen, M.J.; Sharifi, N.; Benn, S.P.; Faghri, A.; Bergman, T.L. Heat pipe heat exchangers and heat sinks: Opportunities, challenges, applications, analysis, and state of the art. *Int. J. Heat Mass Transf.* **2015**, *89*, 138–158. [CrossRef]
3. Chen, X.; Ye, H.; Fan, X.; Ren, T.; Zhang, G. A review of small heat pipes for electronics. *Appl. Therm. Eng.* **2016**, *96*, 1–17. [CrossRef]
4. Ji, X.; Xu, J.; Abanda, A.M.; Xue, Q. A vapor chamber using extended condenser concept for ultra-high heat flux and large heater area. *Int. J. Heat Mass Transf.* **2012**, *55*, 4908–4913. [CrossRef]
5. Naphon, P.; Wongwises, S.; Wiriyasart, S. Application of two-phase vapor chamber technique for hard disk drive cooling of PCs. *Int. Commun. Heat Mass Transf.* **2013**, *40*, 32–35. [CrossRef]
6. Wang, J.-C.; Wang, R.-T.; Chang, T.-L.; Hwang, D.-S. Development of 30Watt high-power LEDs vapor chamber-based plate. *Int. J. Heat Mass Transf.* **2010**, *53*, 3990–4001. [CrossRef]
7. Tang, Y.; Yuan, D.; Lu, L.; Wang, Z. A multi-artery vapor chamber and its performance. *Appl. Therm. Eng.* **2013**, *60*, 15–23. [CrossRef]
8. Ji, X.; Xu, J.; Abanda, A.M. Copper foam based vapor chamber for high heat flux dissipation. *Exp. Therm. Fluid Sci.* **2012**, *40*, 93–102. [CrossRef]
9. Yu, J.-C.; Chien, H.-C.; Chiang, C.-Y.; Liu, E.-C.; Chang, Y.-H.; Huang, H.-H.; Chen, T.-Y.; Kao, C.-L.; Liao, C.-N. High-performance electrodeposited copper wicks for heat-spreading vapor chambers. *Appl. Therm. Eng.* **2023**, *228*, 120495. [CrossRef]
10. Peng, Y.; Liu, W.; Liu, B.; Liu, J.; Huang, K.; Wang, L.; Chen, W. The performance of the novel vapor chamber based on the leaf vein system. *Int. J. Heat Mass Transf.* **2015**, *86*, 656–666. [CrossRef]
11. Wong, S.-C.; Hsieh, K.-C.; Wu, J.-D.; Han, W.-L. A novel vapor chamber and its performance. *Int. J. Heat Mass Transf.* **2010**, *53*, 2377–2384. [CrossRef]
12. Wong, S.-C.; Huang, S.-F.; Hsieh, K.-C. Performance tests on a novel vapor chamber. *Appl. Therm. Eng.* **2011**, *31*, 1757–1762. [CrossRef]
13. Wang, L.; Zhang, D.; Chen, Y.; Sun, L.; Mao, J. Heat dissipation performance of grooved-type and copper foam-type vapor chambers. *Therm. Sci.* **2022**, *26*, 1357–1366. [CrossRef]
14. Yang, Y.; Li, J.; Wang, H.; Liao, D.; Qiu, H. Microstructured wettability pattern for enhancing thermal performance in an ultrathin vapor chamber. *Case Stud. Therm. Eng.* **2021**, *25*, 100906. [CrossRef]
15. Liu, T.; Yan, W.; Wu, W.; Wang, S. Thermal performance enhancement of vapor chamber with modified thin screen mesh wick by laser etching. *Case Stud. Therm. Eng.* **2021**, *28*, 101525. [CrossRef]

16. Yu, J.; Xin, Z.; Zhang, R.; Chen, Z.; Li, Y.; Zhou, W. Effect of spiral woven mesh liquid pumping action on the heat transfer performance of ultrathin vapour chamber. *Int. J. Therm. Sci.* **2022**, *182*, 107799. [CrossRef]
17. Deng, D.; Huang, Q.; Xie, Y.; Huang, X.; Chu, X. Thermal performance of composite porous vapor chambers with uniform radial grooves. *Appl. Therm. Eng.* **2017**, *125*, 1334–1344. [CrossRef]
18. Shaeri, M.R.; Attinger, D.; Bonner, R. Feasibility study of a vapor chamber with a hydrophobic evaporator substrate in high heat flux applications. *Int. Commun. Heat Mass Transf.* **2017**, *86*, 199–205. [CrossRef]
19. Ju, Y.S.; Kaviany, M.; Nam, Y.; Sharratt, S.; Hwang, G.; Catton, I.; Fleming, E.; Dussinger, P. Planar vapor chamber with hybrid evaporator wicks for the thermal management of high-heat-flux and high-power optoelectronic devices. *Int. J. Heat Mass Transf.* **2013**, *60*, 163–169. [CrossRef]
20. Patankar, G.; Weibel, J.A.; Garimella, S.V. Patterning the condenser-side wick in ultra-thin vapor chamber heat spreaders to improve skin temperature uniformity of mobile devices. *Int. J. Heat Mass Transf.* **2016**, *101*, 927–936. [CrossRef]
21. Zhao, Z.; Li, L.; Wang, Y.; Wang, Y.; Hui, Y. Heat transfer of copper mesh–powder composite-based sintered-wick vapor chamber. *AIP Adv.* **2023**, *13*, 015027. [CrossRef]
22. Sudhakar, S.; Weibel, J.A.; Garimella, S.V. Design of an Area-Scalable Two-Layer Evaporator Wick for High-Heat-Flux Vapor Chambers. *IEEE Trans. Compon. Packag. Manuf. Technol.* **2018**, *9*, 458–472. [CrossRef]
23. Damoulakis, G.; Megaridis, C.M. Wick-free paradigm for high-performance vapor-chamber heat spreaders. *Energy Convers. Manag.* **2022**, *253*, 115138. [CrossRef]
24. Liu, W.; Peng, Y.; Luo, T.; Luo, Y.; Huang, K. The performance of the vapor chamber based on the plant leaf. *Int. J. Heat Mass Transf.* **2016**, *98*, 746–757. [CrossRef]
25. Peng, Y.; Liu, W.; Wang, N.; Tian, Y.; Chen, X. A novel wick structure of vapor chamber based on the fractal architecture of leaf vein. *Int. J. Heat Mass Transf.* **2013**, *63*, 120–133. [CrossRef]
26. Peng, Y.; Liu, W.; Chen, W.; Wang, N. A conceptual structure for heat transfer imitating the transporting principle of plant leaf. *Int. J. Heat Mass Transf.* **2014**, *71*, 79–90. [CrossRef]
27. Liu, W.; Luo, Y.; Wang, L.; Luo, T.; Peng, Y.; Wu, L. Water transport in leaf vein systems and the flow velocity measurement with a new method. *J. Plant Physiol.* **2016**, *204*, 74–84. [CrossRef]
28. Luo, Y.; Liu, W.; Wang, L.; Xie, W. Heat and mass transfer characteristics of leaf-vein-inspired microchannels with wall thickening patterns. *Int. J. Heat Mass Transf.* **2016**, *101*, 1273–1282. [CrossRef]
29. Das, A.; Jain, A.; Pant, A. Study on the liquid flow behaviour of cotton wick. *Fibers Polym.* **2008**, *9*, 176–186. [CrossRef]

materials

Article

Kinetics of Catalytic Oxidation of Methylene Blue with La/Cu Co-Doped in Attapulgite

Jianping Shang [1,*], Wei Zhang [1], Zhengliang Dong [2] and Hua-Jun Shawn Fan [1]

[1] College of Chemical Engineering, Sichuan University of Science and Engineering, Zigong 643099, China
[2] Sichuan Institute of Geological Survey Salt Geology Drilling Team, Chengdu 610059, China
* Correspondence: shangjianping@suse.edu.cn; Tel.: +86-15708137030

Abstract: Methylene blue (MB) is a common pollutant in wastewater from the printing and dyeing industries. In this study, attapulgite (ATP) was modified with La^{3+}/Cu^{2+}, using the method of equivolumetric impregnation. The La^{3+}/Cu^{2+} -ATP nanocomposites were characterized using X-ray diffraction (XRD) and scanning electron microscopy (SEM). The catalytic properties of the modified ATP and the original ATP were compared. At the same time, the influence of the reaction temperature, concentration of methylene blue and pH on the reaction rate were investigated. The optimal reaction conditions are as follows: MB concentration is 80 mg/L, the dosage of the catalyst is 0.30 g, the dosage of hydrogen peroxide is 2 mL, the pH is 10 and the reaction temperature is 50 °C. Under these conditions, the degradation rate of MB can reach 98%. The recatalysis experiment was carried out reusing the catalyst, and the experimental results showed that the degradation rate could reach 65% after three uses, indicating that the catalyst could be recycled many times and costs could be reduced. Finally, the degradation mechanism of MB was speculated, and the reaction kinetic equation was obtained as follows: $-dc/dt = 14{,}044\ \exp(-3598.34/T)C(O)^{0.28}$.

Keywords: attapulgite; methylene blue; catalytic; kinetics

1. Introduction

Lanthanum is one of the more common rare earth elements. It is easily oxidized to a white oxide powder in the air [1]. Supported lanthanum oxide and lanthanum-based mixed oxides are effective catalysts for the removal of organics from wastewater [2,3].

Nowadays, printing and dyeing wastewater is characterized by high alkalinity, complex composition and difficulty in treatment [4]. At present, the treatment methods of printing and dyeing wastewater are as follows: adsorption [5–7], catalytic oxidation [8–12] and biological decomposition [13,14]. The catalytic wet oxidation (CWO) process is usually carried out under a high temperature and pressure, which limits its application. Therefore, the development of a new process and the preparation of new catalysts with high efficiency are the main directions for improving catalytic oxidation efficiency.

Attapulgite (or palygorskite, as it is commonly known) is a crystalline magnesium aluminosilicate hydrate, formulated as $Si_8O_{20}(Mg, Al, Fe)_5(OH)_2(OH_2)_4 \cdot 4H_2O$. ATP is a natural nanomaterial with active hydroxyl groups on the surface and zeolite-like channels inside [15]. It has a large specific surface area and is often used as an adsorbent, catalyst and catalyst carrier [16–19]. ATP is suitable as a catalyst carrier because of its large specific surface area, high chemical stability and mechanical stability. Especially for its low price, ATP has practical significance in wastewater treatment. Zhang et al. [20] prepared lanthanum-modified mesoporous adsorption materials using the hydrothermal synthesis method and found that the modified materials greatly improved the adsorption capacity of arsenic and phosphorus in an adsorption experiment. Hamed et al. [21] prepared lanthanum-doped $CuCr_2O_4$ nanoparticles using the sol–gel method. The photocatalytic experiments on organic pollutants showed that the degradation rate of nanomaterials mixed with lanthanum reached 98%, which was 20% higher than that of unmixed catalysts. Wang et al. [22] added lanthanum to a V_2O_5 catalyst to investigate the effect of lanthanum on its catalytic performance. The experimental results showed that the catalyst can participate in the reaction at a low temperature and that a certain amount of lanthanum can promote the formation of pyrosulfate and increase the specific surface area and pore diameter of the catalyst. In this paper, attapulgite (ATP) was modified using La^{3+}/Cu^{2+}, through the method of equivolumetric impregnation. The effect of the reaction temperature, catalyst dosage, pH, and hydrogen peroxide dosage on the degradation rate of MB was investigated.

2. Materials and Methods

2.1. Materials

Attapulgite (ATP) came from Xuyi, Jiangsu. Copper nitrate ($Cu(NO_3)_2{\bullet}3H_2O$) was purchased from the Magnolia town industrial development zone, Xindu District, Chengdu. Lanthanum nitrate ($La(NO_3)_3{\bullet}6H_2O$) was purchased from Tianjin Kemeo chemical reagent Co., Ltd., Tianjin, China. Hydrogen peroxide (H_2O_2, 30%) was purchased from Chengdu cologne chemical Co., Ltd. Methylene blue ($C_{16}H_{18}$ $ClN_3S{\bullet}3H_2O$) was purchased from Chengdu cologne chemical Co., Ltd. These drugs were used for the analysis.

2.2. Methods

The experimental set-up and reagents used in this work were described in detail in the literature [23]. The experimental procedure was as follows: about 40 g of attapulgite with a particle size of about 180 mesh was screened and immersed in 200 mL of 4 mol/L dilute hydrochloric acid for 3 h, then the vacuum filter was used to filter, followed by repeated rinse with deionized water until the filtrate was neutral, and the remainder was put in the drying oven, drying under 80 °C for 2 h. Then the composite catalyst was prepared using the equal volume method, in which the mass ratio (ATP:La) was 20:1 and the molar ratio (La:Cu) was 3:1. The loading time was 12 h. The loaded catalyst was put in the drying oven, drying under 80 °C for 6 h. Products were dried in the tube furnace at 300 °C, 400 °C and 500 °C calcine, respectively, for 2 h. The calcined catalyst was sealed and bagged for later use. After the temperature reached the reaction temperature, the La^{3+}/Cu^{2+}-ATP catalyst and hydrogen peroxide were added, and samples were taken at the times of 0, 1, 3, 5, 7, 10, 15 and 20 min.

Methods for the analysis of MB concentration include colorimetry, iodometry, spectrophotometry and liquid chromatography. For the advantages of simpler operation, lower cost and higher accuracy, spectrophotometry was selected as the MB concentration analysis method in this experiment. By measuring the absorbance or luminous intensity of the substance at a specific wavelength or within a certain wavelength range, a qualitative and quantitative analysis method was adopted for the substance. The Lambert–Beer theorem was applied, in the case of the low concentration solutions.

3. Results and Discussion

3.1. Effect of Hydrogen Peroxide

For the 250 mL MB solution, the experimental conditions are as follows: the concentration of MB is 80 mg/L, the dosage of the catalyst is 0.3 g, the reaction temperature is 50 °C, the pH is 10, the dosage of hydrogen peroxide is 0 mL, 0.5 mL, 1.0 mL, 1.5 mL and 2.0 mL, respectively. The experimental results are shown in Figure 1.

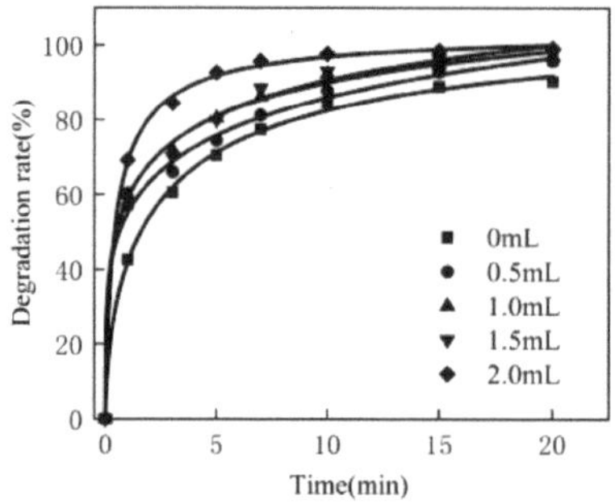

Figure 1. Effect of hydrogen peroxide consumption.

As can be seen from Figure 1, when there was no H_2O_2, MB was adsorbed on the ATP surface, resulting in a decrease in MB and an increase in the degradation rate. When H_2O_2 was added, the degradation rate of MB significantly increased, which was due to the effective collision between H_2O_2 and the catalyst, the Fenton-like reaction and the formation of •OH, which promoted the decomposition of MB. In addition, with the increase in H_2O_2 dosage, the degradation rate of MB accelerated, and the final degradation rate also increased within the same timeframe. Therefore, the dosage of H_2O_2 in subsequent experiments was 2 mL.

3.2. *Effect of Reaction Temperature*

For the 250 mL MB solution, the experimental conditions are as follows: the concentration of MB is 80 mg/L, the dosage of the catalyst is 0.3 g, the pH is 10, the dosage of hydrogen peroxide is 2.0 mL and the reaction temperature is 30 °C, 35 °C, 40 °C, 45 °C and 50 °C, respectively. The experimental results are shown in Figure 2.

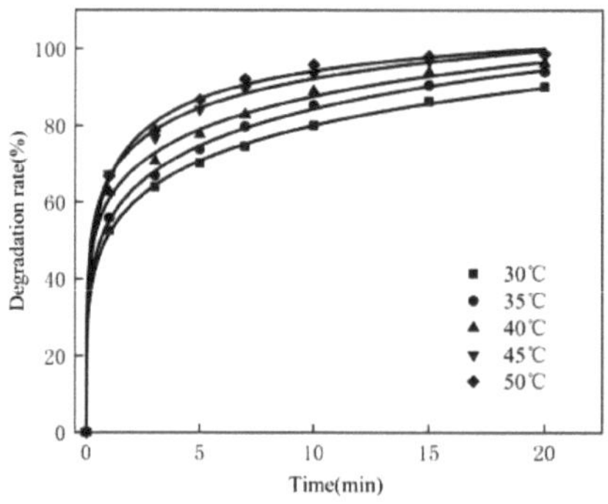

Figure 2. Effect of reaction temperature.

It can be seen from Figure 2 that, in the case of enough hydrogen peroxide, from 30 °C to 50 °C, with the increase in reaction temperature, the degradation rate of MB rises. This may be because as the temperature increases, the activation ratio in the molecules of hydrogen peroxide increases, and the movement will increase the probability of effective collision, produce more •OH and accelerate the decomposition of methylene blue. In the 30 °C to 50 °C temperature range, the degradation rate of MB increases with the rise of temperature. Therefore, in subsequent experiments, the choice reaction temperature is 50 °C.

3.3. *Effect of Dosage of Catalyst*

For the 250 mL MB solution, the experimental conditions are as follows: the concentration of MB is 80 mg/L, the reaction temperature is 50 °C, the pH is 10, the dosage of hydrogen peroxide is 2.0 mL and the dosage of the catalyst is 0.10 g, 0.15 g, 0.20 g, 0.25 g and 0.30 g, respectively. The experimental results are shown in Figure 3b. Meanwhile, the original soil was compared with the modified ATP. The experimental results are shown in Figure 3a.

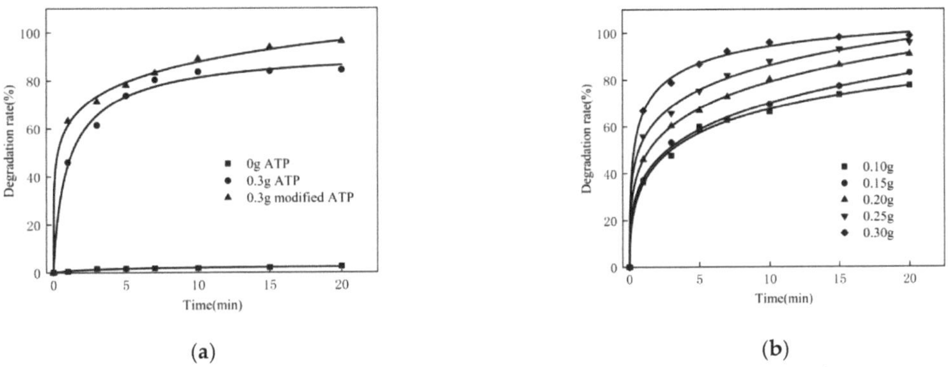

Figure 3. (**a**) Performance of modified ATP and ATP. (**b**) Effect of dosage of catalyst.

As can be seen from Figure 3a, the degradation rate of MB is very low without a catalyst, because hydrogen peroxide hardly reacts with MB. After adding 0.3 g attapulgite, the degradation of MB becomes evident, because ATP is a kind of water-containing crystallization of magnesium aluminum silicate and itself contains certain metal ions. After adding the hydrogen peroxide solution, the Fenton reaction happens, leading to the catalytic decomposition of methylene blue. It has internal holes that can produce adsorption, which has a certain impact. Compared with the original attapulgite, the catalytic degradation effect using the modified ATP is further enhanced because the modified ATP can introduce specific metal ions, which can accelerate the generation of hydroxyl radicals. In the same timeframe, the degradation rate of MB will increase.

It can be seen from Figure 3b that the degradation rate of MB increases with the increase in the amount of modified ATP. Moreover, the higher the amount of catalyst, the faster the initial degradation rate and the higher the final degradation rate. This is due to the fact that, with a certain concentration of hydrogen peroxide and MB, the amount of catalyst increases, and the catalyst can provide more active sites, so that the effective collisions per unit volume increase, the generation rate of •OH increases and the degradation of MB is promoted. Therefore, in subsequent experiments, the dosage of catalyst was determined to be 0.30 g.

3.4. *Effect of pH*

For the 250 mL MB solution, the experimental conditions are as follows: the concentration of MB is 80 mg/L, the dosage of the catalyst is 0.3 g, the reaction temperature is 50 °C, the dosage of hydrogen peroxide is 2.0 mL, and the pH is 4, 7, 8, 9 and 10, respectively. The experimental results are shown in Figure 4.

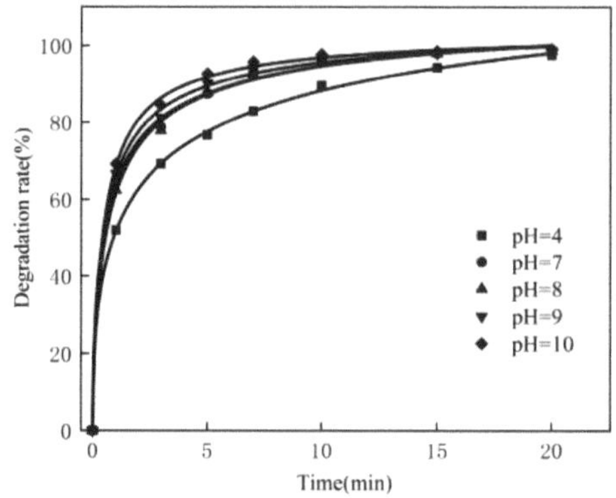

Figure 4. Effect of pH.

It can be seen from Figure 4 that the degradation rate of MB under acidic conditions is lower than that under neutral conditions; the reason is that the degradation rate of MB under acidic conditions is inhibited. Because the MB solution exists in the form of cations in water, this weakens the MB under acidic conditions and weakens the electrostatic attraction to the modified catalyst. In alkaline environments, MB increases with the electrostatic attraction to the modified catalyst [24]. This makes the MB respond more rapidly to the modification of the catalyst and, at the same time, MB generates and reacts to the •OH, which make H_2O_2 more quickly, prompting the metal on the surface of the modified catalyst to release •OH. Therefore, in a follow-up experiment, the pH is 10.

3.5. *Effect of Concentration of MB*

For the 250 mL MB solution, the experimental condition are as follows: the dosage of the catalyst is 0.3 g, the reaction temperature is 50 °C, the dosage of hydrogen peroxide is 2.0 mL, the pH is 10 and the concentration of MB is 60 mg/L, 70 mg/L, 80 mg/L, 90 mg/L and 100 mg/L, respectively. The experimental results are shown in Figure 5.

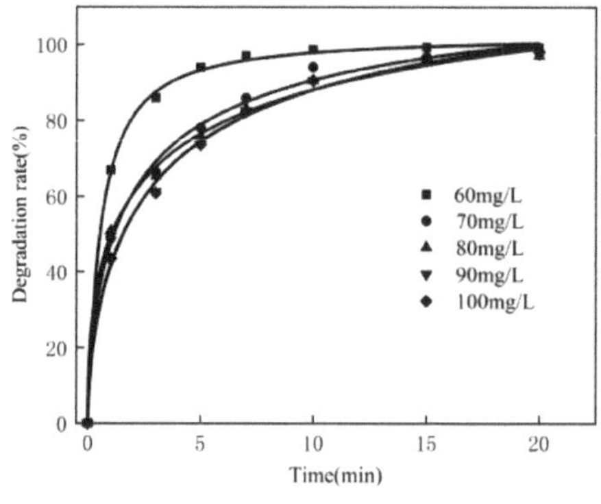

Figure 5. Effect of MB concentration.

As can be seen from Figure 5, with the increase in MB concentration, the trend of the increasing MB degradation rate becomes slower. The higher the concentration of MB, the smaller the final degradation rate. This is because the amount of H_2O_2 is the same; thus, in the unit volume solution, the amount of •OH produced is constant, which means that the amount of MB that can catalyze degradation is constant. The lower the concentration, the smaller the amount of MB in the unit volume solution and the more complete the degradation. The higher the concentration, the more MB per unit volume, and the less complete the degradation. Moreover, H_2O_2 will decompose itself, which means that H_2O_2 is not completely used for the catalytic degradation of MB. Therefore, in subsequent experiments, the concentration of MB was 80 mg/L.

3.6. Effect of Calcination Temperature

According to the above experiments, the best experiment conditions are as follows: the MB concentration is 80 mg/L, 250 mL, the dosage of the catalyst is 0.30 g, the dosage of hydrogen peroxide is 2 mL, the pH is 10 and the reaction temperature is 50 °C. Under these experiment conditions, the effect of calcination temperature on the catalyst was investigated. The experimental results are shown in Figure 6.

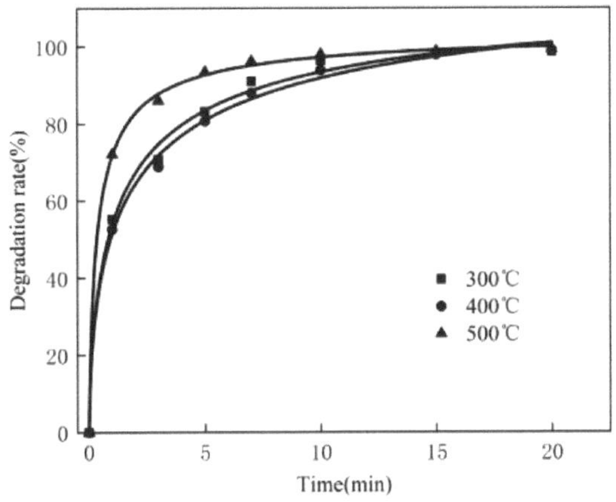

Figure 6. Effect of calcination temperature.

As you can see from Figure 6, a calcination temperature of 500 °C for the modified catalyst is better than that of 300 °C and 400 °C. The calcination temperatures of 300 °C and 400 °C for the modified catalyst have a similar effect. This is because in the 500 °C calcination, the crystallization water almost completely removes the ATP [25], expanding the internal space of the ATP and further increasing the specific surface area. H_2O_2 can interact more quickly and easily with modified metal on the surface of the catalyst when effective collision occurs, so that OH is produced more quickly, accelerating the degradation of MB. Where the calcination temperature is 300 °C or 400 °C, the modified catalyst is removed from the channel of free water by a small amount of crystal water; thus, the ATP interior hole does not get bigger.

3.7. Regeneration Experiment

Under the best experiment conditions, the experiment to reuse the catalyst was conducted.

As can be seen from Figure 7, the catalytic decomposition rate of MB was relatively high in the first place. With the increase in calcination temperature, the degradation rate of MB was 95%, 96% and 97%, respectively. The degradation rate of MB using the reused catalyst is also relatively good. With the increase in calcination temperature, the degradation rate can reach 76%, 78% and 80%, respectively, indicating that the catalyst has good secondary catalytic performance and can be recycled. After the third use, the degradation rate of MB is still around 65%. The results showed that the reuse ability of the catalyst was good.

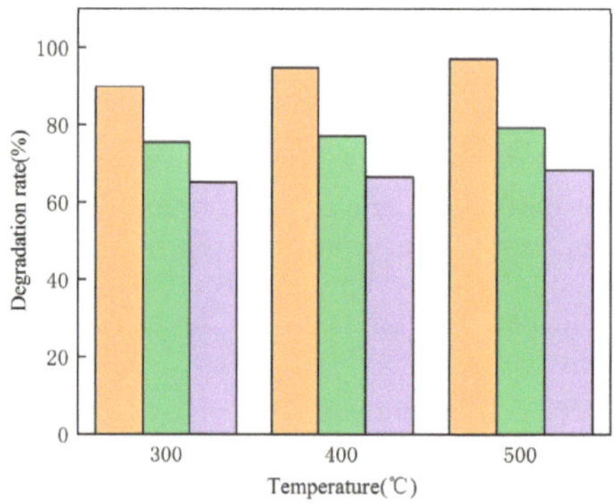

Figure 7. Regeneration experiment.

3.8. Characterization

The ATP and modified ATP were characterized using SEM, with a voltage of 15 KV and a magnification of 2000 and 5000 times, respectively. The obtained images were as follows.

As can be seen from Figure 8a, the original soil is larger than the modified ATP at 2000 times magnification, with many folds and protrusions on the surface. It can be seen from Figure 8b that the size of the La^{3+}/Cu^{2+} -ATP is around 30 μm. Compared with the original ATP, the surface of the modified ATP is relatively smooth, which may damage the outer structure and cause some surface structures to fall off during the acid soaking and calcining stages. As can be seen from Figure 8c, the original ATP soil has internal depressions, but its diameters are all large. As can be seen from Figure 8d, the surface of the modified ATP has a lump, which is presumed to be a metal material loaded on it. In addition, the diameter of the surface of the modified ATP becomes larger towards the inner depression, which may be due to the removal of water from the ATP during calcining.

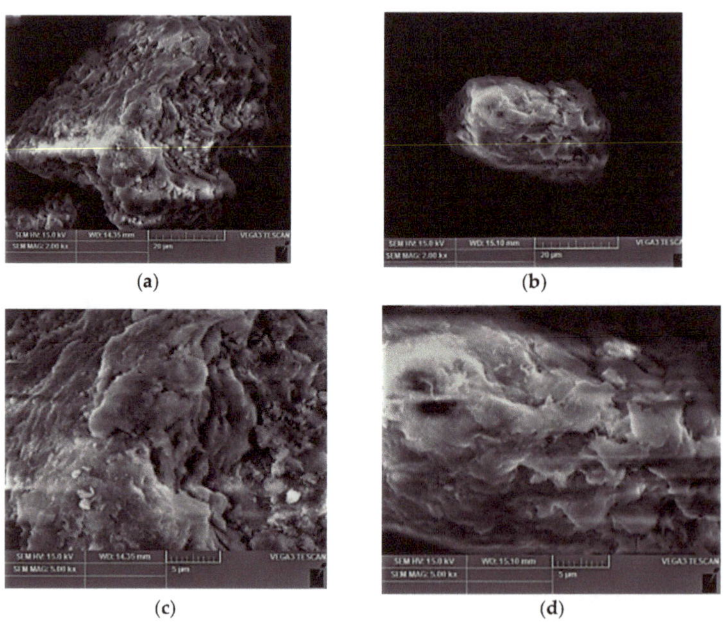

Figure 8. (**a**) SEM image of original soil, (**b**) modified catalyst, (**c**) original soil, (**d**) modified catalyst.

The ATP and modified ATP were characterized using XRD, and the changes of materials in ATP and modified ATP were compared. The XRD results are shown in Figure 9.

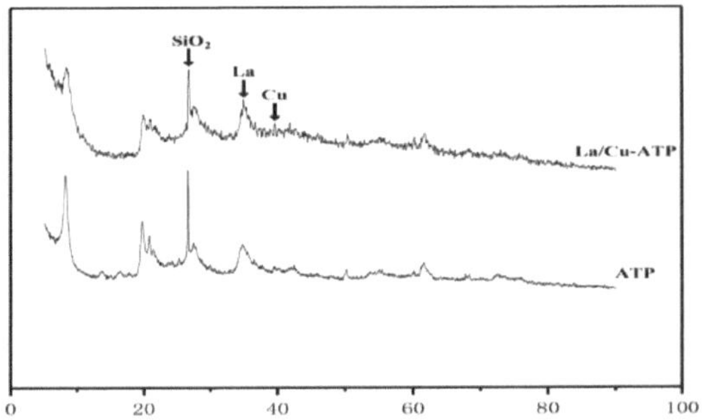

Figure 9. XRD of La^{3+}/Cu^{2+}-ATP.

It can be seen from Figure 9 that a strong diffraction characteristic peak of lanthanum appears when 2θ is at 34.3° and a diffraction characteristic peak of copper appears when 2θ is at 39.4°. Other peaks in the graph are caused by the presence of impurities in the sample itself. The overall crystal lattice of the catalyst is intact and not destroyed. The product has obvious diffraction peaks and the interlayer spacing fluctuates.

3.9. The Reaction Mechanism

Oxidation of the MB solution by the ATP rod–soil occurs through a heterogeneous catalytic oxidation mechanism. With hydrogen peroxide in the solution, the reaction of MB is divided into two kinds. One is the direct reaction of hydrogen peroxide and MB through direct action; the results are shown in Figure 10 (experimental conditions: reaction temperature: 40 °C, pH: 7, H_2O_2: 2 mL, MB: 50 mg/L, 250 mL and the modification of the catalyst: 0 g).

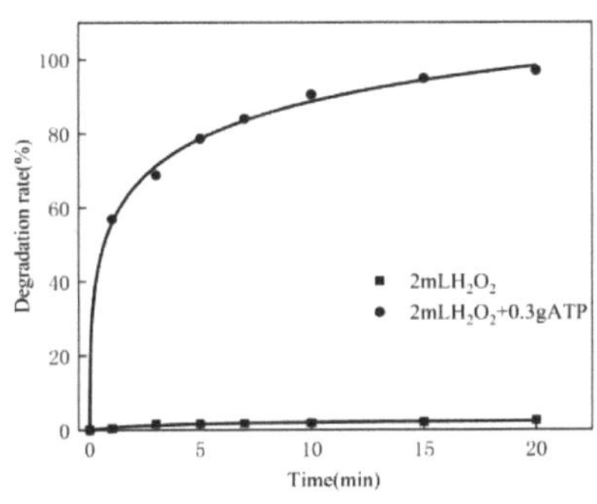

Figure 10. Experimental results of H_2O_2.

The degradation rate of MB is less than 5%. This almost indicates that H_2O_2 does not react directly with MB. The other kind is the chain reaction of H_2O_2 in the solution to form the highly oxidizing •OH, which catalyzes the decomposition of MB. At the same time, CuO is contained in the catalyst and part of the CuO will enter the lattice of La_2O_3 [26]. A small number of lanthanum ions will be replaced by copper ions, resulting in more empty oxygen positions [27,28]. The presence of empty oxygen potential will participate in the decomposition of H_2O_2. According to the mechanism proposed by Deraz [29], the reaction pathway of H_2O_2 decomposition in the La–Cu mixed oxide system can be proposed, and its surface is exposed to the La^{2+}/La^{3+}, Cu^{2+}/La^{2+}, Cu^{+}/Cu^{2+} ion pairs:

$$H_2O^2 \leftrightarrow H^+ + HO_2^- \tag{1}$$

$$Cu^+ \rightarrow Cu^{2+} + e \tag{2}$$

$$H^+ + e \rightarrow H \tag{3}$$

$$HO_2^- \rightarrow HO_2 + e \tag{4}$$

$$La^{3+} + e \rightarrow La^{2+} \tag{5}$$

$$Cu^+ + La^{3+} \rightarrow Cu^{2+} + La^{2+} \tag{6}$$

$$H_2O_2 \leftrightarrow H{+}{+}HO_2 \tag{7}$$

$$La^{2+} \rightarrow La^{3+} + e \tag{8}$$

$$H^+ + e \rightarrow H \tag{9}$$

$$HO_2^- \rightarrow HO_2 + e \tag{10}$$

$$Cu^{2+} + e \rightarrow Cu^+ \tag{11}$$

$$Cu^{2+} + La^{2+} \rightarrow Cu^+ + La^{3+} \tag{12}$$

Based on the above information, it is speculated that the degradation of MB is as follows:

$$La^{2+} + H_2O_2 \rightarrow La^{3+} + \bullet OH + OH^- \tag{13}$$

$$Cu^+ + H_2O_2 \rightarrow Cu^{2+} + \bullet OH + OH^- \tag{14}$$

$$\bullet OH + H_2O_2 \rightarrow HO_2\bullet + H_2O \tag{15}$$

$$La^{3+} + HO_2\bullet \rightarrow La^{2+} + HO_2^- \tag{16}$$

$$Cu^{2+} + HO_2\bullet \rightarrow Cu^+ + HO_2^- \tag{17}$$

$$Cu^+ + La^{3+} \rightarrow Cu^{2+} + La^{2+} \tag{18}$$

$$MB + \bullet OH \rightarrow CO_2 + H_2O \tag{19}$$

As the main active substance, La^{2+} firstly initiates the REDOX reaction to produce •OH (Equation (13)). It is reported that Cu+ also has Fenton-like properties and can catalyze the activation of H_2O_2 to generate •OH (Equation (14)). At the same time, Cu^{2+} can react with the $HO_2\bullet$ generated in the reaction to form Cu^+, achieving the purpose of the Cu^{2+} and Cu+ cycle (Equation (17)). In addition, as previously mentioned, La^{3+} was reduced by Cu^+ to La^{2+} (Equation (18)). Thus, the reaction, to some extent, increased the content of La^{2+} in the Fenton reaction, promoting the generation of •OH in the reaction process, which is likely to be a key step in improving the catalytic activity of the catalyst.

3.10. Reaction Kinetic

The catalytic oxidation of hydrogen peroxide has many applications in the oxidation and decomposition of organic impurities in wastewater. In the catalytic oxidation system, temperature, time, catalyst, substrate and the amount of oxidant are the main factors affecting the oxidation rate.

The reaction rate equation is as follows:

$$-dC/dt = k(C_{MB})^m \tag{20}$$

Here, k represents the reaction rate constant, C represents the concentration of methylene blue in solution, mol/L and m represents the reaction series.

The reaction conditions are as follows: the temperature is 50 °C, the concentration of MB is 80 mg/L and the reaction rate is simulated using MATLAB software. According to the results of the fitting theory, the dynamics equation is as follows:

$$-\,dC/dt = 6.027 \times (C)^{2.45} \quad (21)$$

According to the experimental data on the effect of reaction temperature on MB degradation mentioned above, MATLAB software was used for data fitting, and the results are shown in Table 1.

Table 1. k values at different reaction temperatures.

Temperature/°C	k
30	3.112
35	3.698
40	4.267
45	5.159
50	6.027

According to Arrhenius' formula:

$$k = k0\ \exp(-Ea/RT) \quad (22)$$

where k_0 represents the preexponential factor and dimension is related to the reaction series m; Ea represents the reaction activation energy, J/mol; R represents the ideal gas constant, 8.314 J/(mol•K); and T represents the reaction temperature, K. The results are shown in Figure 11.

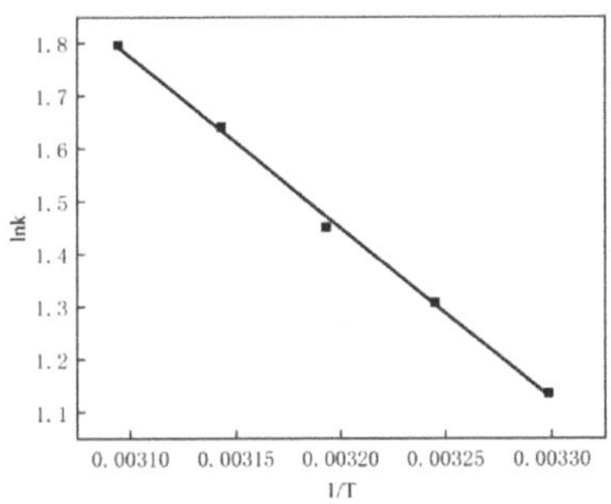

Figure 11. lnk-T^{-1}.

The intercept is 11.82. That is, $\ln k_0 = 11.82$, $k_0 = e^{11.82} = 1.36 \times 10^5$ and the slope is −3241.32, (−Ea/R) = −3241.32, Ea = 26.95 kJ/mol. Therefore, k = 1.36×10^5 exp (3241.32/T).

The kinetic equation of methylene blue catalytic oxidation is as follows:

$$-dC/dt = 1.36 \times 10^5 \exp(3241.32/T)(C_{MB})^{2.45} \quad (23)$$

The final degradation rate can reach 98% under optimum conditions. This result indicates that the above kinetic equation is suitable for most reactions. Meanwhile, the degradation rate of MB in this process is greater than the 73.8% rate obtained by other researchers under the optimum conditions [30].

4. Conclusions

Through preparing the modified ATP, the influencing factors of MB degradation were investigated, examining various aspects. ATP is an excellent catalyst carrier with capacities for adsorption and catalytic oxidation. Compared with ATP, the modified ATP improved its catalytic performance. The optimum experimental conditions obtained from the experiment are as follows: reaction temperature: 50 °C, pH: 10, hydrogen peroxide dosage: 2 mL and modified catalyst: 0.3 g. This technology will provide certain theoretical support for the industry.

Author Contributions: Methodology, W.Z.; Writing—original draft, J.S.; Writing—review & editing, Z.D. and H.-J.S.F. All authors have read and agreed to the published version of the manuscript.

Funding: This research was funded by Guangxi Key Laboratory of Petrochemical Resource Processing and Process Intensification Technology (2020K005), The Key Laboratory of Fine Chemical Application Technology of Luzhou (HYJH-2102-B) and The Opening Foundation of The State Key Laboratory of Molecular Engineering of Polymers (K2020-24).

Institutional Review Board Statement: Not applicable.

Informed Consent Statement: Not applicable.

Data Availability Statement: The authors confirm that the data supporting the findings of this study are available within the article.

Conflicts of Interest: The authors declare no conflict of interest.

References

1. Zhang, J.H.; Tang, D.X.; Zhang, H.J. The role and application of rare earth elements in magnesium alloy. *Rare Met.* **2008**, *32*, 659–667.
2. Bi, J.; Wu, Y.B.; Wei, B.B. The preparation and photocatalytic performance of perovskite-type $LaCoO_3$. *J. Dalian Jiaotong Univ.* **2014**, *35*, 78–81.
3. Yang, L.P.; Zhao, X.H.; Dong, C. Synthesis of $LaCoO_3$ by microwave irradiation and its photocatalytic degradation of reactive brilliant blue. *Chin. J. Anal. Chem.* **2009**, *37*, 99–104.
4. Chen, B. Status quo and development direction of printing and dyeing wastewater treatment. *Resour. Conserv. Environ. Prot.* **2018**, *11*, 96–97.
5. Huang, J.; Liu, Y.; Jin, Q.; Wang, X.; Yang, J. Adsorption studies of a water soluble dye, Reactive Red MF-3B, using sonicationsurfactant—Modified attapulgite clay. *J. Hazard. Mater.* **2007**, *143*, 541–556. [CrossRef] [PubMed]
6. Alkan, M.; Doan, M.; Turhan, Y.; Demirbas, Ö.; Turan, P. Adsorption kinetics and mechanism of maxilon blue 5G dye on sepiolite from aqueous solutions. *Chem. Eng. J.* **2008**, *139*, 213–221. [CrossRef]
7. Doan, M.; Abak, H.; Alkan, M. Adsorption of methylene blue onto hazelnut shell: Kinetics, mechanism and activation parameters *J. Hazard. Mater.* **2009**, *164*, 172–179. [CrossRef]
8. Malik, P.K.; Saha, S.K. Oxidation of direct dyes with hydrogen peroxide using ferrous ion as catalyst. *Sep. Purif. Technol.* **2003**, *31*, 241–248. [CrossRef]
9. Liu, Y.; Sun, D. Development of Fe_2O_3-CeO_2-TiO_2/γ-Al_2O_3 as catalyst for catalytic wet air oxidation of methyl orange azo dye under room condition. *Appl. Catal. B* **2007**, *72*, 205–210. [CrossRef]
10. Rodríguez, A.; Ovejero, G.; Romero, M.D.; Díaz, C.; Barreiro, M.; García, J. Catalytic wet air oxidation of textile industrial waste water using metal supported on carbon nanofibers. *J. Supercrit. Fluids* **2008**, *46*, 163–174. [CrossRef]
11. Ikram, M.; Haider, A.; Bibi, S.T.; Ul-Hamid, A.; Haider, J.; Shahzadi, I.; Nabgan, W.; Moeen, S.; Ali, S.; Goumri-Said, S.; et al. Synthesis of Al/starch co-doped in CaO nanoparticles for enhanced catalytic and antimicrobial activities: Experimental and DFT approaches. *RSC Adv.* **2022**, *12*, 32142–32155. [CrossRef] [PubMed]
12. Moeen, S.; Ikram, M.; Haider, A.; Haider, J.; Ul-Hamid, A.; Nabgan, W.; Shujah, T.; Naz, M.; Shahzadi, I. Comparative Study of Sonophotocatalytic, Photocatalytic, and Catalytic Activities of Magnesium and Chitosan-Doped Tin Oxide Quantum Dots. *ACS Omega* **2022**, *7*, 46428–46439. [CrossRef]
13. Walker, G.M.; Weatherley, L.R. Biodegradation and biosorption of acid anthraquinone dye. *Environ. Pollut.* **2000**, *108*, 219–223 [CrossRef] [PubMed]
14. El-Sheekh, M.M.; Gharieb, M.M. Biodegradation of dyes by some green algae and cyanobacteria. *Int. Biodeterior. Biodegrad.* **2009**, *63*, 699–710. [CrossRef]
15. Xu, J.; Li, W.; Yin, Q.; Zhong, H.; Zhu, Y.; Jin, L. Direct electron transfer and bioelectrocatalysis of hemoglobin on nano-structural attapulgite clay-modified glassy carbon electrode. *J. Colloid Interface Sci.* **2007**, *315*, 170–182. [CrossRef]
16. Zhang, J.; Xie, S.; Ho, Y.S. Removal of fluoride ions from aqueous solution using modified attapulgite as adsorbent. *J. Hazard. Mater.* **2009**, *165*, 218–224. [CrossRef]
17. Zhao, D.; Zhou, J.; Liu, N. Characterization of the structure and catalytic activity of copper modified palygorskite/TiO_2(Cu2+-PG/TiO_2) catalysts. *Mater. Sci. Eng. A* **2006**, *431*, 256–263. [CrossRef]

18. Zhao, D.; Zhou, J.; Liu, N. Surface characteristics and photoactivity of silver-modified palygorskite clays coated with nanosized titanium dioxide particles. *Mater. Charact.* **2007**, *58*, 249–255. [CrossRef]
19. Liu, Y.; Liu, P.; Su, Z.; Li, F.; Wen, F. Attapulgite-Fe_3O_4 magnetic nanoparticles via co-precipitation technique. *Appl. Surf. Sci.* **2008**, *255*, 2020–2031. [CrossRef]
20. Zhang, H.T.; Zhang, Y.M.; Wang, T.H.; Wang, P. Adsorption of phosphorus and arsenic on La-modified mesoporous materials. *Chin. J. Environ. Eng.* **2019**, *13*, 1791–1799.
21. Ramezanalizadeh, H.; Peymanfar, R.; Khodamoradipoor, N. Design and development of a novel lanthanum inserted $CuCr_2O_4$ nanoparticles photocatalyst for the efficient removal of water pollutions. *Optik* **2019**, *180*, 113–124. [CrossRef]
22. Wang, X.; Kang, Y.; Cui, D.; Li, J.; Li, D. Influence of lanthanum promoter on vanadium catalyst for sulfur dioxide oxidation. *Catal. Commun.* **2019**, *118*, 39–45. [CrossRef]
23. Shang, J.; Jiang, Y.; Qin, X.; Zhao, B.; Li, X. Catalytic Oxidation of Methylene Blue by Attapulgite/TiO_2. *Front. Environ. Sci. Water Wastewater Manag.* **2021**, *9*, 783313. [CrossRef]
24. Shi, Y. Preparation of titanium and lanthanum mesoporous bioglass and its dye-removal performance in water. *Tianjin Univ. Technol.* **2016**, *20*, 67–68.
25. Fan, G.; Shen, M. Research and application of attapulgite clay. *Prog. Chem. Ind.* **2009**, *28*, 100–101.
26. Wang, J.B.; Tsai, D.H.; Huang, T.J. Synergistic catalysis of carbon monoxide oxidation over copper oxide supported on samaria-doped ceria. *J. Catal.* **2002**, *208*, 370–377. [CrossRef]
27. Wu, X.; Liang, Q.; Weng, D.; Lu, Z. The catalytic activity of $CuOCeO_2$ mixed oxides for diesel soot oxidation with a NO/O_2 mixture. *Catal. Commun.* **2007**, *8*, 2110–2118. [CrossRef]
28. Tabakova, T.; Idakiev, V.; Papavasiliou, J.; Avgouropoulos, G.; Ioannides, T. Effect of additives on the WGS activity of combustion synthesized CuO/CeO_2. *Catal. Catal. Commun.* **2007**, *8*, 101–112. [CrossRef]
29. Deraz, N.M. Characterization and catalytic performance of pure and Li_2O-doped CuO/CeO_2 catalysts. *Appl. Surf. Sci.* **2009**, *255*, 3884–3891. [CrossRef]
30. Yin, L.; Lu, X.; Ai, F. Effect of Ti-Attapulgite Catalyst on Ozonation Degradation of Dye Wastewater. *J. Chin. Ceram. Soc.* **2003**, *31*, 66–69.

materials

MDPI

Article

Optimization of Hexagonal Structure for Enhancing Heat Transfer in Storage System

Natalia Raźny [1,*], Anna Dmitruk [1], Artur Nemś [2], Magdalena Nemś [2] and Krzysztof Naplocha [1]

[1] Department of Lightweight Elements Engineering, Foundry and Automation, Faculty of Mechanical Engineering, Wrocław University of Science and Technology, Wybrzeże Wyspiańskiego 27, 50-370 Wrocław, Poland

[2] Department of Thermodynamics and Renewable Energy Sources, Faculty of Mechanical and Power Engineering, Wrocław University of Science and Technology, Wybrzeże Wyspiańskiego 27, 50-370 Wrocław, Poland

* Correspondence: natalia.razny@pwr.edu.pl

Abstract: Thermal performance was tested during cycling work for latent heat storage systems based on KNO_3 and $NaNO_3$ (weight ratio 54:46). For heat transfer improvement, cast aluminum honeycomb-shaped structures were produced via 3D printing of polymer model and investment casting. Different wall thicknesses were tested at 1.2 mm and 1.6 mm. The obtained results were compared to working cycles of pure PCM bed. The use of enhancers is reported to improve the rate of charging and discharging of the deposit. In the next step, the structures were examined with numerical simulation performed with ANSYS Fluent software. The wall thicknesses taken into consideration were the following: 0.8, 1.2, 1.6, and 2.0 mm. An insert with a greater wall thickness allows for smaller dT/dt and better heat distribution in the vessel. The investment casting process enables the manufacturing of complex structures of custom shapes without porosity and contamination.

Keywords: heat storage; heat transfer; PCM; spatial structures; honeycomb; metal structures

1. Introduction

Due to the currently growing worldwide demand for energy and increasing environmental constraints, the key issue appears to be developing new, safe, and sustainable energy sources, storage, and transport technologies. In this case, renewable energy sources (RES) such as solar, wind, or geothermal deserve special attention. An important aspect of thermal energy storage (TES) is the ability to gather large amounts of energy in relatively small volumes, which can be guaranteed by the high phase transition enthalpy, typical for some of the Phase Change Materials (PCMs). This feature allows them to accumulate large amounts of latent heat during a phase change (melting, evaporating) [1]. For solar energy storage, salt hydrates and their eutectic mixtures are the most commonly used PCMs. This choice is connected to the fact that the operating temperature of solar energy storage oscillates around 250 °C, in which organic materials can degrade. Nevertheless, the bottleneck limiting the wider application of PCMs in heat storage tanks is their low thermal conductivity, prolonging the charging period of such units.

Numerous materials and structures are being used to support the transport of heat in PCMs. Such, so-called heat transfer enhancers directly improve the thermal conductivity of the deposit, creating composite PCM. The following examples can be highlighted among the particles' additives [2,3]: carbon-based fillers or nanofillers (e.g., expanded graphite, carbon nanotubes (CNTs)), 2D materials (e.g., graphene or MXenes), and metallic or ceramic powders (e.g., Al, Cu, SiO_2, Al_2O_3, Fe_2O_3, ZnO). Variously shaped and manufactured by casting or plastic working, metal heat exchanging surfaces are also commonly applied as follows: tubes (plain, helical, or finned [2,3]), fins [4,5], metal foams (e.g., Cu, Al, Ni) [6] and biomimetic fractal-like (e.g., tree-, Y-, snowflake-shaped [7–10]), and pin-fin [11] or honeycomb structures [12]. Another method is micro- or macroencapsulation in organic or inorganic shells of superior thermal conductivity [5,6,13].

The main focus of this work will be put on the application of honeycomb structures in heat exchange, especially in PCM heat storage units. Lu et al. analyzed the factors such as wall thickness, cell size, and orientation and stated that the heat transfer characteristics of micro-cell Al honeycombs are alike the ones of open-cell Duocel® foams [14]. Kong et al. tested and simulated numerically the thermal behavior of gradient honeycomb heat enhancer 3D-printed from 316 L stainless steel. Comparing the results with uniform structures the 14–17% overall thermal performance enhancement was observed [15]. In the work of Andreozzi et al., a ceramic cordierite honeycomb system with paraffin wax as PCM was subjected to numerical simulation for different pores per unit of length (PPU) values. It was found that higher PPU benefits with faster charging [16]. The influence of different geometry of cores (triangular, trapezoidal, rectangular, hexagonal, circular) on the thermal cycling of aluminum honeycomb structures with paraffin was researched by Duan and co-authors. It was stated that cells with a smaller geometrical factor (GF), such as triangular or quadrilateral, ensure the shortening of melting times of PCM, compared to a hexagonal structure. Moreover, the cell orientation appears to matter, as turning the hexagonal cell corner down resulted in a 9.9% reduction in melting duration [17]. Egolf et al. studied the possibility of the utilization of inorganic or organic (e.g., glass, polycarbonate (PC)) honeycombs in the construction of translucent solar collectors integrated with PCM ($CaCl_26H_2$) heat storage modules to be used in buildings façades. Longitudinal honeycomb cells forming panels were considered with various degrees of filling with PCM (e.g., 33.3%, 66.6%—the rest of the hexagons were left empty) resulting in physical models, numerical simulation results, determination of honeycombs absorption properties, and solar impacts [18]. The aluminum honeycomb wallboard elaborated by Lai et al. was filled with microencapsulated PCM (mPCM, here paraffin–pure or with expanded graphite or ironwire, covered with polymer shell). A honeycomb sample (10 cm × 10 cm × 2.54 cm, core cell 8 mm) was tested experimentally in terms of thermal performance in PV panels providing sufficient heat conduction enhancement and unifying the interior temperatures of the module with no thermal stratification [19]. A subsequent study on mPCMs by similar authors confirmed the benefit of Al honeycomb working as a structural support and heat transfer channel facilitating thermal energy dissipation [20]. Another study reported an investigation of the effect of honeycomb core on latent heat storage (LHS) in PCM solar air heater (SAH). Three SAHs were compared, containing the following: a flat absorber plate, a heat storage panel with PCM (paraffin), and the one with PCM and a honeycomb core. The main findings were that, due to the utilization of aluminum honeycomb structure, the charge-discharge times were significantly reduced and the daytime temperature was increased [21]. A similar honeycomb was introduced also to the paraffin by Xie et al., resulting in a thermal conductivity increase up to 2.08 W/m·K [22]. Kant et al. considered n-octadecane as a PCM reinforced with aluminum honeycomb in numerical simulation investigating the effect of cells length, thickness, and inclination angle. The analyzed composite PCM exhibits an improved charging rate with the increasing volume of a metal insert, while the tilt influence is rather marginal. Nevertheless, it has to be taken into account that replacing the PCM volume with a honeycomb structure reduces the total energy stored, and therefore a compromising solution must be identified [12].

In this paper, the use of an efficient hexagonal cast aluminum structure to enhance heat transfer in the salt PCM deposit is proposed. In the experimental part, three arrangements are taken into consideration: the deposit filled only with salt PCM, and two deposits enhanced with honeycomb inserts with differing wall thickness of 1.2 mm and 1.6 mm. The inserts are designed by 3D modeling and cast by investment casting. The charging and discharging of the deposits supported by the structures are described. Thermal and physical phenomena occurring during operations are depicted. During the numerical part of the research, the influence of the wall thickness on the performance of the unit is studied.

The analysis is performed for the following dimensions: 0.8 mm, 1.2 mm, 1.6 mm, and 2.0 mm.

2. Materials and Methods

2.1. Experimental Approach

Aluminum honeycomb inserts were manufactured by the means of the investment casting method, consisting of the design and execution of the model, molding with plaster, the burn-out cycle of the mold, and metal pouring under low pressure. Models were designed in Autodesk Inventor Professional 2018, then the Simplify3D software (Simplify3D 4.1.2, Simplify3D Inc., Cincinnati, OH, USA) was used to create the G-code file. Spatial models from polylactide (PLA) were produced with the use of HBOT 3D printer F300 (3D Printers, Wrocław, Poland). Thus, prepared patterns with attached wax gating systems after molding and initial hardening were burned out at the maximum temperature of 730 °C in the furnace. During this stage, the combustion and gasification of the polymer model occur and the plaster mold hardens and obtains its full strength. The evaporated pattern leaves the internal precise cavity of its shape in the plaster. Ceramic plaster Randolph Ransom-type R&R® ARGENTUM™ (Ransom & Randolph, Maumee, OH, USA) (quartz < 50%, cristobalite < 50%, $CaSO_4$ binder) was utilized. Mold, still at the elevated temperature, was mounted in autoclave and under the pressure of 0.04 MPa, the chosen aluminum alloy (AC 44200, Si-10.5%-13.5%; Fe-0.55%; Cu-0.05%; Mn-0.35%; Zn-0.10%; Ti-0.15%; Al-rest) was cast.

Cast inserts, varying in wall thicknesses (1.2 and 1.6 mm), were alternately located horizontally in the center of the isolated PCM-based accumulator, fully immersed in the chosen PCM—the eutectic composition of KNO_3 and $NaNO_3$ (Archem) with a weight ratio of 54:46 and a melting point of 222 °C. Both of the salts were purchased separately, weighed carefully, mixed, and melted together. Subsequently, the thusly prepared heat storage unit was subjected to multiple working cycles of charging and discharging. The temperature distribution in the unit was controlled with the use of three thermocouples located on the bottom (12 mm), in the center (50 mm), and on the top (87 mm) of the heat storage tank, and in the middle of the honeycomb cells. The scheme of the stand is presented in Figure 1. Charging and discharging cycles were conducted by the hot plate (400 °C) used as a heat source, in order to achieve the semi-directional heat flux in the height direction of the accumulator. The heating–cooling thermocouple placed in the cooling plate indicates the temperature of the heating source while charging the accumulator. When the deposit is completely charged, the salt is melted, and water cooling is activated. The temperature shown then by the thermocouple corresponds to the temperature of the water during cooling. Thus, it is possible to observe, by the changes in the water temperature, how the accumulator is discharged by transferring the heat from the PCM to the cooling plate and, then, to the water.

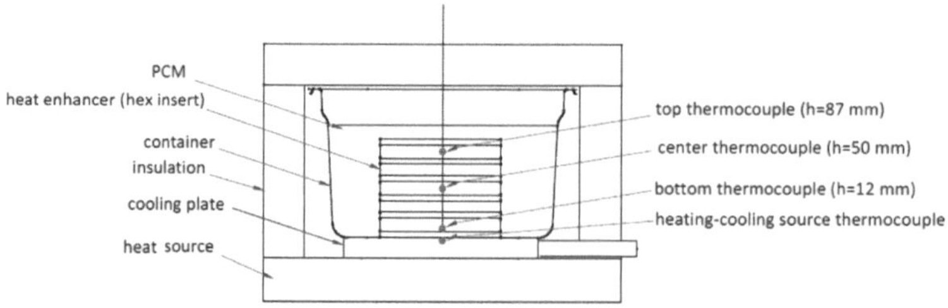

Figure 1. Test stand scheme: the thermocouples are placed in the center of the cells, in the middle of the structure.

The deposit was heated up with the use of a heating plate (Hot Plate, Model SH-II-5B, heating power 1200 W, temperature accuracy ±1%). The experimental data, the temperature at different heights in the bed, were registered by K-type NiCr-NiAl thermocouples, class 1 (±1.5 °C) according to the PN-EN 60584 standard, collected, and recorded by Adam 4018 type adapter (American Advantech Corporation, Sunnyvale, CA, USA) with 8 channels, and the VisiDAQ program (VisiDAQ Professional Version 3.1, American Advantech Corporation, Sunnyvale, CA, USA) The loading step's duration was dependent on the time needed to melt the entire deposit. The first trials established the performance of the accumulator filled only with the PCM material (11 h charging/discharging cycle), while the next concerned the composite PCM with the immersed aluminum honeycomb structures (5 h charging/discharging cycle).

2.2. Numerical Analysis—Geometry, Boundary, Initail Conditions, and Numercial Schemes

In the next step, the influence of the wall thickness of the insert on the melting process of PCM was analyzed. The heat flow in PCM can be described using standard equations [23,24]. Partial differential equations are used to solve the mass, momentum, and energy equations in complex systems [25]. For this purpose, the Finite Volume Method (FVM) was used [17], and the calculations were performed with the commercial Ansys® Academic Research Fluent, 2021R1 (ANSYS, Inc., Canonsburg, PA, USA).

The meshes for all 12 domains shown in Figure 2 were carefully prepared to meet the high regime for phase change calculations. Their quality allowed for a high accuracy of calculations, which is described later. In addition, the mesh was densified near the walls to reduce the calculation error on the walls and in places of lower mesh quality, as shown in Table 1. The smallest thickness of adjacent elements was 0.05 mm.

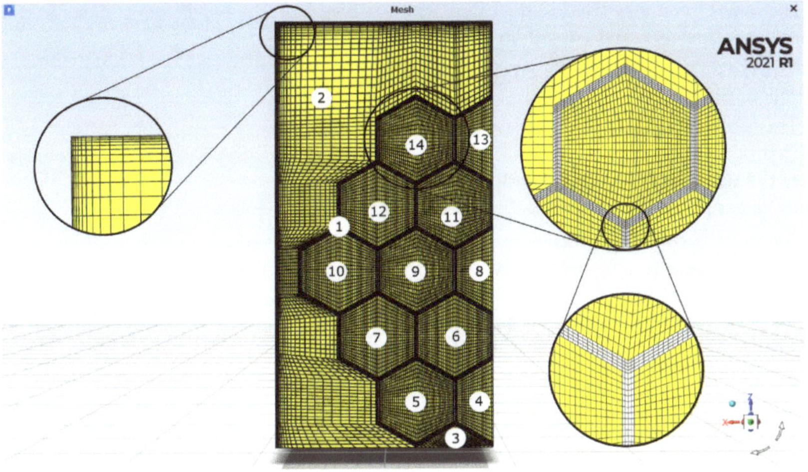

Figure 2. Mesh with details used during numerical calculations. Images used courtesy of ANSYS, Inc.

Table 1. Mesh quality in domains.

Domain	Hexahedral Elements	Quality		Thickness of Elements Adjacent	
		Min	Max	Min	Max
1	3168	1	1	0.400148	1.84752
2	3528	0.758	1	0.050000	1.47029
3, 8, 13	243	0.692	0.956	0.133432	1.15556
4–7, 9–12, 14	361	0.95	0.974	0.607737	1.21547

In the numerical calculations, the geometry used in the experimental studies was used. The boundary conditions used in the simulations are shown in Figure 3.

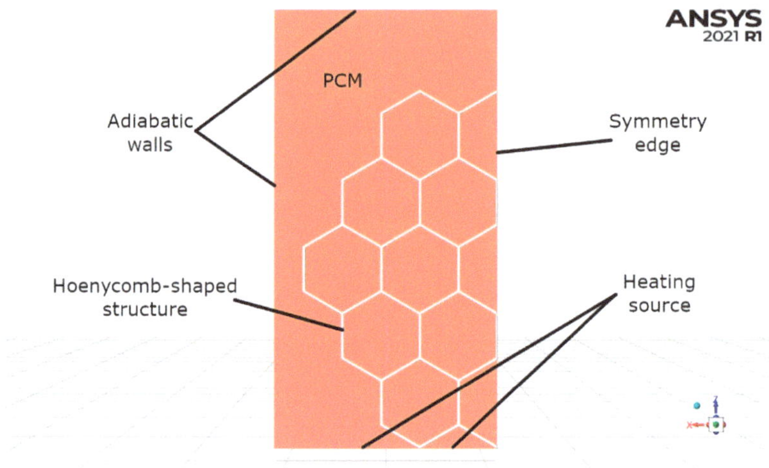

Figure 3. Boundary conditions employed during calculations. Images used courtesy of ANSYS, Inc.

The insert geometry used in the calculations is based on the experiments described in the previous chapters, and considers the following conditions:

1. The vessel geometry has been simplified;
2. The bottom surface temperature was 400 °C;
3. The remaining walls were adiabatic;
4. In order to speed up the calculations, the symmetry condition was used and only half of the geometry was calculated;
5. PCM properties have been defined identically for each domain (2 ÷ 14).

The physicochemical properties of PCM and aluminum are presented in Table 2. Due to the small differences in specific heat and thermal conductivity of the liquid and solid PCM phases, the same values were set. The convective movement of the liquid phase of PCM is caused by the difference in density; therefore, it was dependent on the temperature and in the tested range, as described by Equation (1).

$$\rho(T) = -0.8958T + 2462.5 \tag{1}$$

Table 2. Thermophysical properties of materials used during calculations [26–29].

Material	Density, kg/m^3	Specific Heat, J/(kg·K)	Thermal Conductivity, W/(m·K)	Dynamic Viscosity, kg/(m·s)	Melting Heat, J/kg	Melting Temperature, K
PCM	Equation (1)	1492	0.457	0.0063	108,000	495
Aluminum	2644	960	160	-	-	-

The geometry and numerical mesh were made in Ansys® Academic Research ICEM CFD, 2021R1 (ANSYS, Inc., Canonsburg, PA, USA). Each of the closed spaces inside the insert was a separate domain filled with PCM. The PCM was also outside the insert. All 14 domains were discretized by using hexahedral elements. Additionally, on the outer surfaces, inflation layers were applied. The thickness of the first element adjacent to the wall was set to 0.07 mm. A laminar model for PCM was used. Calculations were performed in a transient mode with a time step of 0.05 s. The solution in each time-step was considered as converged if the residuals were less than 10^{-3}, except for the energy equation, where it was 10^{-6}. However, a maximum of 30 iterations per time-step was completed.

In order to determine the influence of the wall thickness of the insert on the rate of heat flow inside its cells, calculations were made for several wall thicknesses: 0.8, 1.2, 1.6, and 2.0 mm. A separate geometry was made for each thickness and a mesh was generated in accordance with the above rules. The boundary conditions were the same for each case.

3. Results and Discussion

This section presents the results of both research paths undertaken in the article. In the first part, the experimental observations are described and concluded, while in the second one, the discussion of the numerical simulation approach is depicted.

3.1. Experimental Results and Discussion

A detailed examination of the heating and cooling processes of the chamber filled with PCM and enhanced with prepared inserts was carried out and the best performing enhancer was selected. It should be taken into account that the duration of charging and discharging the storage unit was affected by error, due to heat losses in the environment, despite the use of insulation. Nevertheless, all measurements were conducted under the same conditions; therefore, the error can be treated as systematic and the results obtained can be compared.

3.1.1. Investment Casting of Honeycomb Structures

Figure 4a presents the real cast insert placed in the horizontal position, in the same way as in the experiment, while in Figure 4b the designed model with the most important dimensions is presented. An appropriate selection of parameters based on the experience and observations allowed for obtaining a structure without visible surface defects.

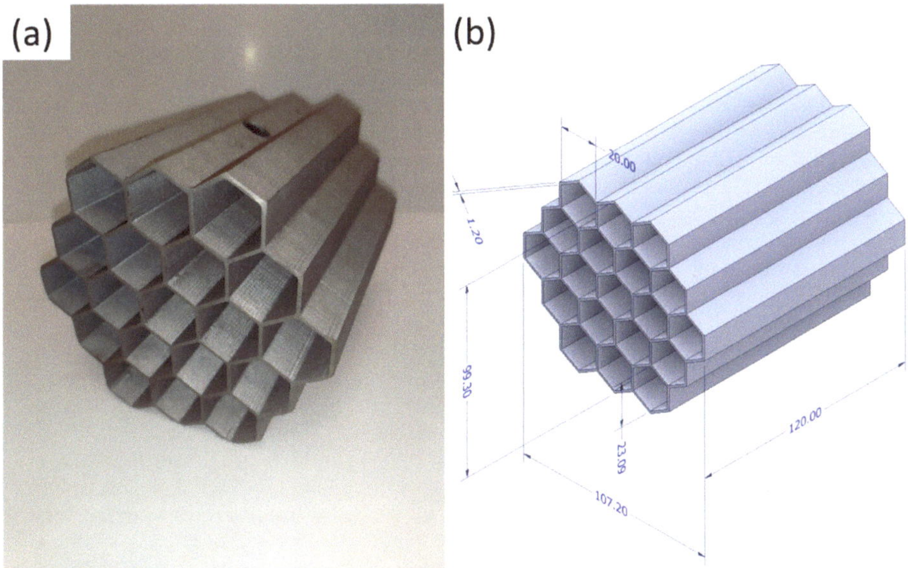

Figure 4. The honeycomb structure: (**a**) cast element made of EN AC 44200 Al-Si alloy; (**b**) dimensions of 1.2 mm wall-thick structure (in experiment and simulation).

3.1.2. Experimental Heat Transfer Performance

Figure 5 shows the temperature variation over time for the thermocouples placed in the system. The units were heated for different periods due to the assumption that the heating is stopped approximately when the whole deposit is molten. In the case of the insert with a 1.2 mm thick wall, even in a significantly shorter time of charging, the deposit reaches a considerably higher temperature (285 °C) than for the pure salt deposit (235 °C).

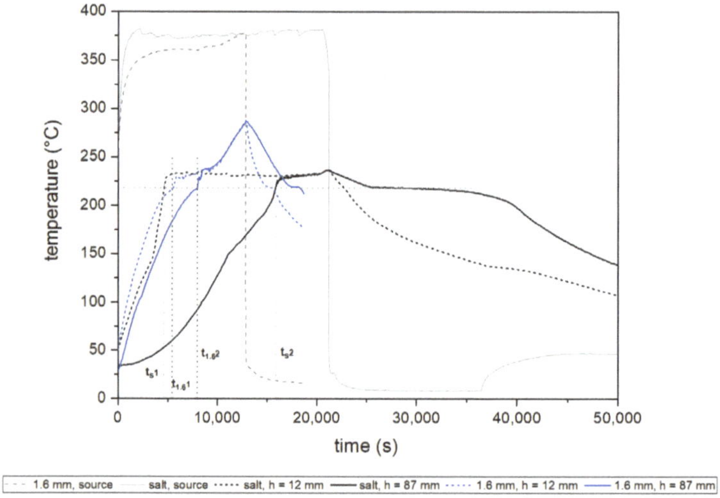

Figure 5. The temperature change in the deposit during the heating and cooling of the store: containing pure PCM and the one with honeycomb heat transfer enhancer with a wall thickness of 1.6 mm.

It can be noticed that the use of the heat transfer enhancer has significantly accelerated the charging period and furthered the melting of the deposit. The influence is especially pronounced for the top thermocouple in both cases, and can be seen based on the different lengths of flattening in the graph. The plateau is related to the melting of the part of the salt PCM in the deposit, which is located outside the volume of the insert structure. The time required to reach the melting temperature is two times higher for the pure PCM deposit than for the one with the elaborated cast structure. Similar observations were reported in [21]. Abuşka et al. tested solar air heaters based on PCM with and without metal hexagonal enhancers placed in the deposit. Tests were driven for six different mass flow rates of air. They claimed that the use of honeycomb structures as heat enhancers can substantially shorten the time of charging and discharging the deposit. A structure applied in the deposit allowed the PCM temperature to increase by 8.8 °C (about 10%) in comparison to a deposit with only PCM.

In the case of crystallization, the long continuous line can be observed, especially in the case of the top (h = 87 mm) thermocouple for the salt deposit, while for the deposit with the insert crystallization, time is much lower. The discharging of the deposit can be successfully supported by the use of such structures.

The dashed reference lines in the graph indicate the time to reach the phase transition temperature in the bottom cell of the honeycomb (referred to as t_1 in Table 3), and in the top cell (referred to as t_2 in Table 3). Furthermore, lower source temperatures were recorded for the PCM deposit with the metallic structure than for the one without it. As mentioned in the section before, in all cases the target heating source temperature was set to 400 °C. At the beginning of each measurement, the source had a temperature of approx. 50 °C. The target temperature was never reached due to the absorption of the heat by the unit. In the case of pure salt units, the reached temperature of the source was higher than for the units with metal enhancers. It can be concluded that the insert is absorbing heat and transferring it into the deposit, so the source temperature slowly increases, but the phase transition process and temperature equalization in the heat storage tank are facilitated. The reduction in the surface temperature using encapsulated PCM was also highlighted in [19]. Lai and Hokoi presented a concept of a honeycomb wallboard filled with microencapsulated PCM, which can be used in building construction to control the temperature in the rooms. Paraffin was used as the PCM. The system was tested with a series of different heat fluxes. Results obtained suggest that the honeycomb structure placed inside the wall allows good heat conduction and the surface of the wall is characterized by the lowered temperature. Likewise, in the case of observations presented hereby, the temperature of the source while using the honeycomb-enhancing structure is lower than for the only-PCM unit since the heat is absorbed by the cast enhancer and transported to PCM located further in the deposit.

Table 3. Starting time of the deposit's melting at the height of the bottom (h = 12 mm) thermocouple (t_1) and the top (h = 87 mm) thermocouple (t_2) and the calculated duration of the deposit's melting (delta) for the tested cases: a pure mixture of salts, PCM, and an insert with a wall thickness of 1.2 mm, and a PCM insert with a wall thickness of 1.6 mm.

Time	PCM	1.2 mm	1.6 mm
t_1 [s]	4710–5590	5480–7540	5220–7920
t_2 [s]	15,960–16,850	10,570–12,010	7850–10,400
Δ ($t_2 - t_1$) range [s]	10,950–11,900	4640–5700	2420–2620
Δt average [s]	11,425	5170	2520
Δt average [min]	190.42	86.17	42.00

On the basis of the data presented in Table 3, one can point out the significant variation between the average derivative difference of each studied case. The time of gaining the melting temperature by the bottom thermocouple (h = 12 mm) t_1 was similar in all three units. By analyzing the average Δt, it is observed that a cast enhancer with a wall of 1.2 mm thickness can shorten the time of charging the deposit twice, while the insert with a 1.6 mm thick wall can shorten the time of charging by over four times.

Figure 6 includes the time differential temperature plot for the measurements of the top thermocouple for the three variants tested. Positive values of the derivative indicate the absorption of heat energy by the system, while the negatives are responsible for releasing heat into the environment.

The first peak (A) is responsible for heat conduction through the insert. A sudden rise and subsequent drop are seen primarily for the dT/dt 1.6 mm curve, while a similar tendency but of lower values is observed for the dT/dt 1.2 mm dependence. In the case of an only-salt deposit, no such behavior was present as the temperature of the hotplate stabilizes. It could be concluded that the insert conducts heat throughout the deposit, while salt PCM by itself has an insulating nature and does not efficiently transport heat. Mihalka and Matiasovsky drew a parallel conclusion in [30], which they have proved by CFD simulations. Better conductivity was obtained for PCM enriched with metallic honeycomb enhancing structure than for the pure PCM deposit.

A slow decline in the value of the temperature derivative (B, characteristic for 1.2 mm) is connected to the phase change (melting) of PCM. The temperature equalizes in the whole deposit when the entire volume of PCM has undergone phase change. The stage ends with a sharp peak, indicating the start of the heating of the liquid phase. The peak falls quickly as the deposit is already heating evenly. Similar behavior can be observed earlier for the 1.6-thick-wall insert (at approximately 10,000 s), and, later, for the only-salt deposit (at approximately 17,000 s).

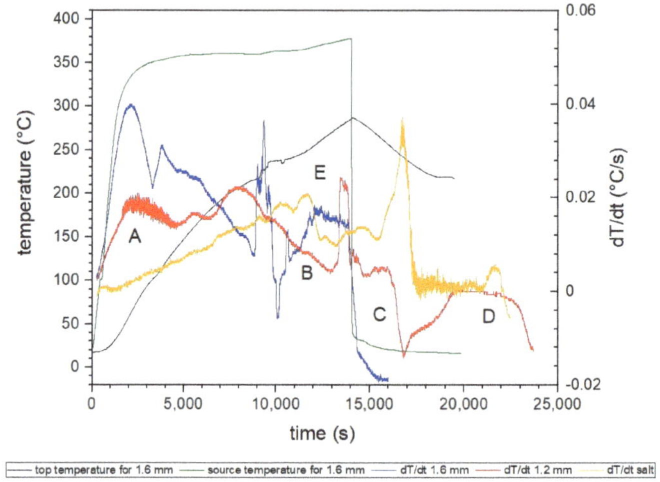

Figure 6. Derivative curves for the investigated cases and the temperature course of the heat source and the top (h = 87 mm) thermocouple for the 1.6 mm insert. Letters symbolize areas as follows: A—the first peak responsible for heat conduction through the insert, B—phase change area, C—the beginning of water cooling, D—crystallization process, E—change of crystalline system of KNO_3.

The following decrease in the derivative (C on the graph, for 1.2 mm insert) and its change to negative values are connected to switching off the heating plate and the start of water cooling. Water used in the experiment for cooling the system was at room temperature, approx. 17 °C. When the water reached the cooling system placed under the deposit, the system started to cool immediately. In the beginning, water was leaving the system in the form of steam, but then it quickly changed to liquid and started to decrease its temperature. The most rapid temperature decrease can be observed for the 1.6 mm-thick-wall honeycomb. Slightly varying behavior (less distinct decrease) was obtained for the 1.2 mm-thick-wall insert, while for pure salt deposit, only the stopping of heating was recorded. The derivative is oscillating around 0, which corresponds to keeping the constant temperature.

At the D point (1.2 mm), a crystallization process occurs. The derivative is close to 0 value, which indicates that no temperature changes are going on in the system as the PCM releases latent heat stored during solidification. Similar processes can be seen earlier for 1.6 mm honeycomb due to the facilitated heat transport.

Other peaks are observed for temperatures in the region of 130–150 °C (E, characteristic of the yellow curve). In this range, the transition from the rhombic to the trigonal solid phase (change of crystalline system) of KNO_3 takes place [31]. The location of the peaks may be slightly changed due to the effect of the heating rate.

Figure 7 presents the temperature difference (ΔT) between the top (h = 87 mm) and bottom (h = 12 mm) thermocouple measurements over time for all tested configurations. The solely PCM deposit exhibits the highest temperature gradient, indicating limited heat flow. In the case of 1.6 and 1.2 mm-thick-wall honeycombs, the differences are significantly lower. The highest value for a pure salt deposit (175 °C) is three times bigger than the peak value for the insert with a 1.2 mm-thick wall (approx. 55 °C), and for 1.6 mm—approx. 30 °C. Those are the difference occurring during the charging of the accumulator. While discharging, the significant differences can be still observed—when the peak for pure salt deposit reaches approx. −73 °C, the lowest values for 1.2 mm is −26 °C, and the lowest for 1.6 mm is −14 °C.

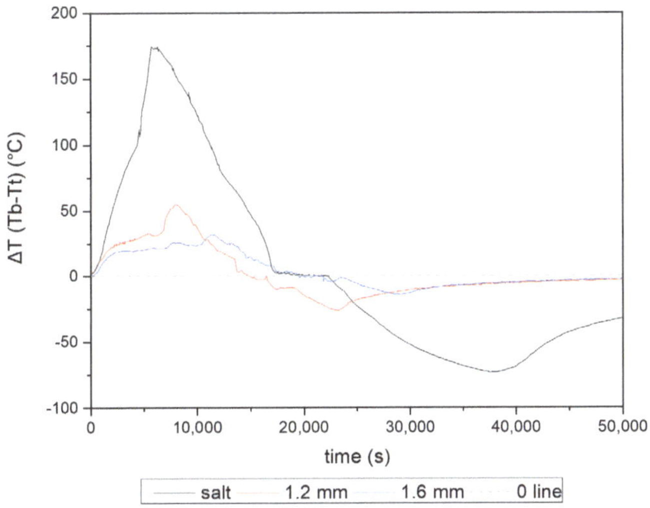

Figure 7. The temperature gradient between the top and bottom thermocouple during the experiment for all tested cases.

The huge negative ΔT visible for only PCM deposit is strictly connected to the heat transfer rate in the bed. Salt mixture in solid state is characterized by low thermal conductivity (for KNO_3-$NaNO_3$ mixture 0.457 W/(m·K)); thus, even if the part of the deposit is already melted, the heat provided to the rest of the material is significantly lower than the amount enabling phase change. In case of enhanced structures, the heat is transported through the aluminum structure, a material with considerably higher thermal conductivity of 160 W/(m·K); therefore, the melting process, as well as that which is strictly connected to the material temperature, is more evenly distributed in the system.

As shown in Table 4, the temperature gradient for different deposit heights, at the point when the PCM in the bottom cell reaches its melting point (h = 12 mm), is reduced with the increase in the honeycomb's wall thickness, resulting in improved heat dissipation. For all tested cases, the melting process starts at a similar temperature at all heights. However, a vast difference can be noticed between the temperature of top thermocouples in time t_1, where pure PCM ΔT amounts to 165.1 °C, while for a 1.2 mm-thick-wall insert the temperature is 48 °C, almost four times less, and for a 1.6 mm-thick-wall insert—32.3 °C. At t_2, the temperatures at all heights are close because the majority of the deposit is already melted.

Table 4. Temperature values for used thermocouples (b—bottom h = 12 mm, c—center h = 50 mm, t—top h = 87 mm) at the time t_1 and t_2 (from Table 1) for the application of different enhancers.

Temperature Gradient		PCM	1.2 mm	1.6 mm
	T_b [°C]	218.3	218.8	218
t_1	T_c [°C]	98.1	191.5	204
	T_t [°C]	53.2	170.8	185.7
	T_b [°C]	230	231.1	232.6
t_2	T_c [°C]	229.5	232.3	230.7
	T_t [°C]	218.3	218.8	218
ΔT for t_1		165.1	48	32.3

The heat transfer enhancer absorbs energy from the heat source and transports it evenly through the deposit; thus, vast differences in the temperature at different heights of the accumulator were compensated. The thicker the walls of the cast structure, the more heat is received and transported through the store, and, in consequence, the shorter the time needed to melt the entire volume of PCM. The difference in the temperature value during the charging of the accumulator was also reported in [32]. The article aimed to present a numerical prediction of solid–liquid interface location and an understanding of temperature distribution in the deposit. Basing on a numerical one-dimensional model of the phase change material storage with a fin, they calculated and presented temperature distribution. The tank filled with paraffin PCM was heated by an aluminum fin structure. As the distance of the measurement point from the heat source increases, a lower PCM temperature is visible. The resulting gradient depends not only on the source temperature and the heat flux rate provided into the deposit, but on the thickness of the enhancing element as well. The increase in the diameter of the fin or, in our case, in the thickness of the wall, will result in lower gradient values and faster charging of the accumulator.

These conclusions are also verified in Figure 8, where the increase in the liquid phase of PCM over time is shown. Initially, there are no relevant differences between the cases considered; however, after melting of the bottom of the deposit, about 25% of the PCM volume, the process significantly accelerates for the tanks with metal inserts. The heat flux supported by metal enhancers moves more quickly through the deposit, which results in a notably shorter time required for a fully charged accumulator. In the case of a 1.2 mm-thick-wall insert, the entirely melted deposit was observed after 10,650 s (2.95 h), while for 1.6 mm, a complete transition to liquid phase takes place after 9190 s (1.5 h). In comparison with the only-salt deposit, in which full remelting occurs after 17,000 s (4.7 h), in which the differences are 1.75 h and 3.2 h, respectively.

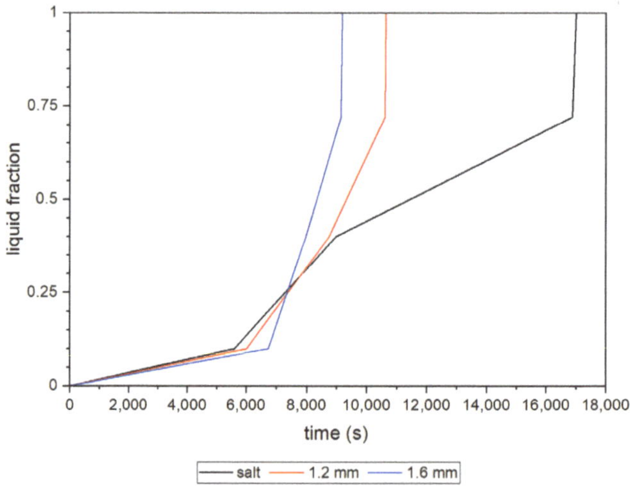

Figure 8. Liquid phase formation estimated on the basis of the excess temperature of the phase transition at the measuring points.

3.2. Numerical Analysis Results and Discussion

Performing numerical calculations allowed the determination of the influence of the wall thickness of the insert on the change of PCM temperature and the heat transfer rate. Figure 9 shows the dependence of temperature in time at three heights corresponding to the height of thermocouples in experimental studies.

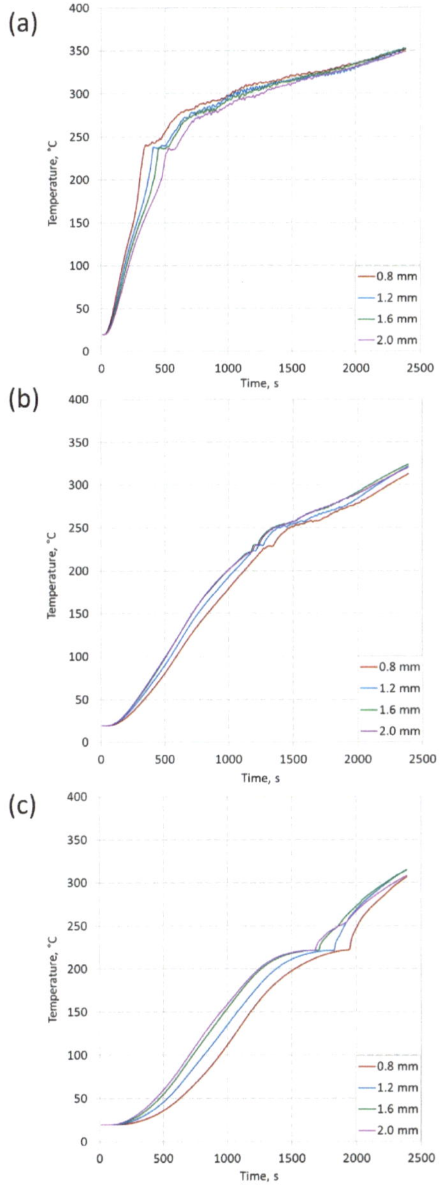

Figure 9. Calculated temperatures evolution on three different heights, for four thicknesses of the insert: (**a**) h = 12 mm; (**b**) h = 50 mm; (**c**) h = 87 mm.

The time of the phase change is clearly visible for each height. Figure 9a shows a rapid temperature increase in the first minutes of the process. It is caused by the largest temperature gradient. This increase is not so visible for h = 50 mm (Figure 9b). This is due to the slow transport of heat in the vertical direction. The slowest temperature increase was recorded for the height was 87 mm. There is no clear difference between the temperature rise before and after the phase change at this height. This may be due to heat transport in the insert in multiple directions simultaneously. There is a noticeable influence of the insert thickness on the rate of temperature increase. It is worth noting that at the height of mm, PCM heats up the fastest for the insert with the smallest wall thickness. This is due to the slower transport of heat through the insert in the vertical direction. This causes the PCM at the bottom of the vessel to heat up faster (close to the heat source) and the phase change occurs faster. However, the temperature distribution changes with the height, and for h = 87 mm, the phase change occurs the fastest for an insert with a thickness of 2.0 mm. This proves as a better heat transport in the vertical direction than for an insert with a smaller wall thickness. At the end of the process, the differences are small as the PCM temperature tends to the same value, equal to the heat source temperature.

In order to better define the dynamics of the process, the temperature change in time dT/dt was determined, presented in Figure 10, for three different heights.

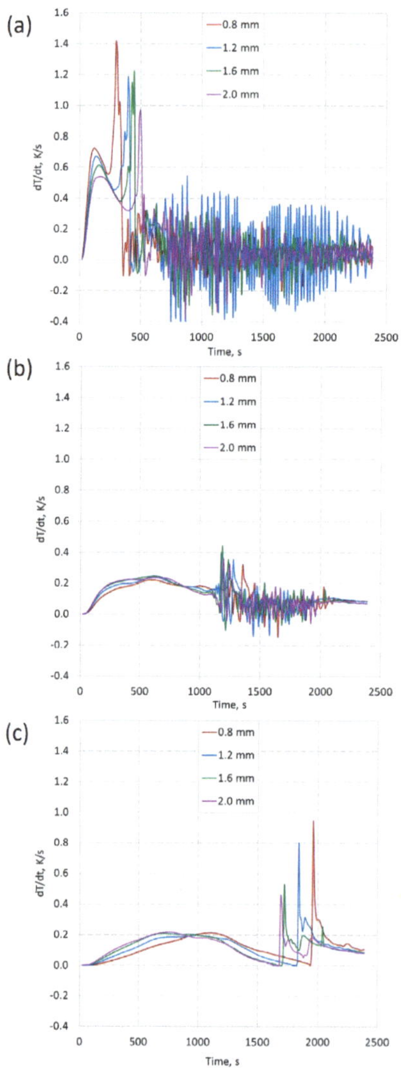

Figure 10. Derivation evolution on three different heights, for four thicknesses of the insert: (**a**) h = 12 mm; (**b**) h = 50 mm; (**c**) h = 87 mm.

As shown in Figure 10, the heat transport rate is very high at the beginning of the process, causing large changes in temperature in time, mainly at the height of 12 mm, where the phase change begins the fastest. The temperature gradient is the highest among the three analyzed heights and exceeds 0.5 K/s. The rapidly increasing temperature over time is the result of a short distance from the heat source. Moreover, the highest temperature increase at h = 12 mm was obtained for the insert thickness of 0.8 mm. This may be due to slower heat transfer in the vertical direction. After this stage, there is a noticeable decrease in dynamics, especially in Figure 10b. It is caused by reaching the temperature at which the phase change takes place. At h = 12 mm, the dynamics decrease, but increase at other heights. It is worth noting that after the described phase change, the dT/dt values oscillate close to zero. This oscillation is also visible in Figure 10b, after the phase change. This may be due to the convective motion of the PCM and the sinking of the solid phase. This would correspond to the non-uniform temperature changes visible, especially in Figure 9a. The influence of the wall thickness at the height of 50 mm and 87 mm is visible, where the dynamics increase with increasing wall thickness. This confirms the observations described for h = 12 mm, for which the phase change occurs later, with the increase in wall thickness. This suggests an increased heat transfer along the insert to the top of the vessel. Figure 10b shows the effect of using a thicker insert. The phase change process is the slowest there for a wall thickness of 0.8 mm. However, for the other thicknesses, the phase change time is similar. Figure 10c shows the best effect of using inserts with a larger wall thickness. The phase change process at the height of h = 87 mm (Figure 10c) occurs the fastest for a wall thickness of 2.0 mm. It should be noted that the rate of temperature change in time does not increase linearly with increasing wall thickness. It is also visible in the maximum values of dT/dt, i.e., right after the PCM has been melted. For a wall thickness of 2.0 mm, there is the smallest increase in dT/dt during the phase change (shown in Figure 9c). For thicknesses 0.8, 1.2, 1.6, and 2.0, the maximum values were 0.94, 0.80, 0.52, and 0.45, respectively. This suggests a more even heat distribution through the insert. This also reduces the temperature gradient throughout the vessel, which is a desirable effect.

4. Conclusions

Based on the research and results presented, the conclusions can be drawn as follows:

- Additive manufacturing followed by investment casting allowed for obtaining customized complex metal structures;
- The use of heat enhancer can notably increase the rate of charging. In the case of studied systems, melting temperature was reached two times faster in case of the system enhanced with a 1.2 mm-thick structure in comparison to the pure deposit of salt PCM. The presence of the casting influences the behavior of the heat source.
- Due to the nonconductive properties of used salt mixture, the application of the insert with 1.2 mm walls improved the time of charging twice, while improving the time of 1.6 mm at a rate of four times. Additionally, the use of the thicker insert lowered the temperature gradient at the beginning of phase change in the deposit five times, compared to pure PCM.
- Even for salt mixture, phase change from rhombic to triagonal structure of KNO_3 in the temperature range of 130–150 °C is observed in the form of flattering the temperature-in-time derivative plot.
- According to the numerical analysis, which confirmed the beneficial effect of using inserts with higher wall thicknesses, it can be concluded that the higher the thickness of the enhancer wall, the better the distribution of heat throughout the vessel volume is observed.
- The main innovation presented in the article is a manufacturing method which allows new, thin-walled, fully customized complex structures, with the possibility of rapid modifications, to be fabricated. The elements produced via investment casting method are characterized by a high precision of execution, repeatability, and extensively developed surface area, allowing good contact between PCM and the material of the

enhancer. Before that, the enhancers were commonly produced by stainless steel by the means of extrusion, which was a huge limitation when it comes to the use of complex shapes. Moreover, the use of stainless steel in molten salt environment carries a high risk of corrosion of the system, so the service life of the system will be in danger of being drastically reduced. The use of a customized aluminum enhancer reduces the chances of such phenomena and allows the insert to be optimally adapted to the case study. In addition, the developed method allows the production of any metal shape quickly, and is much cheaper than, e.g., selective laser melting (SLM).

Author Contributions: Conceptualization, N.R., A.D., K.N., A.N. and M.N.; methodology, N.R., A.D., K.N., A.N. and M.N.; validation, N.R., A.D., K.N., A.N. and M.N.; investigation, N.R., A.D., K.N., A.N. and M.N.; writing—original draft preparation, N.R., A.D. and A.N.; writing—review and editing, N.R., A.D., K.N., A.N. and M.N.; supervision, K.N. All authors have read and agreed to the published version of the manuscript.

Funding: This research received no external funding.

Institutional Review Board Statement: Not applicable.

Informed Consent Statement: Not applicable.

Data Availability Statement: Not applicable.

Conflicts of Interest: The authors declare no conflict of interest.

References

1. Sharma, A.; Tyagi, V.V.; Chen, C.R.; Buddhi, D. Review on Thermal Energy Storage with Phase Change Materials and Applications. *Renew. Sustain. Energy Rev.* **2009**, *13*, 318–345. [CrossRef]
2. Khaled, A.R.A.; Siddique, M.; Abdulhafiz, N.I.; Boukhary, A.Y. Recent Advances in Heat Transfer Enhancements: A Review Report. *Int. J. Chem. Eng.* **2010**, *2010*, 106461. [CrossRef]
3. Saxena, A.; Agarwal, N.; Cuce, E. Thermal Performance Evaluation of a Solar Air Heater Integrated with Helical Tubes Carrying Phase Change Material. *J. Energy Storage* **2020**, *30*, 101406. [CrossRef]
4. Maradiya, C.; Vadher, J.; Agarwal, R. The Heat Transfer Enhancement Techniques and Their Thermal Performance Factor. *Beni-Suef Univ. J. Basic Appl. Sci.* **2018**, *7*, 1–21. [CrossRef]
5. Shriwas, D.; Saini, J. Heat Transfer Enhancement Technique in Heat Exchanger: An Overview. *Int. J. Res. Trends Innov.* **2018**, *3*, 2456–3315.
6. Huang, X.; Alva, G.; Jia, Y.; Fang, G. Morphological Characterization and Applications of Phase Change Materials in Thermal Energy Storage: A Review. *Renew. Sustain. Energy Rev.* **2017**, *72*, 128–145. [CrossRef]
7. Konan, H.C.; Cetkin, E. Snowflake Shaped High-Conductivity Inserts for Heat Transfer Enhancement. *Int. J. Heat Mass Transf.* **2018**, *127*, 473–482. [CrossRef]
8. Zheng, J.; Wang, J.; Chen, T.; Yu, Y. Solidification Performance of Heat Exchanger with Tree-Shaped Fins. *Renew. Energy* **2020**, *150*, 1098–1107. [CrossRef]
9. Zhang, C.; Li, J.; Chen, Y. Improving the Energy Discharging Performance of a Latent Heat Storage (LHS) Unit Using Fractal-Tree Shaped Fins. *Appl. Energy* **2020**, *259*, 114102. [CrossRef]
10. Ghalambaz, M.; Mohammed, H.I.; Mahdi, J.M.; Eisapour, A.H.; Younis, O.; Ghosh, A.; Talebizadehsardari, P.; Yaïci, W. Intensifying the Charging Response of a Phase-Change Material with Twisted Fin Arrays in a Shell-and-Tube Storage System. *Energies* **2021**, *14*, 1619. [CrossRef]
11. Karthikeyan, R.; Arul Kumar, R.; Manikandan, P.; Senthilnathan, A.K. Investigation of Solar Air Heater with Phase Change Materials Using Packed Bed Absorber Plate. *Mater. Today Proc.* **2021**, *45*, 1360–1365. [CrossRef]
12. Kant, K.; Biwole, P.H.; Shukla, A.; Sharma, A.; Gorjian, S. Heat Transfer and Energy Storage Performances of Phase Change Materials Encapsulated in Honeycomb Cells. *J. Energy Storage* **2021**, *38*, 102507. [CrossRef]
13. Khan, Z.; Khan, Z.; Ghafoor, A. A Review of Performance Enhancement of PCM Based Latent Heat Storage System within the Context of Materials, Thermal Stability and Compatibility. *Energy Convers. Manag.* **2016**, *115*, 132–158. [CrossRef]
14. Lu, T.J. Heat Transfer Efficiency of Metal Honeycombs. *Int. J. Heat Mass Transf.* **1999**, *42*, 2031–2040. [CrossRef]
15. Kong, D.; Zhang, Y.; Liu, S. Convective Heat Transfer Enhancement by Novel Honeycomb-Core in Sandwich Panel Exchanger Fabricated by Additive Manufacturing. *Appl. Therm. Eng.* **2019**, *163*, 114408. [CrossRef]
16. Andreozzi, A.; Buonomo, B.; Ercole, D.; Manca, O. Phase Change Materials (PCMs) in a Honeycomb System for Solar Energy Applications. *Int. J. Heat Technol.* **2017**, *35*, S472–S477. [CrossRef]
17. Duan, J.; Xiong, Y.; Yang, D. Melting Behavior of Phase Change Material in Honeycomb Structures with Different Geometrical Cores. *Energies* **2019**, *12*, 2920. [CrossRef]

18. Egolf, P.W.; Amacker, N.; Gottschalk, G.; Courret, G.; Noume, A.; Hutter, K. A Translucent Honeycomb Solar Collector and Thermal Storage Module for Building Façades. *Int. J. Heat Mass Transf.* **2018**, *127*, 781–795. [CrossRef]
19. Lai, C.M.; Hokoi, S. Thermal Performance of an Aluminum Honeycomb Wallboard Incorporating Microencapsulated PCM. *Energy Build.* **2014**, *73*, 37–47. [CrossRef]
20. Shih, H.M.; Lin, Y.-P.; Lin, L.P.; Lai, C. Thermal Characterization of a Heat Management Module Containing Microencapsulated Phase Change Material. *Energies* **2019**, *12*, 2164. [CrossRef]
21. Abuşka, M.; Şevik, S.; Kayapunar, A. A Comparative Investigation of the Effect of Honeycomb Core on the Latent Heat Storage with PCM in Solar Air Heater. *Appl. Therm. Eng.* **2019**, *148*, 684–693. [CrossRef]
22. Xie, B.; Cheng, W.L.; Xu, Z.M. Studies on the Effect of Shape-Stabilized PCM Filled Aluminum Honeycomb Composite Material on Thermal Control. *Int. J. Heat Mass Transf.* **2015**, *91*, 135–143. [CrossRef]
23. Mazzeo, D.; Oliveti, G.; Arcuri, N. A Method for Thermal Dimensioning and for Energy Behavior Evaluation of a Building Envelope PCM Layer by Using the Characteristic Days. *Energies* **2017**, *10*, 659. [CrossRef]
24. Zastawna-Rumin, A.; Kisilewicz, T.; Berardi, U. Novel Simulation Algorithm for Modeling the Hysteresis of Phase Change Materials. *Energies* **2020**, *13*, 1200. [CrossRef]
25. Zhang, P.; Xiao, X.; Ma, Z.W. A Review of the Composite Phase Change Materials: Fabrication, Characterization, Mathematical Modeling and Application to Performance Enhancement. *Appl. Energy* **2016**, *165*, 472–510. [CrossRef]
26. Bauer, T.; Laing, D.; Tamme, R. Overview of PCMs for Concentrated Solar Power in the Temperature Range 200 to 350 °C. *Adv. Sci. Technol.* **2010**, *74*, 272–277.
27. D'Aguanno, B.; Karthik, M.; Grace, A.N.; Floris, A. Thermostatic Properties of Nitrate Molten Salts and Their Solar and Eutectic Mixtures. *Sci. Rep.* **2018**, *8*, 10485. [CrossRef]
28. Kamimoto, M.Y. Thermodynamic Properties of 50 Mole % $NaNO_3$-50% KNO_3 (HTS_2). *Thermochim. Acta* **1981**, *48*, 319–331. [CrossRef]
29. Zhao, Q.-G.; Hu, C.-X.; Liu, S.-J.; Guo, H.; Wu, Y.-T. The Thermal Conductivity of Molten $NaNO_3$, KNO_3, and Their Mixtures. *Energy Procedia* **2017**, *143*, 774–779. [CrossRef]
30. Mihalka, P.; Matiasovsky, P. Numerical Simulation of Photovoltaic Cells Cooling by Honeycomb PCM Structure in Real Climatic Conditions. *AIP Conf. Proc.* **2020**, *2305*, 020011. [CrossRef]
31. Li, C.; Yan, N.; Ye, Y.; Lv, Z.; He, X.; Huang, J.; Zhang, N. Thermal Analysis and Stability of Boron/Potassium Nitrate Pyrotechnic Composition at 180 °C. *Appl. Sci.* **2019**, *9*, 3630. [CrossRef]
32. Sushant, S.; Ankit, Y. Numerical Analysis of Latent Heat Storage with an Internal Fin Using Flexpde. In Proceedings of the World Congress on Engineering, London, UK, 5–7 July 2017; Volume 2.

MDPI AG
Grosspeteranlage 5
4052 Basel
Switzerland
Tel.: +41 61 683 77 34

Materials Editorial Office
E-mail: materials@mdpi.com
www.mdpi.com/journal/materials

www.ingramcontent.com/pod-product-compliance
Lightning Source LLC
Chambersburg PA
CBHW041350210926
51850CB00013B/3039
* 9 7 8 3 7 2 5 8 4 4 1 3 5 *